A Primer of Genome Science
Third Edition

A Primer of
Genome Science THIRD EDITION

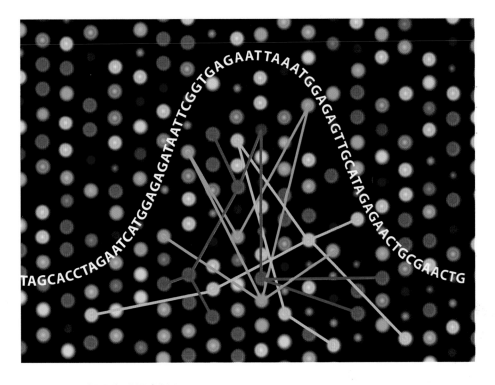

GREG GIBSON   Georgia Institute of Technology

SPENCER V. MUSE   North Carolina State University

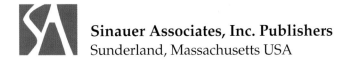

Sinauer Associates, Inc. Publishers
Sunderland, Massachusetts USA

The Cover

The cover represents each of the five major areas of genome science covered in the text. DNA sequencing and the study of genomic variation are represented by the white letters in the form of a normal distribution. These white letters spell out "A Primer of Genome Science" in codons corresponding to the single-letter amino acid abbreviations. The red, yellow, and green spots derive from a stylized image of a gene expression microarray, based on Illumina bead technology. The radiating lines connecting spots with similar colors represent an interaction network such as can be inferred using proteomics and investigated by computational genomicists.

A Primer of Genome Science, Third Edition

Sinauer Associates Inc., 23 Plumtree Road, Sunderland, MA 01375 U.S.A.
FAX 413-549-1118
publish@sinauer.com; orders@sinauer.com
www.sinauer.com

Book Web Site: http://www.sinauer.com/genomics

Trademarks and Photo Credits

All trademarked names appearing in this book are used in an editorial fashion, to the benefit of the trademark owners, and with no intent to infringe upon these trademarks.

Credits for photos, p. 29: *Chimp*: Gerry Ellis/DigitalVision. *Macaques*: Claire George/istockphoto.com. *Dog*: National Human Genome Research Institute. *Wallaby*: State of Victoria, Department of Innovation, Industry and Regional Development. *Pufferfish*: Exotic Fish and Pet World, MA. *Sea urchin*: Kirt L. Onthank. *Sea squirt*: Rich Mooi/California Academy of Sciences. *Platypus*: Nicole Duplaix, Getty Images. *Mosquito*: Jim Gathany/CDC. *Honeybee*: David McIntyre.

Library of Congress Cataloging-in-Publication Data
Gibson, Greg.
  A primer of genome science / Greg Gibson, Spencer V. Muse. — 3rd ed.
      p. ; cm.
  Includes bibliographical references and index.
  ISBN 978-0-87893-236-8 (paperbound)
1. Genomics.   I. Muse, Spencer V., 1966- II. Title.
  [DNLM: 1.   Genomics. 2.   Computational Biology—methods. 3.   Gene Expression.
4. Genome.   QU 58.5 G448p 2009]

QH447.G534 2009
  572.8'6—dc22   2008049294

Printed in China
6 5 4 3

To Ann and Alan Gibson, and to John Goebel
for opening our minds to science

Contents

Preface   xi

1 Genome Projects: Organization and Objectives 1

The Core Aims of Genome Science 1

Mapping Genomes 4

Genetic Maps 4

EXERCISE 1.1 *Constructing a genetic map* 7

Physical Maps 8

Cytological Maps 8

Comparative Genomics 10

The Human Genome Project 13

Objectives 13

The Content of the Human Genome 16

BOX 1.1 The Ethical, Legal, and Social Implications of the Human Genome Project 18

EXERCISE 1.2 *Use the NCBI and Ensembl genome browsers to examine a human disease gene* 22

Internet Resources 22

BOX 1.2 GenBank Files 26

Animal Genome Projects 28

Primate Genome Projects 28

Rodent Genome Projects 30

EXERCISE 1.3 *Compare the structure of a gene in a mouse and a human* 33

Other Vertebrate Biomedical Models 34

Animal Breeding Projects 35

Invertebrate Model Organisms 36

BOX 1.3 Managing and Distributing Genome Data 40

Plant Genome Projects 40

Arabidopsis thaliana 40

Grasses and Legumes 44

Other Flowering Plants 46

Microbial Genome Projects 48

The Minimal Genome 48

Sequenced Microbial Genomes 51

EXERCISE 1.4 *Compare two microbial genomes using the CMR* 53

Yeast 54

EXERCISE 1.5 *Examining a gene in the Saccharomyces Genome Database* 56

Metagenomics 59

Summary • Discussion Questions • Literature Cited 60

2 Genome Sequencing and Annotation 65

Automated DNA Sequencing 65

The Principle of Sanger Sequencing 65
High-Throughput Sequencing 68
Reading Sequence Traces 68
EXERCISE 2.1 *Reading a sequence trace* 71
Contig Assembly 71
BOX 2.1 Pairwise Sequence Alignment 74
EXERCISE 2.2 *Computing an optimal sequence alignment* 78
Emerging Sequencing Methods: The Next Generation 79

Genome Sequencing 83

Hierarchical Sequencing 84
Shotgun Sequencing 88
BOX 2.2 Searching Sequence Databases Using BLAST 90
Sequence Verification 94

Genome Annotation 95

EST Sequencing 95

Ab Initio Gene Discovery 98
BOX 2.3 Hidden Markov Models and Gene Finding 100
Regulatory Sequences 103
Non-Protein Coding Genes 104
Structural Features of Genome Sequences 107

Functional Annotation and Clusters of Gene Families 113

EXERCISE 2.3 *Perform a BLAST search* 114
Clustering of Genes by Sequence Similarity 114
Clusters of Orthologous Genes 116
Phylogenetic Classification of Genes 119
BOX 2.4 Phylogenetics 120
EXERCISE 2.4 *A simple phylogenetic analysis* 123
Gene Ontology 124
BOX 2.5 Gene Ontologies 126

Summary • Discussion Questions • Web Site Exercises • Literature Cited 128

3 Genomic Variation 133

The Nature of Single Nucleotide Polymorphisms 133

Classification of SNPs 133
Distribution of SNPs 136

Linkage Disequilibrium and Haplotype Maps 138

BOX 3.1 Disequilibrium between Alleles at Two Loci 138
EXERCISE 3.1 *Quantifying heterozygosity and LD* 143

Applications of SNP Technology 146

Population Genetics 146
BOX 3.2 The Coalescent 148
Recombination Mapping 152
EXERCISE 3.2 *Inferring haplotype structure* 154
QTL Mapping 155

Linkage Disequilibrium Mapping 158
BOX 3.3 Case-Control Association Studies 164
BOX 3.4 Family-Based Association Tests 167
EXERCISE 3.3 *Perform a case-control association test* 170
BOX 3.5 Genome-Wide Association Studies 173

SNP Genotyping 177

SNP Discovery 177
SNP Genotyping 178
EXERCISE 3.4 *Designing a genotyping assay for a double polymorphism* 183
High-throughput genotyping platforms 183
Haplotype phasing methods 185

Summary • Discussion Questions • Web Site Exercises • Literature Cited 186

4 Gene Expression and the Transcriptome 191

Parallel Analysis of Gene Expression: Microarrays 191
Applications of Microarray Technology 192
Experimental Design 194
EXERCISE 4.1 *Design a microarray experiment 196*
Microarray Technologies 198
Labeling and Hybridization of cDNAs 205
Statistical Analysis of cDNA Microarray Data 207
EXERCISE 4.2 *Calculate which genes are differentially exposed 209*
  BOX 4.1 Microarray Image Processing 211
  BOX 4.2 Basic Statistical Methods 214
EXERCISE 4.3 *Evaluate the significance of the following gene expression differences 217*
Microarray Data Mining 220
  BOX 4.3 Clustering Methods 221
EXERCISE 4.4 *Perform a cluster analysis on gene expression profiles 224*
ChIP Chips and Gene Regulation 225

DNA Applications of Microarrays 227
  BOX 4.4 Motif Detection in Promoter Sequences 228
Parallel Analysis of Gene Expression: RNA Sequencing 231
Serial Analysis of Gene Expression 231
RNA-Seq 234
Single-Gene Analyses 236
Northern Blots 236
Quantitative PCR 237
Properties of Transcriptomes 239
Microbial Transcriptomics 239
Cancer and Clinical Applications 243
Development, Physiology, and Behavior 246
Evolutionary and Ecological Functional Genomics 248
Gene Expression Databases 252
Summary • Discussion Questions • Web Site Exercises • Literature Cited 253

5 Proteomics and Functional Genomics 259

Functional Proteomics 259
Protein Annotation 259
EXERCISE 5.1 *Structural annotation of a protein 262*
  BOX 5.1 Hidden Markov Models in Domain Profiling 264
Protein Separation and 2D-PAGE 267
Mass Spectrometry 270
EXERCISE 5.2 *Identification of a protein on the basis of a mass spectrometry profile 273*
Immunochemistry 276
Protein Microarrays 277
Protein Interaction Maps 280
EXERCISE 5.3 *Formulating a network of protein interactions 281*

  BOX 5.2 Biological Networks in Genome Science 283
Structural Proteomics 286
Objectives of Structural Proteomics 286
Protein Structure Determination 288
Protein Structure Prediction and Threading 291
Functional Genomics 294
Saturation Forward Genetics 295
High-Throughput Reverse Genetics 300
  BOX 5.3 Transgenic Animals and Plants 304
Fine-Structure Genetics 308
EXERCISE 5.4 *Designing a genetic screen 309*
Genetic Fingerprinting 314
Summary • Discussion Questions • Web Site Exercises • Literature Cited 317

6 Integrative Genomics 323

Metabolomics 325
Analysis of Cellular Constituents 325
Metabolic Profiling 328
Metabolic and Biochemical Databases 331

In Silico Genomics 333
Metabolic Control Analysis 333
Systems-Level Modeling of Gene Networks 338

Summary • Discussion Questions • Literature
Cited 342

Glossary 345

List of Abbreviations 355

Index 357

Preface

This book is an introduction to genome science intended for senior undergraduates or for graduate students who are new to the field. Eight years ago, we were inspired to write this text as a result of our experiences teaching introductory graduate classes in functional genomics (GG) and bioinformatics (SM) at North Carolina State University. The general features of the field of genome science seem to have settled down around genome-scale sequencing, analysis of variation, gene expression profiling, proteomics, and metabolomics. Increasingly, computational biology and statistical analysis play a central role in genome research, and we endeavor to capture this trend toward a merger of molecular and theoretical biology. Another major theme is that genome research is not restricted to human biology or just a handful of biomedical model organisms: agricultural, ecological, behavioral and evolutionary genomics are emerging as dynamic subdisciplines. By combining the basics of bioinformatics and experimental methods in one text, we hope to encourage students to develop a familiarity with genomics that will be useful whether they pursue careers in science, law, journalism, marketing, or wherever their interests may lead.

This third edition follows the same sequence and style as the second edition, but includes substantial revisions that reflect some dramatic technological innovations that became integral components of contemporary genome analysis in 2007 and 2008. These are most notable in relation to genome sequencing and "NextGen" resequencing (Chapter 2), whole-genome genotyping and genome-wide association studies (Chapter 3), and commercial gene chips and bead arrays for whole-transcriptome profiling (Chapter 5). We have included several new illustrations to help explain these advances and have discarded a surprising number of methods that are now obsolete. Chapter 6 has been updated to include some exciting new developments in the field of metabolomics (also known as metabonomics) and systems biology.

The exercises associated with each chapter are for the most part the same as in the previous edition, as are the Boxes that explain statistical and bioinformatic principles. Regrettably, we have found that many of the Web links mentioned in the text become inactive or move to new addresses with a half-life of a couple of years—which is also the time it takes the major genome databases to overhaul their content and appearance. There is little we can do about this continual metamorphosis, and we apologize to students in advance for the frustration it will surely cause them. Without a doubt new methods and concepts will arise as soon as this book appears in print, if not sooner; such is the frenetic pace of genomics. We hope, however, that this edition at least captures the major trends of the past few years.

We assume that the reader is familiar with the content of typical "300 level" undergraduate courses in genetics, and so have eschewed an opening chapter that introduces basic concepts and terminology such as exons and introns, enhancers, cDNA, the central dogma, and the chromosomal basis of inheritance. Most readers will probably benefit from having a copy of a standard genetics text available for reference, and some will wish to graduate quickly to any of the many detailed treatments of the main topics covered in single chapters of this book.

We recognize that students will come to the discipline with vastly different educational backgrounds. Some will know how to program neural networks but be unaware of the distinction between SNPs and indels; some will have used microarrays but never have heard of a mapping function, let alone a Markov chain. Our strategy is to describe the core experimental methods side-by-side with the analytical approaches. We emphasize throughout the importance of statistics and analysis in the design and interpretation of genomic data. Some readers will undoubtedly be frustrated by our decision not to concentrate on the biological impact of genomic data, but that decision was made for two reasons. The first is that there is so much in flux that such coverage would be premature; the second is that this task is best left to individual teachers, who can put their own emphasis on the field in the context of class discussions.

In addition to the exercises in the text, some much more complicated exercises involving large datasets are provided online. Use of the internet is an essential part of instruction in genomics, which also depends on computers for laboratory information management, database establishment, data sharing, and statistical analysis. Consequently, the Web site for this book provides, among other things, a series of related DNA sequence, microarray, and SNP datasets that can be downloaded and analyzed at will. The descriptive titles for these exercises can be found at the conclusions of Chapters 2 through 5.

In the final analysis, there is no substitute for reading the primary literature. It is unlikely that an undergraduate course will cover all of the topics, or that most graduate instructors will want to. This *Primer* is not meant to be used as a reference so much as an aid, in conjunction with classroom instruction, for the building of a foundation from which forays can be made into specific topics, from psychogenomics to motif detection, from comparison of genome sequences to modeling metabolism. There are undoubtedly topics we have missed and some discussions that will be out of date before long. But given an understanding of the basic approaches in genome science, there is no reason why these cannot be explored independently.

It has been our opinion that genomics is too broad a topic to be accommodated by simply adding a chapter to standard textbooks. Furthermore, to do so misses the central reality that the view of biology is different from the perspective of the whole genome as opposed to that from individual genes. We imagine that some students drawn to the genomic way of doing biology will seek to incorporate mathematics and computer science into their research from the beginning of their training, while others will seek to integrate classical molecular genetics and cell biology into their work. If this book helps make both of these tasks easier, it will have done its job.

Acknowledgments

We would both like to acknowledge the contributions of our students in helping to bring the structure of this book together, as well as the input from instructors who have used the *Primer* in their classes over the years. A number of colleagues reviewed and otherwise helped shape specific chapters, including Lisa Goering, Trisha Wittkopp, Erika Zimmerman, David Pollock, Sergey Nuzhdin, Steven Carr, David Bird, Dahlia Nielsen, Matt Rockman, Greg Wray, Andy Clark, Patricia Hurban, Alison Motsinger-Reif, and several anonymous reviewers (none of whom, of course, bears any responsibility for errors of omission or commission). Greg Gibson would particularly like to thank Walter Gehring, David Hogness, Cathy Laurie, Trudy Mackay, and Julian Adams, who helped build scaffolds and bridge gaps. Spencer Muse adds his special appreciation to Bruce Weir for a decade of guidance and advice.

It was an absolute pleasure working with S. Mark Williams at Pyramis Studios in Durham, NC, who created the artwork for the first two editions, and with Jack Haley, who did the new illustrations in this edition. Both rendered the figures with care and creativity. Andy Sinauer, Carol Wigg, Chris Small, Janice Holabird, Jefferson Johnson, Susan McGlew, and everyone else at Sinauer Associates make writing a textbook a pleasure, and their efficiency and enthusiasm quickly turn possibilities into reality.

Finally, we thank our wives, Diana and Cindy, from the bottom of our hearts, for understanding and so much more. Sydney, Paddington, Marcy, Quinn, Cooper, Incense, Calvin, Bernie, and Oliver played their supporting roles superbly as well.

Greg Gibson and Spencer Muse
Brisbane, Australia and Raleigh, North Carolina
November, 2008

1 Genome Projects: Organization and Objectives

Genome science is the study of the structure, content, and evolution of genomes. Initially driven by high-volume nucleotide sequencing, technological advances have enabled the field to grow rapidly, and genome science, or "genomics," now also encompasses analysis of the expression and function of both genes and proteins. The related disciplines of bioinformatics and computational biology are increasingly integrated with empirical analyses. Genome projects are now in place for well over 300 different organisms. This chapter provides an overview of these projects as well as an introduction to some of the central concepts in genome science.

The Core Aims of Genome Science

All genome projects share a common set of aims, subsets of which are emphasized according to available resources, economic and scientific goals, and the biological attributes of the organism under study. The major aims of individual genome projects are:

1. **To establish an integrated Web-based database and research interface.** Access to the enormous volume of data generated by genome researchers is facilitated by ever-evolving Web interfaces. Many Web sites initially grew out of the efforts of a small group of leaders at a single research institution, but such sites quickly demand millions of dollars for upkeep and innovation, leading to the creation of mirror sites and organizations dedicated to collecting and linking data, to quality control, and to presenting data in a useful manner. Rather than merely serving as storage bins for sequence and other data, most sites are now built on state-of-the-art relational databases. Many of these utilize a common resource of innovative software for data searches and online analysis.

2. **To assemble physical and genetic maps of the genome.** The location of genes in a genome can be specified both according to physical distance and relative position defined by recombination frequencies. This information is crucial for comparing the genomes of related species and for putting together phenotypic and genetic data. Genetic maps are used in animal and plant breeding, and in numerous areas of basic biological research. The art of map making has been honed by geneticists for the better part of a century, but genomicists are extending it to more and more species, with greater and greater accuracy.

3. **To generate and order genomic and expressed gene sequences.** High-volume sequencing is one of the defining features of genome science. The basic procedure used is the same one developed by Fred Sanger two decades ago, except that it has been heavily automated both at the levels of running reactions and reading sequences. Sequencing of whole genomes is now generally performed in a "shotgun" fashion: millions of short pieces of DNA sequence are computationally assembled into contigs, scaffolds, and eventually whole chromosomes. The term **contig** refers to a set of sequence fragments that have been ordered into a contiguous, linear stretch on the basis of sequence overlaps at the fragment ends. A set of contigs constitutes the **scaffold** of a whole-genome sequence. Because much of a genome consists of noncoding DNA, genomicists are also interested in sequencing large numbers of **cDNA clones**. Such sequences are cloned from mRNA transcripts (see Chapter 2) and thus identify expressed genes and exon boundaries. Only one end of a cDNA need be sequenced to identify a clone, and these fragments are called **expressed sequence tags**, or **ESTs**. Because of alternative splicing and errors in the construction of cDNA libraries, there is not a one-to-one correspondence between ESTs and genes, but EST collections are thought to give a good first approximation of the diversity of genes expressed in a tissue.

4. **To identify and annotate the complete set of genes encoded within a genome.** Having the complete sequence of a genome is only the first step toward characterizing its gene content. The genes encoded within the sequence must then be identified using a combination of experimental and bioinformatic strategies. These include alignment of cDNA and genomic sequences, looking for sequences that are similar to those already identified in other genomes (both of these procedures rely on DNA alignment and comparison algorithms such as BLAST), and applying gene-finding software that recognizes DNA features that are associated with genes, such as open reading frames (ORFs), transcription start and termination sites, and exon/intron boundaries. Once a gene has been identified, it must be **annotated**, which entails linking its sequence to genetic data about the function, expression, and mutant phenotypes of the protein associated with the locus, as well as to comparative data from homologous proteins in other species.

5. **To characterize DNA sequence diversity.** It has been known for some time that all genomes are full of polymorphisms: sequence sites at which two or more variants are found in natural populations. **Single-nucleotide polymorphisms** are called **SNPs**, and it is generally assumed that most quantitative genetic variation—that is, the heritable component of variation in characters such as size, shape, yield, and disease susceptibility—should be traceable to SNPs or to insertion/deletion polymorphisms. Characterization of the distribution of SNPs is a crucial first step in efforts to find associations between SNP variation and phenotypic variation. Another fundamentally important quantity that remains to be characterized is the level of **haplotype** structure due to linkage disequilibrium (LD). A haplotype is a set of adjacent polymorphisms found on a single chromosome, while LD refers to nonrandom associations between sites. The farther apart two sites are, the more they tend to assort independently (i.e., randomly), but there is great variation in the distances involved, from tens of bases to tens of kilobases. Disease locus mapping now generally utilizes detailed knowledge of LD. Further, SNP variation, along with variation in repetitive sequences such as microsatellites, can be an invaluable tool for inferring relationships between individuals, in forensics, in evolutionary studies of the history of a species, and in studies of population structure.

6. **To compile atlases of gene expression.** Important clues to gene function can be gleaned from analyzing profiles of transcription and protein synthesis. Traditional methods for characterizing gene expression include Northern blots and in situ hybridization (and, where an antibody exists, Western blots and immunohistochemistry). Genomic methods aim to contrast the expression of thousands of genes simultaneously. EST sequencing and **serial analysis of gene expression** (**SAGE**) rely on the detection of transcript tags in libraries of cDNA fragments. These are being displaced by new high-throughput, "next-generation" sequencing of tens of millions of fragments. Such methods are used to contrast gene expression in different tissues, at different stages of development or infection, or in the presence of toxins and other environmental agents.

   Transcription profiling is used to compare gene expression across multiple treatments or conditions. A microarray is a collection of gene probes that have been spotted onto a glass slide or synthesized as oligonucleotides on a silicon chip or collection of beads. These are hybridized to fluorescently labeled cDNA, and the relative intensity of signals between samples provides a measure of the abundance of each transcript. Analysis of gene expression profiles provides information about the regulation of transcription, can yield clues to gene function, and may identify biomarkers for disease processes. Gene expression is also a powerful tool in the identification of candidate genes and genetic pathways for any given process, since a gene must (usually) be expressed in a tissue in order to have an effect on it.

7. **To accumulate functional data, including biochemical and phenotypic properties of genes.** The term **functional genomics** refers to a panoply of approaches under development to ascertain the biochemical, cellular, and/or physiological properties of each and every gene product. These include *near-saturation mutagenesis* (screening hundreds of thousands of mutants to identify genes that affect traits as diverse as embryogenesis, immunology, and behavior), *high-throughput reverse genetics* (methods to systematically and specifically inactivate individual genes), and elaboration of genetic tools. **Proteomics**, a core element in functional genomics, includes methods for detecting protein expression and for detecting protein-protein interactions. Another subfield, **structural genomics**, seeks to elucidate the tertiary structure of each class of protein found in cells. **Pharmacogenomicists** are particularly interested in studying the interactions between small molecules (i.e., potential drugs) and proteins, both in vitro and in the context of living organisms. Research on model organisms such as mice, fruitflies, nematodes, various plants, and yeast is a crucial component of functional genomics.

8. **To provide the resources for comparison with other genomes.** Just as nothing in biology makes sense except in the light of evolution, nothing in genomics makes sense except in the light of comparative data. Comparative maps allow genetic data from one species to be used in the analysis of another, because local gene order along a chromosome tends to be conserved over millions of years—a phenomenon known as **synteny.** Even without synteny, the conservation of gene function is now known to be extensive enough that studies of the genetics of neuronal signaling or heart development in fruit flies can tell us much about the same processes in a primate. The development of online resources is being prioritized in order to enhance an individual researcher's ability to use data being generated anywhere in the world, from any organism.

This chapter will describe the core elements of each of these objectives for the major genome projects in animals, plants, and microbes of medical, agricultural, and basic biological interest. Before doing so, it is essential to define the concepts of physical and genetic maps, as these two resources are central to all of genome science.

Mapping Genomes

Genetic Maps

A **genetic map** is a description of the relative order of genetic markers in linkage groups in which the distance between markers is expressed as units of recombination. The genetic markers are most often physical attributes of the DNA (such as single-nucleotide polymorphisms, simple repeats, or variable restriction enzyme sites), but may include phenotypes associated with Mendelian loci. In diploid organisms, genetic maps are typically assembled from data on the co-segregation of genetic markers either in pedigrees or in the progeny of controlled crosses. The standard unit of genetic distance is the **centiMorgan (cM)**, an expression of the percentage of progeny in which a recombination event has occurred between two markers. Named for Thomas Hunt Morgan, 1 cM is an arbitrary unit equivalent to a recombination frequency of 0.01. In human autosomal euchromatin, 1 cM is approximately 1,000 kb, which is twice as long as the equivalent parameter for *Drosophila*.

While there is a high variance, individual chromosomes in many animal species tend to be on the order of 100 cM in length, indicating that one crossover occurs per chromosome per generation. Markers on different chromosomes have a 50-50 chance of co-segregating, and thus are 50 cM apart—which is the threshold for assigning markers to the same linkage group. Two markers that recombine 50% of the time may nevertheless be on the same linkage group, so long as they are joined by a third marker that shows less than 50% recombination with each of the markers (Figure 1.1). With sufficient markers, the number of linkage groups and the number of chromosomes should be the same for any given organism.

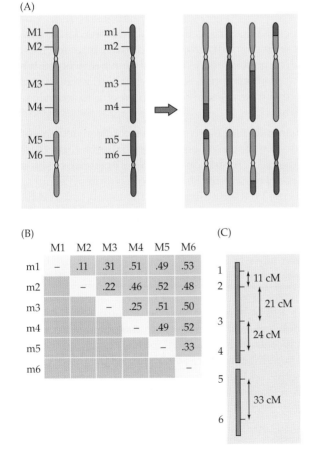

Figure 1.1 Assembling a genetic map. (A) Starting with a pair of different parental genomes, represented as green and blue chromosomes, a series of recombinant chromosomes are generated in controlled crosses. (B) The genotypes of multiple molecular or phenotypic markers in the recombinant individuals are determined, from which a table showing the frequency of recombinants between each marker is drawn up. (C) Software is then used to calculate the most likely genetic map from the data. In this hypothetical example, two linkage groups are inferred, one of which (top) is longer than 50 cM. The best estimate of the genetic distance in cM between each pair of markers is indicated on the right. Genetic maps can be assembled from pedigree data using similar principles.

Software for the assembly of genetic maps is freely available from a number of sources (see http://linkage.rockefeller.edu/soft/list.html for a comprehensive listing of genetic analysis software), perhaps the most popular of which is the Mapmaker/QTL program (Lander et al. 1987). Linkage data are converted to map distances by virtue of a **mapping function** that adjusts for the observation that the probability of crossovers leading to recombination does not increase linearly with physical distance. The two most common mapping functions bear the names of their developers, Haldane and Kosambi. The Kosambi mapping function also adjusts for the phenomenon of **interference**, whereby the presence of one crossover reduces the likelihood of another crossover in the vicinity.

Multiple factors affect the correspondence between physical and genetic distances, including the variability of recombination rate along a chromosome (in most species, centromeres and telomeres are less recombinogenic than general euchromatin); genome-wide recombination rate variation within and between species; gain and loss of repetitive DNA; and the low resolution of genetic maps due to small sample sizes of progeny used to establish linkage.

Several methods have emerged for generating precise genetic maps in humans and most model organisms:

- **Recombinant inbred lines (RIL)** are collections of highly inbred lines generated by sib-mating or selfing of individual lines derived from the cross of a pair of genetically divergent inbred parents. RIL have proven particularly useful in mapping plant genomes. Seed for each line can be stored and disseminated to large numbers of workers, who are able to add markers to the existing maps at will.

- Construction of human, mouse, and many other vertebrate genetic maps has been greatly facilitated by the development of **radiation hybrid (RH) mapping**, in which fragments of chromosomes from the organism of study are incorporated into a panel of hamster fibroblast cell cultures (Deloukas et al. 1998). Species-specific PCR amplification is then used to ascertain which loci are present in each line, and the frequency with which markers co-segregate is an indicator of the physical distance between the markers. There is no need for the markers to be polymorphic in the species under study. Mapping functions unique to RIL and RH mapping have been developed that allow precise ordering of loci, as well as comparison with physical data.

- **Single-sperm typing** is a PCR-based method for actually counting the number of crossovers between two markers, with the potential to accurately define recombination fractions as small as 0.001 over just tens of kilobases. Evidence is emerging that there is variation in the distribution of crossovers along a gene, so that recombination may be focused on hotspots that could have a major effect on the structure of polymorphism (Jeffreys et al. 2004).

EXERCISE 1.1 *Constructing a genetic map*

Suppose that a breeder of orange trees begins to assemble a genetic map based on four recessive loci—thickskin, reddish, sour, and petite—named after the fruit phenotypes of homozygotes. After identifying two true-breeding trees that are either completely wild-type or are mutant for all four loci, the breeder crosses them and plants an orchard of the resultant F_2 trees. Based on the following frequencies of mutant classes, determine which loci are likely to be on the same chromosome and which are the most closely linked.

Normal	402	Thick	18
Petite	127	Red, sour, and petite	12
Sour	115	Red	11
Thick and red	108	Thick and petite	10
Thick, red, and petite	42	Thick, red, sour, and petite	8
Thick, red, and sour	41	Thick and sour	7
Sour and petite	38	Red and petite	3
Red and sour	24	Thick, sour, and petite	2

ANSWER: *First confirm that each mutant segregates with approximately Mendelian ratios by adding up the total number of each phenotype. One-quarter of the total number of 968 trees, or around 242, should show each recessive phenotype. This seems to be the case, since there are 242 petite, 249 reddish, 247 sour, and 236 thickskin orange-producing trees. Given these ratios, we would expect that unlinked loci would segregate independently, such that just 60 or so trees ($1/4 \times 1/4 \times 968$) produce each double-mutant class, but that completely linked loci would produce close to 240 double mutants. Adding the actual numbers of double mutants produced gives the following results:*

Thick and red	199	Thick and petite	62	Thick and sour	58
Red and petite	65	Red and sour	85	Sour and petite	60

These numbers indicate that the thickskin *and* reddish *loci are quite closely linked, and that the* reddish *and* sour *loci are also probably linked, though less closely than* thickskin *and* reddish. *Since* thickskin *and* sour *appear to segregate independently, the* reddish *locus must lie between them.* Petite *maps to a different linkage group.*

• Recombination rates can also be estimated indirectly from population genetic data, using advanced statistical algorithms to analyze the pattern of variation in the level of linkage disequilibrium along a locus in a population of sequences (McVean et al. 2004).

Physical Maps

A physical map is an assembly of contiguous stretches of chromosomal DNA—**contigs**—in which the distance between landmark sequences of DNA is expressed in kilobases. The ultimate physical map is the complete DNA sequence, which allows physical distances to be defined in nucleotides, notwithstanding variance among individuals due to insertion/deletion polymorphisms. Physical maps provide a scaffold upon which anonymous polymorphic markers can be placed, thereby facilitating finer scale linkage mapping; they confirm linkages inferred from recombination frequencies, and resolve ambiguities about the order of closely linked loci; they enable detailed comparisons of regions of colinearity between genomes; and they can be a first step in the assembly of a complete genome sequence.

Two general strategies can be used to assemble contigs:

1. *Alignment of randomly isolated clones based on shared restriction fragment length profiles.* Clones range in size from 1,000s of kilobases (yeast artificial chromosomes, YACs) to 100s of kilobases (bacterial artificial chromosomes [BACs], or P1 clones) to kilobases (plasmid or bacteriophage clones). Restriction profiling (see Figure 2.10) has been automated using a combination of robotics, which reduces human error while increasing throughput, and continually improving software for efficient analysis of profiles. DNA fragments that have been separated by gel electrophoresis and visualized using standard dyes are imaged digitally, then placed in size bins chosen to optimize resolution and facilitate probabilistic comparison of profiles.

2. *Hybridization-based approaches.* In these techniques, a common probe is used to identify which of an arrayed set of clones are likely to be contiguous. In **chromosome walking**, a series of adjacent clones are isolated using the terminal sequence of one clone as a probe to identify a set of overlapping clones that share the probe sequence. One of these is then chosen to identify a new terminal probe for the second step (Figure 1.2), and so forth. The extent of overlap of sets of clones identified by hybridization is also determined by comparing the restriction enzyme profiles of each clone. Contigs can be extended by end-sequencing particular clones, leading to the identification of **sequence-tagged sites (STSs)**, which in turn can be used as hybridization probes to either extend the chromosome walk or fill in gaps in the contig. The assembly of contigs and even whole genome physical maps is discussed in Chapter 2.

Cytological Maps

One historically prevalent aid in the alignment of physical and genetic maps is the use of cytological maps, as shown in Figure 1.3. **Cytological maps** are the banding patterns observed through a microscope on stained chromosome spreads. Traditional cytological preparations have included the salivary gland polytene chromosomes of insects and Giemsa-banded mammalian metaphase karyotypes. The demonstration that certain mutant

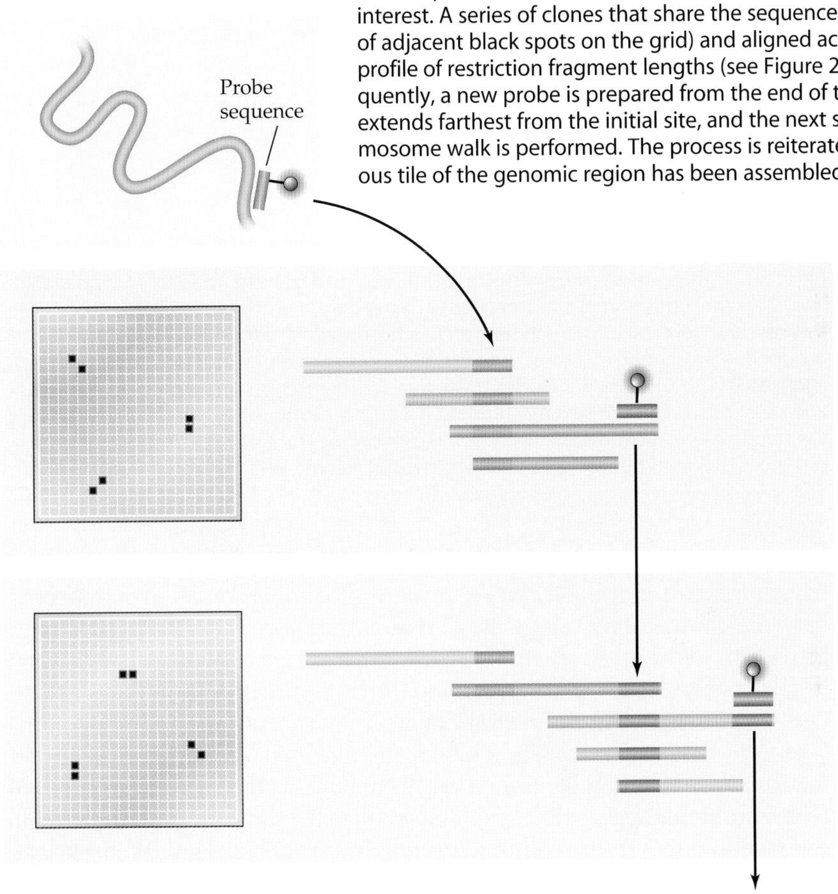

Figure 1.2 Chromosome walking.   A library of random genomic DNA fragments, typically up to 250 kb in length, is cloned into bacterial artificial chromosomes, or P1 vectors. The clones are arrayed in pairs on a grid so that they can be screened by hybridization to a labeled probe from a known sequence fragment in the region of interest. A series of clones that share the sequence are isolated (pairs of adjacent black spots on the grid) and aligned according to their profile of restriction fragment lengths (see Figure 2.10). Subsequently, a new probe is prepared from the end of the clone that extends farthest from the initial site, and the next step in the chromosome walk is performed. The process is reiterated until a contiguous tile of the genomic region has been assembled.

phenotypes or medical conditions correlate with the deletion or rearrangement of chromosome sections provided some of the first evidence that chromosomes are the genetic material, and human karyotype mapping remains a vital tool in diagnosing a range of human disorders.

In situ hybridization of cloned DNA fragments to the chromosomes allows alignment with the physical map as well. Human chromosomes are divided into bands, with numbering on the small (petite, p) and long (nonpetite, q) arms from the centromere to the telomere, giving rise, for example, to the nomenclature that yields 7p15 as the signifier for the location of the *Hox* complex in the middle of the short arm of chromosome 7. Genome browsers give researchers a wide range of options for the visual display of cytological, physical, and genetic map data.

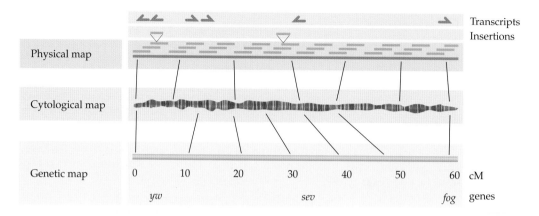

Transcripts
Insertions

Physical map

Cytological map

Genetic map

| 0 | 10 | 20 | 30 | 40 | 50 | 60 | cM |

yw   *sev*   *fog*   genes

Figure 1.3 Alignment of cytological, physical, and genetic maps.   A cytological map is a representation of a chromosome based on the pattern of staining of bands, in this case a schematic of the polytene X chromosome of the fruit fly *Drosophila melanogaster*. Genetic (bottom), cytological (middle), and physical (top) maps can be aligned. The physical map includes the location of transcripts and sites of insertions and deletions as determined by standard molecular biology. Since recombination rates vary along a chromosome, typically being reduced near the telomere and centromere, distances between genetic, physical, and cytological markers are not uniform.

Comparative Genomics

In genomic parlance, the term **synteny** refers to the conservation of gene order between chromosome segments of two or more organisms. Physical maps provide the most direct means for characterizing the extent of synteny, because highly conserved loci that can confidently be regarded as **homologs** (that is, derived from a common ancestral locus) provide anchoring landmarks. However, appropriate phylogenetic methods (discussed in Chapter 2) must be used to distinguish true homologs—properly known as **orthologs**—from **paralogs**, which are similar genes that arose as a result of duplication in one or both lineages subsequent to an evolutionary split. In many cases, entire complexes of genes have undergone multiple rounds of duplication, which complicates efforts to distinguish orthology from paralogy. The distinction is not just of academic interest; it can be crucial when comparative physical maps are used to suggest the location of a locus with medical or other phenotypic effects based on its linkage to a cluster of syntenic loci.

The extent of conservation of gene order is an inverse function of time since divergence from the ancestral locus, since chromosomal rearrangement is required to break up linkages. Rates of divergence vary considerably at all taxonomic levels. A notable example is the Japanese pufferfish, *Fugu rubripes*. Chosen for study largely on the basis of its unusually small genome size for a vertebrate (7.5 times smaller than the human genome), *Fugu* appears to have achieved this state by virtue of extensive DNA loss accompanied by rearrangements that probably shortened the average length of syntenic regions (Aparicio et al. 2002). *Fugu* and humans have a similar number of genes, and even short blocks of synteny are usually disrupted by the insertion of one or more genes between the conserved ones. Monocotyledenous grasses and dicotyledenous plant species also retain segments of conserved gene order, covering hundreds of kilobases, that are likely to be common to all plants (Mayer et al. 2001), but extensive gene duplication, gene loss, and local rearrangement disrupts synteny and complicates efforts to use comparative physical maps to clone genes of interest (Delseny 2004).

Chromosome fusions and splits, reciprocal translocations, and inversions all contribute to rearrangement of gene order (Ferguson-Smith and Trifonov 2007). Within the great apes, each chromosome is thought to have experienced an average of two or three rearrangements, many of which can be observed directly in karyotypes inferred from the banding patterns of mitotic chromosome spreads. This method has been enhanced dramatically by the development of the **chromosome painting** technique (Figure 1.4),

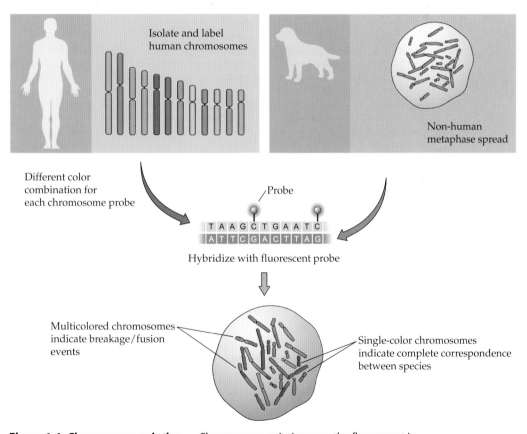

Figure 1.4 Chromosome painting.   Chromosome painting uses the fluorescent in situ hybridization (FISH) technique to detect DNA sequences in metaphase spreads of animal cells. The fluorescently labeled hybrid karyotype (bottom) shows some chromosomes that are a single color, indicating complete correspondence with the human chromosomes, while multicolored chromosomes indicate that a chromosome breakage or fusion event occurred and has contributed to chromosomal evolution between the species.

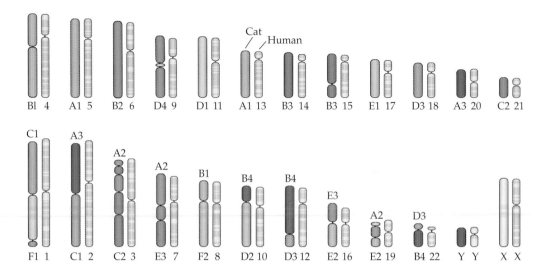

Figure 1.5 Synteny between cat and human genomes.   Ideograms for each of the 24 human chromosomes (22 autosomes plus the two sex chromosomes; lavender) are aligned against color-coded representations of corresponding cat chromosomes, as determined by a combination of FISH and RH mapping. Cat chromosomes are assigned to six groups (A–F) of 2–4 chromosomes each. The top row shows 12 autosomes that are essentially syntenic along their entire length, except for some small rearrangements that are not shown. The bottom row shows 10 autosomes that have at least one major rearrangement. The two sex chromosomes are essentially syntenic between cat and human. (After Murphy et al. 2000.)

in which each chromosome of one species is separately labeled with a set of fluorescent dyes that produce a unique hue, and hybridized to chromosome spreads of the other genome. Chromosome painting has been used to define regions of synteny covering segments of the order of one-tenth of a chromosome arm between dog or cat and human (Figure 1.5; Murphy et al. 2000). In addition to anchoring sequence scaffolds to cytological maps, this method is an important component of efforts to perform comparative mapping of species for which well-resolved physical maps are unavailable.

High-resolution comparative physical mapping has shown that even within megabase-long stretches of synteny between human and mouse that may contain hundreds of genes, local inversions and insertions/deletions involving one or a few genes may not be uncommon. Families of genes organized in tandem clusters are particularly likely to diverge in number, organization, and coding content between species (Dehal et al. 2001). Breakpoints between syntenic blocks often also harbor local duplicated segments of DNA, though it is not clear whether these are the cause of, or occur in conjunction with, genomic rearrangements (Bailey et al. 2004). Considerable size variation in intergenic "junk" DNA is also apparent.

The Human Genome Project

Genome science originated with a set of objectives conceived primarily to facilitate molecular biological research. With technological advances and increased involvement of private industry, the pursuit of these objectives emerged as a discipline in its own right. Historians and philosophers of science will undoubtedly debate the relationship between the acquisition of information versus knowledge, and critics will continue to lament the absence of hypothesis-driven research in genomics, but the field is unique in having a clearly defined set of criteria by which its impact can be judged. By the objective criteria of goal attainment, the Human Genome Project in particular has been one of the most successful publicly funded scientific endeavors in history. In the remainder of this chapter we explore its objectives and achievements, as well as those of mouse, invertebrate, plant, and microbial genome projects.

Objectives

The objectives of the publicly funded **Human Genome Project** (**HGP**) were established in a series of 5-year plans authored by leading figures of the U.S. National Institutes of Health and Department of Energy research planning groups (Collins and Galas 1993; Collins et al. 1998, 2003). These objectives are summarized in Tables 1.1 and 1.2 and can be found on the HGP Web site http://www.ornl.gov/sci/techresources/Human_Genome/home.shtml. Most of the goals of the first and second plans were achieved ahead of time and under budget, including completion of the first-draft human genome sequence. Achievements beyond those originally included in the plans, such as the emergence of structural genomics and association mapping of disease loci, will help shape the discipline over the next decade.

The initial objectives of the HGP can be summarized briefly as follows. *First*, the generation of high-resolution genetic and physical maps that have become an invaluable component of efforts to localize disease-associated genes. *Second,* the attainment of sequence benchmarks, leading to generation of a complete genome sequence by the year 2005. (Two draft versions were achieved in 2000, one by privately and one by publicly financed initiatives, but completion of the "finished" sequence—namely, an error rate of less than 1 in 10,000 nucleotides—of each chromosome took several more years.) *Third,* identification of each and every gene in the genome by a combination of bioinformatic identification of open reading frames (ORFs), generation of voluminous expressed sequence tag (EST) databases, and collation of functional data including comparative data from other animal genome projects. And *fourth,* compilation of exhaustive polymorphism databases, in particular of single-nucleotide polymorphisms (SNPs), to facilitate integration of genomic and clinical data, as well as studies of human diversity and evolution.

TABLE 1.1 *Initial Goals of the Human Genome Project*

From the First 5-Year Plan: 1993–1998

1. THE GENETIC MAP
   Complete 2 to 5 cM map by 1995
   Develop new technology for rapid and efficient genotyping

2. THE PHYSICAL MAP
   Complete STS map to 100 kb resolution

3. DNA SEQUENCING
   Develop approaches to sequence highly interesting regions on Mb scale
   Develop technology for automated high throughput sequencing
   Attain sequencing capacity of 50 Mb per year; sequence 80 Mb by 1998

4. GENE IDENTIFICATION
   Develop efficient methods for gene identification and placement on maps

5. TECHNOLOGY DEVELOPMENT
   Substantially expand support for innovative genome technology research

6. MODEL ORGANISMS
   Finish STS map of mouse genome to 300 kb resolution
   Obtain complete sequence of biologically interesting regions of mouse
     genome
   Finish sequences of *E. coli* and *S. cerevisiae* genomes
   Substantial progress on complete sequencing of *C. elegans* and
     D. melanogaster

7. INFORMATICS
   Continue to create, develop, and operate databases and database tools
   Consolidate, distribute, and develop software for genome projects
   Continue to develop tools for comparison and interpretation of genome
     information

8. ETHICAL, LEGAL, AND SOCIAL IMPLICATIONS (ELSI)
   Continue to identify and define issues and develop policy options
   Develop and disseminate policy regarding genetic testing
   Foster greater understanding of human genetic variation
   Enhance public and professional education programs on sensitive issues

9. OTHER
   Training of interdisciplinary genome researchers
   Technology transfer into and out of genome centers
   Outreach

A list of ancillary but no less important objectives includes: support for and fostering of technological innovation in the domains of sequencing methodology and gene expression analysis; bioinformatic research and application; database establishment and standardization; and support for

TABLE 1.2 *A Blueprint for the Human Genome Project*

Grand Challenges in the Third 5-Year Plan: 2003–2008

I. GENOMICS TO BIOLOGY

1. Comprehensively identify the structural and functional components encoded in the human genome
2. Elucidate the organization of genetic networks and protein pathways and establish how they contribute to cellular and organismal phenotypes
3. Develop a detailed understanding of the heritable variation in the human genome
4. Understand evolutionary variation across species and the mechanisms underlying it
5. Develop policy options that facilitate the widespread use of genome information in both research and clinical settings

II. GENOMICS TO HEALTH

1. Develop robust strategies for identifying the genetic contributions to disease and drug response
2. Develop strategies to identify gene variants that contribute to good health and resistance to disease
3. Develop genome-based approaches to prediction of disease susceptibility and drug response, early detection of illness, and molecular taxonomy of disease states
4. Use new understanding of genes and pathways to develop powerful new therapeutic approaches to disease
5. Investigate how genetic risk information is conveyed in clinical settings, how that information influences health strategies and behaviors, and how these affect health outcomes and costs
6. Develop genome-based tools that improve the health of all

III. GENOMICS TO SOCIETY

1. Develop policy options for the uses of genomics in medical and non-medical settings
2. Understand the relationships between genomics, race, and ethnicity, and the consequences of uncovering these relationships
3. Understand the consequences of uncovering the genomic contributions to human traits and behaviors
4. Assess how to define the ethical boundaries for uses of genomics

model systems genome science. A significant portion of the HGP budget in the United States has also been set aside for research on the ethical, legal, and social implications of genetic research (the ELSI project; see Box 1.1).

The current blueprint for the future of the Human Genome Project focuses more on conversion of genomic data into biological knowledge than achievement of specific objectives. The visual metaphor in Figure 1.6 pictures

Figure 1.6 The architecture of the Human Genome Project in the twenty-first century.   This visual metaphor shows how the three major themes for future genome research are founded on six pillars of genome resources. (After Collins et al. 2003; courtesy of Darryl Leja, National Human Genome Research Institute, NIH.)

genomics informing general biology, public health, and social issues, utilizing a foundation of six research pillars: genomic resources; technology development; computational biology; training; ELSI; and education. The fifteen grand challenges listed in Table 1.2 range from elucidation of the structure of genetic networks through prediction of susceptibility to disease to the development of policy options for ensuring equal access to the benefits of genome research.

The Content of the Human Genome

Completion of the first draft of the human genome sequence was announced at a press conference in May of 2000, but publication of this milestone was delayed until February of 2001 (IHGSC 2001; Venter et al. 2001). The intervening months were used to initiate a phase of refinement of the sequence assembly, including gap closure and verification of ambiguous aspects, along with gene annotation and prediction. The published draft genome fell well short of a definitive statement of the total gene content of the human genome, primarily because identifying coding sequences within random higher eukaryotic DNA sequences is not a trivial process. Estimates of the number of genes in the human genome initially ranged from a low of 30,000 to in excess of 120,000 genes. The current estimate centers around a consensus figure of 22,000 genes (IHGSC 2004).

This figure is not much greater than the gene content of two invertebrate multicellular eukaryotes—the fruit fly *Drosophila melanogaster* and the nematode worm *Caenorhabditis elegans*—and is less than four times greater than that of a unicellular eukaryote, the yeast *Saccharomyces cerevisiae* (Table 1.3). Much of the increase in size can be attributed to two rounds of whole-genome duplication, although independent expansion of certain gene families is also apparent. The complement of metabolic enzymes is essentially no different from that of other eukaryotes, while the numbers of genes involved in regulatory functions such as transcription, signal transduction, and intercellular signaling has expanded in proportion to the whole genome increase. Further analyses have confirmed that there are no dramatic differences in gene content between humans and other mammals, such as the mouse; in fact the human genome is notable for a *reduction* in some gene classes, such as olfactory receptors.

The first high-resolution genetic map of the complete human genome was published in September of 1994. This map established the existence of 23 linkage groups (one per chromosome) with 1,200 markers at an average of 1 cM intervals (Gyapay et al. 1994). This map was quickly followed by a physical map generated from 52,000 sequence tag sites (STS) at approximately 60 kb intervals that formed the scaffold for the top-down public sequencing effort (Hudson et al. 1995). A database of 30,000 expressed sequence tags, initially thought to represent the first collection of unique genes, was reported by Adams et al. in 1995, and several commercial consortia were also assembling orders-of-magnitude greater EST databases by that time. A collection of more than 3,000 SNPs was described by Wang et al. in 1998, but by 2004 this figure had exceeded 1.8 million mapped SNPs—in essence providing polymorphic markers at 2 kb intervals and placing 85%

TABLE 1.3 *Comparison of Gene Content in Some Representative Genomes*

	E. coli	S. cerevisiae	Drosophila[a]	C. elegans	A. thaliana[a]	H. sapiens[a]
Genome size[a] (Mb)	4.6	12.0	120+	97	115+	3,000+
Number of genes[b]	4,300	6,250	13,600	18,425	25,500	22,000
Average gene density (kb)	1.1	1.9	8.8	5.3	4.5	135
Number of gene families[c]	2,500	4,500	8,000	9,500	11,000	10,000

[a]*Drosophila*, human, and *Arabidopsis* genome sizes indicate sequenced euchromatin, excluding heterochromatin.
[b]Approximate number of genes estimated from original annotation, rounded to 25.
[c]Number of gene families rounded to 500.

BOX 1.1 The Ethical, Legal, and Social Implications of the Human Genome Project

The ELSI Program was established in January of 1990 after a combined working group of the National Institutes of Health and the Department of Energy recommended that research and education relating to the ethical, legal, and social implications of human genome research be incorporated as an essential component of the entire project from its inception. The National Human Genome Research Institute (NHGRI) now commits 5% of its annual budget to ELSI, funding three types of activity: regular R01 research grants, R25 education grants, and intramural programs at the NIH campus in Bethesda, Maryland. Web sites describing the program can be found at http://www.genome.gov/10001618 and http://www.ornl.gov/sci/techresources/Human_Genome/research/elsi.html, the latter documenting activities funded by the DOE.

The original mission statement of the project had four major objectives:

- Anticipation of the implications for individuals and society of sequencing the human genome.
- Examination of the ethical, legal, and social implications of obtaining the sequence.
- Stimulation of public discussion of the issues.
- Development of policy options that would ensure beneficial use of HGP information.

More practically, research is centered around four main subject areas:

- Privacy and fairness in the use and interpretation of genetic information (by insurers, employers, courts, schools, adoption agencies, and the military, among others).
- Clinical integration of new genetic technologies.
- Issues relating to design and implementation of clinical research, including consent, participation, and reporting.
- Public and professional education.

One pressing social and individual issue that has attracted wide attention and generated great concern is the privacy and confidentiality of genetic information. This topic is particularly prominent in countries such as Iceland and Estonia, where government-sponsored databases of medical records (including both genetic histories dating back centuries and newly obtained genotype data) have been supplied to medical research companies. Another area of concern is the psychological impact and potential for stigmatization inherent in the generation of genetic data, particularly in the context of racial mistrust and socioeconomic differences in gathering of and access to genetic information. Reproductive issues can arise relating to informed consent and the rights of parents to know versus their fear of knowing, as well as potential moral (and possibly even legal) obligations once data has been obtained.

The program works toward implementing clear and uniform standards for informed consent and the conduct of clinical research, against a backdrop of federal reluctance to regulate and public uncertainty about the new technology. Another goal is the education of self-help groups, educators, and the media about distinguishing between very complex statistical associations and oversimplified assertions about the nature-nurture debate, biological determinism, and heritability. Philosophical discussions cover issues ranging from the basis of human responsibility, to the human right to "play God" with genetic material, to the meaning of free will in relation to genetically influenced behaviors.

In the more "practical" arena are studies of economic, safety, and environmental issues in relation to the release of genetically modified organisms (GMOs) at a time when public mistrust of science is increasing, particularly in Europe; the commercial and legal issues associated with the patenting of genetic material, procedures, and data, including international treaties and obligations as well as the right to free access to published data; and the forensic implications of DNA profiles in legal issues from paternity testing to the presumption of innocence.

Web pages and documents providing more detail on each of these issues can be accessed through Web links at the ORNL site. In addition, at least two major university institutes have established useful Web sites relating to their involvement in the ELSI project. The University of Kansas Medical Center site (http://www.kumc.edu/gec/prof/geneelsi.html) provides a series of policy papers put out by the American Society of Human Genetics and the American College of Medical Genetics on issues as diverse as fetal screening, cancer testing, testing for late-onset psychological disease, genetics and adoption, population screening of at-risk populations, and eugenics. It also provides an extensive set of links to legislative acts dealing, for example, with health insurance portability, Americans with disabilities, and birth defects prevention, as well as links to other internet resources established by public policy institutes and journals. Medical genetics courses and education opportunities, including the 30 or so genetic counseling programs in the United States, are also listed.

The Lawrence Berkeley Laboratory's ELSI Project site (http://www.lbl.gov/Education/ELSI/ELSI.html) is more concerned with the development of educational materials and is developing a series of teaching modules on topics such as breast cancer screening, genetic patents and intellectual property, and personal privacy and medical databases.

In 1998, the ELSI Research Planning and Evaluation Group issued a new set of goals as part of the NHGRI reevaluation of the future of the Human Genome Project. The five new major aims are:

1. Examine issues surrounding the completion of the human genome sequence and the study of human genetic variation: How will SNP mapping affect our understanding of race and ethnic diversity? How can we balance individual rights with ongoing research needs? Are there new concerns relating to the commercialization of human genetics? and, How can we best educate professionals and the general public alike about the implications of human genetic variation?

2. Examine issues raised by the integration of genetic technologies and knowledge into health care and public health activities: Will genetic testing promote risky behavior or intolerance? What are the social implications of pharmacogenomics, the tailoring of treatments for complex conditions to genotype? Will genetic factors be overemphasized merely because they can be more objectively defined than environmental ones? and, What will be the impact of genomics on health care provision and insurance issues, morbidity and mortality, and reproductive behavior?

3. Examine issues raised by the integration of knowledge about gene-environment interactions in nonclinical settings: Are there conditions under which genetic testing should not be considered? What issues are raised by the storage of blood and other tissue samples by the military and police, among other groups? What legal issues arise in relation to adoption and child custody, and are they affected by prior exposure to cultural and environmental variables and pathogens? and, Can we identify potential abuses of genetic information in the workplace, in classrooms, and by the media?

4. Explore interactions between new genetic knowledge and philosophy, theology, and sociology: Will our appreciation of the place of humans in relation to other living creatures change? What are the implications of behavioral genetics for traditional notions of self, responsibility, and spirituality? and, Is lengthening the human life span likely and/or desirable?

5. Explore how socioeconomic factors and concepts of race and ethnicity influence the use and interpretation of genetic information, the utilization of genetic services, and development of policy: How are individual views about the impact of genetics influenced by ethnic and social factors? Will particular communities be more vulnerable to abuse or more likely to benefit from genomics? and, What are the most effective strategies to ensure that genetic counseling is provided in a culturally sensitive and relevant manner?

of all exons within 5 kb of a SNP. Today there are estimated to be in excess of 10 million common SNPs, and the latest haplotype map (based on 3.1 million SNPs genotyped in 270 individuals from Africa, Europe, and Asia) captures the vast majority of human genetic polymorphism (IHMC 2007).

The first draft chromosome sequence was also published in 2000 for the smallest human chromosome, chromosome 21 (Hattori et al. 2000), and finished quality whole chromosomes were published at regular intervals between 2003 and 2007. New builds of the genome sequence are updated on the public Web sites almost monthly, ensuring constant integration of the latest sequence information.

One question that is often asked in relation to the content of the human genome is: Whose genome was sequenced? The answer is that the sequence was derived from a collection of several libraries obtained from a set of anonymous donors. In the mid 1990s, ethical concerns were raised over the ramifications of one individual contributing the complete sequence. The issue was resolved to general satisfaction by reconstructing the core libraries. Both the International Human Genome Sequencing Consortium (IHGSC) and the private firm Celera Genomics reported that they assembled their sequence from multiple libraries of ethnically diverse individuals, although one particular individual's DNA contributed three-quarters and two-thirds of the raw sequence, respectively, as shown in Figure 1.7. Both groups adhered to strict privacy and consent guidelines when enrolling donors, whose identities remain unknown to researchers and are unavailable to the general public.*

The IHGSC made initial libraries from individuals of both sexes but, as reported in their paper, by chance only males were used in the assembly of the first draft. The ethnic identity of the eight individuals whose samples were included is unknown, since the final samples were chosen at random. By contrast, the Celera sample included at least one individual from each of four ethnic groups, as well as both males and females. Race is not regarded as a concern, both because ongoing genome diversity surveys have been far more informative with regard to identification of racial differences, and because it is well known that the overwhelming majority of human sequence variation is shared across the human races.

Extensive tracts of heterochromatin, which may account for as much as 20% of the total genome and are mostly located adjacent to the centromeres, will probably never be sequenced. Since the completion of the first draft, much of the sequencing focus of the HGP has been on characterizing human diversity. The International HapMap project (described in Chapter 3) is intended to map all of the major haplotypes in the human genome and to characterize their distribution among populations, as a step toward identification of human disease susceptibility factors. Agencies in Canada, China,

---

*In May of 2002, J. Craig Venter admitted to CBS' *60 Minutes* that in fact his own DNA contributed substantially to the Celera sequence. A finished version of his diploid genome was published in September of 2007 (Levy et al. 2007). He and his then-wife, Claire Fraser, continued the family tradition by sequencing the genome of their poodle, Shadow, as the first draft canine genome sequence. See Venter's 2007 autobiography *A Life Decoded* for a personal view of the Human Genome Project.

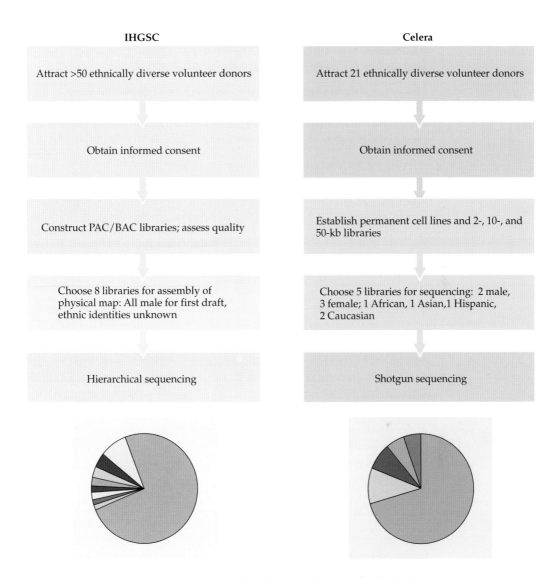

Figure 1.7 Whose genome was sequenced?   The contributions of individual donors to the draft human genome sequences, as reported in February of 2001 by IHGSC and Celera Genomics. The size of each shaded sector in the pie charts is in proportion to the amount of sequence contributed by a single individual. Note that 8% of the IHGSC sequence was derived from clones of unpublished origin. Identification of SNP diversity, discussed in Chapter 3, uses a much wider sample of human variation.

Japan, Nigeria, the United Kingdom, and the United States have developed a public resource (http://www.hapmap.org) that allows researchers around the world to access the latest information concerning the structure of variation in the human genome. Plans to generate 1,000 human genome sequences from all over the world were announced in early 2008.

EXERCISE 1.2 *Use the NCBI and Ensembl genome browsers to examine a human disease gene*

Choose a human disease of interest to you, and then use the OMIM site to identify a gene that is implicated in the etiology of the disease. Then use the NCBI, UCSC, and Ensembl genome browsers to answer the following questions about the gene:

a. *What are the various identifiers (aliases) for your gene?*

b. *Where is the gene located on the chromosome (cytologically and physically)?*

c. *What is the Reference Sequence (RefSeq) for the gene?*

d. *How many exons are there in the major transcript, and how long is it?*

e. *What is known about the function of the gene?*

f. *Do the three annotations agree? Which browser do you prefer, and why?*

SAMPLE ANSWER: *Go to http://www.ncbi.nlm.nih.gov, click on OMIM, and type "asthma" into the search field. Scroll through the genes until you find one of interest, for example Interleukin-13 (IL13), and click on the blue number (*147683). This will bring up a page with a lot of textual information, but you can also link out to various sites, including Entrez Gene, which tells you that the LocusID for IL13 is 3596 and that its cytological location is 5q31 (chromosome 5, long arm, position 132.1 Mb). The gene is also called ALRH, BHR1, and P600. The RefSeq is NM_002188, an mRNA sequence. There appear to be four exons in the major transcript, which is 1,282 bp long and encodes a protein of 146 amino acids, 34 of which are for the signal peptide. The protein is an immunoregulatory cytokine that regulates B-cell maturation and differentiation and the inflammatory response in the presence of allergens.*

All of this information can also be obtained by linking out to the Primary Source from the Human Genome Organization Gene Nomenclature Committee (HGNC: 5973), which links to the http://www.ensembl.org and http://genome.ucsc.edu pages. Alternatively, you can access these sites directly and search for the genes by typing "IL13" into the search engines. The Ensembl gene identifier is ENSG00000169194, and the gene view indicates that the transcript spans almost 3 kb. A number of gene ontology functions are listed as well. The UCSC site displays polymorphisms and conservation plots as well as a large number of other tracks.

Internet Resources

There are three major Web browsers for accessing genome data: (1) the U.S. National Institute of Health's Entrez Life Sciences Search Engine; (2) the European Bioinformatics Institute's Ensembl Genome Browser; and (3) the

University of California at Santa Cruz's UCSC Genome Bioinformatics Site. Each of these is each built around the human genome, but all of them also provide comparable access to annotated genome sequences of a wide range of other species. Each of the major model organisms described later in this chapter also has dedicated browsers that include other attributes such as genetic data, literature links, and scientific images. Exercise 1.2 provides an opportunity for you to compare these genome browsers.

The Entrez site, http://www.ncbi.nlm.nih.gov/sites/gquery, is a service of the U.S. National Center for Biotechnology Information (NCBI). There are now 26 core search pages that coordinate information relating to such topics as gene and mRNA sequences, proteins, gene expression, chemical screens, and polymorphism. The Human Genome Resources site links to data and resources including a convenient map viewer and a tutorial with step-by-step instructions for various types of search. As of June 2008, there are 148 animal, 42 plant, and 152 protist or fungal genomes in various states of completion that can be viewed either by scrolling along a schematic chromosome map, by searching for sequence matches, or in most cases via a genome resource page dedicated for a specific organism.

The NIH (www.ncbi.nlm.nih.gov; Figure 1.8) also maintains GenBank and PubMed. **GenBank** is a genetic sequence database that provides an annotated collection of all genomic and cDNA sequences that are in the public domain (Box 1.2; Benson et al. 2007). These can be searched using a program called **BLAST** in subsets of the whole database that include only nonredundant (NR) sequences; only highly curated reference sequences for well-described genes (RefSeq); expressed sequence tags (ESTs, short fragments of cDNAs); genome survey sequences (GSS, short reads of genomic DNA); HTGS (high-throughput genomic sequences, corresponding to nonassembled reads of whole genomes); and assembled genomes. **PubMed** is a literature search engine providing access to all published medically related abstracts and journals, and PubMed Central is an open-source repository where all NIH-funded research papers must be made freely available within a year of publication.

The Ensembl *e!* Browser (www.ensembl.org) is a joint project between the European Bioinformatics Institute (part of the European Molecular Biology Organization) and the Sanger Institute in Cambridge, England, and is primarily funded by the Wellcome Trust (Birney et al. 2004). There are at least 39 animal species represented, including such unusual characters as the platypus, lesser hedgehog, and Caribbean lizards. Most of the genome sequences are processed automatically through annotation pipelines, but the sequences of five species (human, mouse, zebrafish, pig, and dog) are highly annotated by manual curation at the VEGA (Vertebrate Genome Annotation) database.

Many users find the Ensembl interface friendlier than Entrez. A nice feature is a customizable home page that allows you to keep track of searches and browsing over multiple sessions from different computers. BioMart is a tool for extracting and downloading user-specified aspects of the data in various formats suitable for high end bioinformatics analysis. Gene specific

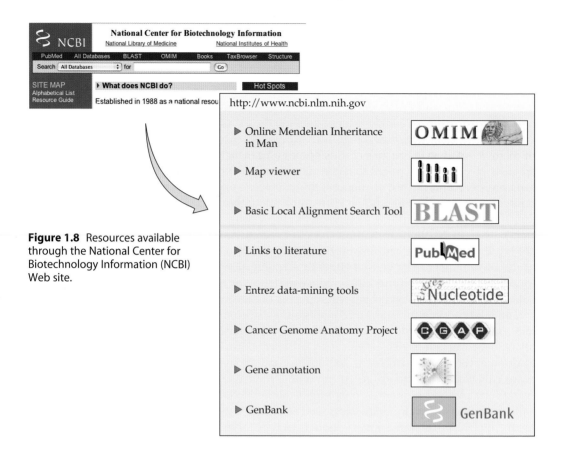

Figure 1.8 Resources available through the National Center for Biotechnology Information (NCBI) Web site.

searches are conducted directly from the home page by selecting your species of interest and then entering the gene name. The resulting page links out to various attributes and a ContigView map that allows you to browse the gene from chromosomal interval all the way to the sequence.

The UCSC Genome Bioinformatics site (http://genome.ucsc.edu) includes a highly configurable browser that provides users with an immediate visualization of hundreds of different genome features (Kuhn et al. 2007). The main page asks you to select the "tracks" that you wish to display, and experienced researchers can also add their own tracks. These tracks are grouped into seven main categories of data plus another six categories relating to the ENCODE project (described in Chapter 2). Tracks can be hidden or displayed in short or expanded views. The categories are: mapping and sequencing; phenotype and disease association; genes and gene predictions; transcript evidence; gene expression data; comparative genomics; and sequence variation. These choices are displayed behind the main graphical display of a region of the genome under consideration and can be updated at will. All of the data behind the tracks can be retrieved in text format using the Table Browser. Sequence searches are performed using a program called BLAT that looks for very high conservation in short 25 nucleotide blocks, rather than by alignment of the whole query sequence, as in BLAST. Another original feature is the Gene Sorter, which provides a table of genes related to the query based on a number of attributes chosen by the researcher, such as sequence, expression, or functional similarity.

Medical and clinical data is integrated with genome data through the **Online Mendelian Inheritance in Man** (**OMIM**) Web site (Figure 1.9), which grew out of an exhaustive catalog of human genetic disorders (McKusick 1998; Hamosh et al. 2000). OMIM is a searchable database that typically provides text summarizing recent genetic research in response to a query about a particular disease. This data is primarily intended for physicians and human geneticists, but a less technical "Genes and Disease" site also exists within the NCBI resource that compiles information according to disease types such as muscle, metabolism, cardiovascular, and psychological disorders. OMIM lists in excess of 19,000 entries, including 4,400 known Mendelian disorders, of which almost 2,800 are linked to genes.

A major objective of the NCBI is to allow researchers in remote locations to be able both to submit their own data for inclusion in the growing databases, or simply to query and search data submitted by others. Increasingly, software for performing complex bioinformatic analyses is incorporated into Web sites so that virtual experiments can be conducted directly over the internet without the need to download voluminous databases. For example, the GEO BLAST tool allows a researcher to search for all genes in the

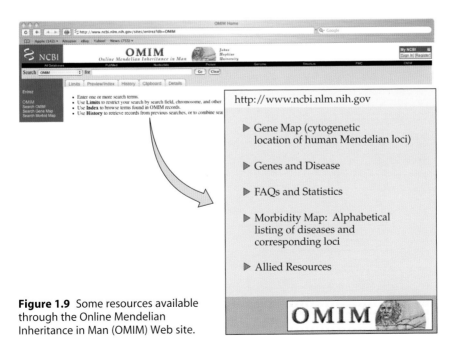

Figure 1.9 Some resources available through the Online Mendelian Inheritance in Man (OMIM) Web site.

BOX 1.2 GenBank Files

There are as many ways to present the structure and annotation of a gene or sequence as there are genomics-related Web sites. A basic problem is that there is often no single correct structure for a given gene, due to alternate splicing and transcription start sites, errors that may occur during cloning, and the fact that algorithms underlying gene-structure prediction software are imperfect. Furthermore, all genomes are full of polymorphism. Thus the same gene may be represented by multiple different sequences or annotations in the genome databases.

The National Center for Biotechnology Information (NCBI) has adopted a policy of accepting all sequences that are submitted to GenBank, but has developed a standard form of annotation that carries the implication "user beware." Hand curation by experts is required before any one sequence is elevated to the status of "RefSeq," or Reference Sequence; all other sequences should be regarded as supporting evidence.

All Entrez nucleotide files in GenBank share a number of features, which we illustrate using the human *HOXA1* file. Access the file by typing the identifier 11421562 into the search field (select "Nucleotide" in the Entrez pulldown menu) at http://www.ncbi.

nlm.nih.gov, and then click on the accession number (for example, XM_004915), which will bring up the view shown below.

1. The top few lines define the **locus**, including the length of the sequence; whether the sequence is derived from mRNA or genomic DNA; the date of submission; the **definition**, including the species from which the sequence was obtained and the common name of the gene; the complete taxonomic classification of the organism; and the source tissue.

2. This information is followed by the **reference**, with the names of the **authors**, the **journal**, and the **title** of the article in which the sequence was or will be published, and any historical notes on updates that have been submitted directly to the NCBI.

3. The meat of the file is the **features** section, which has subheadings describing the known extent of the gene; the coding sequence (CDS) including the predicted protein sequence; and miscellaneous features (**misc_feature**) such as intron-exon boundaries, identified protein domains, variations,

Partial Entrez nucleotide file (GenBank).

mutations (if the sequence is derived from a mutant strain), and alternate transcripts. All of these are associated with links—for example, to OMIM, the HGNC GeneID page, or the PFAM protein domain database file. Longer-sequence "scaffold files" span hundreds of kilobases (for example, *HOXA1* is contained within PAC file AC_004079) and annotate each predicted gene. Scaffold files also include pointers to sequence-tagged sites, repeats, and miscellaneous other features.

4. Next comes the sequence itself, in blocks of 10 bases, 60 bases per row, with a running tally of the site number at the beginning of each row. Users have the option of displaying the sequence in FASTA format (which is simply an uninterrupted sequence of letters following a header) or downloading it as a text file by clicking on the "Display" and "Send to" boxes at the top of the page. XML and ASN.1 files can be used to export the file to certain bioinformatics applications.

For a thorough description of all the elements of a sample GenBank record, see http://www.ncbi.nlm.nih.gov/Sitemap/samplerecord.html; an exhaustive description of the features table is found at http://www.ncbi.nlm.nih.gov/collab/FT.

There are numerous ways to view the structure of a gene graphically. Starting from the GenBank nucleotide file, link out to the EntrezGene file by clicking on the GeneID (in this case 3198) link. This will display a very low-resolution image of the transcripts associated with the gene (in the case of human *HOXA1*, two transcripts are seen). Next, click the MapViewer link on the right-hand side of the GeneID page, which displays various models of the gene in vertical orientation and allows you to zoom in or out along the chromosome. A highlighted set of links point to sequence view (sv), protein reports (pr), download (dl), evidence view (ev), model maker (mm), homologene (hm), and other pages.

Other databases and genome browsers can be accessed directly from the HGNC GeneID page; the link is found near the top of the Entrez Gene page, on the line labeled Primary Source. In the "Database Links" section of the HGNC page one finds a link to the Ensembl GeneView report, the top of which is shown below. GeneView presents a great deal of useful information on a single page, including links to orthologs across many species, transcript data, gene and protein structure information, and graphical representations of the gene. Also from the HGNC Database Links one can reach the UCSC Browser, which is particularly useful for comparing alignments across multiple mammals. In HGNC's "Gene Symbol Links" section one finds GeneCards, a useful site that provides access to an ever growing array of data about any particular gene.

Partial Ensembl GeneView report.

gene expression database that have similar sequences, and then compare levels of expression of the genes across species and experimental conditions. Clearly, the utility of these resources is heavily dependent on the quality of the data they contain. Thus an important new area of bioinformatic research is the development of procedures for efficient database quality control.

Animal Genome Projects

The International Sequencing Consortium (ISC; http://www.intlgenome.org) maintains a database of animal and plant genome sequencing projects supported mainly by funding agencies in the United States, England, France, Japan and Canada. In addition to the projects described below, sequencing is well advanced for the gallery of organisms shown in Figure 1.10, including mosquitoes, honeybees, water fleas, a frog, a fruit bat, sea squirts and urchins, cows, a possum and a wallaby, and several primates. The wide phylogenetic representation of organisms reflects the notion that analysis of patterns of sequence conservation and evolution will be an indispensible tool in making sense of the human genome. Draft genome sequences can be produced for most animals within 3 to 6 months, and may soon only take days. The NHGRI has a policy of rapid public dissemination of the data so that each community can immediately begin the task of genome annotation.

Primate Genome Projects

Rhesus macaques (*Macaca mulatta*) diverged from humans 25 million years ago, yet are sufficiently similar in so many aspects of their physiology, immunology, and neurobiology that they have become an essential model for infectious disease and vaccine research. The average sequence identity with human genes is 93%, compared with 99% for the chimpanzee, so the complete genome sequence (RMGSAC 2007) provides a convenient reference point for defining constrained sequences, for polarizing the identity of gains and losses of genes in higher primates, and for inference of natural selection on components of the immune system and other genes. Levels of polymorphism are similar to those found among humans, but there appears to be much greater population structure, with Indian and Chinese macaque populations diverging considerably throughout the genome. One particularly striking observation is that 229 amino acid substitutions that lead to severe human diseases (including mental retardation and cystic fibrosis) appear to be the normal wild-type allele in macaques.

The draft sequence of Clint, a member of our closest relative species, the chimpanzee (*Pan troglodytes*), has also been used to inform our understanding of human evolution. For example, Clint's genome sequence revealed far more divergence due to insertion-deletion polymorphism than had previously been appreciated (CSAC 2005). Although less common than SNPs, there are approaching 5 million indels that collectively affect approximately 3% of the euchromatin of humans and chimps. Overall, purifying selection and genetic drift are inferred to dominate the landscape of primate genome

Chimpanzee
Pan troglodytes

Rhesus macaque
Macaca mulatta

Domestic dog
Canis familiaris

Tammar wallaby
Macropus eugenii

Pufferfish
Tetraodon nigroviridis

Purple sea urchin
Strongylocentrotus purpuratus

Sea squirt
Ciona savignyi

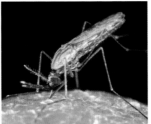

Platypus
Ornithorhynchus anatinus

Mosquito
Anopheles gambiae

Honeybee
Apis mellifera

Figure 1.10 A gallery of some animal genome sequencing projects.

evolution, with almost 30% of all chimp protein sequences being identical to that of their human orthologs. Most other proteins differ by only a few amino acids. Those that are more divergent have probably experienced positive selection, including proteins that offer protection from the malaria parasite and pathogenic microbes, sperm proteins, and pain receptors. A major insight offered by the chimp sequence is that it helps geneticists to infer whether disease-associated variants in the human genome are derived or ancestral, since the chimp allele is assumed to have been inherited from our common ancestor.

Two research groups have announced first-pass, high-throughput sequences of Neanderthal genomic DNA from the leg bone of a 38,000 year old Croatian fossil (Green et al. 2006; Noonan et al. 2006). Both studies conclude—not without controversy—that modern humans had diverged from Neanderthals by 350,000 years ago (considerably before humans arrived in Europe), and that little if any admixture between the primitive hominids occurred after that time. Further whole-genome sequences of multiple Neanderthals will soon add significantly to our perspective on contemporary diversity.

Low-quality draft sequences are available for several other primates, including the gorilla, gibbon, and orangutan. Those of two primitive primates, the bushbaby or galago (*Otolemur garnettii*) and the mouse lemur (*Microcebus murinus*), are available through the Ensembl genome browser.

Rodent Genome Projects

Almost a century of genetic research on mice and rats has ensured that these small mammals will occupy a central place in genome research. Three major advantages of rodent research are:

1. The existence of a large number of mutant strains that, combined with the potential for whole-genome mutagenesis, will lead eventually to genetic analysis of every identified locus in the genomes.

2. The existence of a panel of approximately 100 commonly used laboratory mouse strains with well-characterized genealogy—a formidable resource for the study of genetic variation and complex quantitative traits.

3. The evolutionary position of rodents, which are sufficiently divergent at the DNA sequence level from humans that the existence of conserved sequence blocks is generally an indicator of functional constraint, yet are sufficiently close to humans that many aspects of development, physiology, and the genetics of disease are shared.

The first public draft of the mouse genome sequence was published in late 2002 (MGSC 2002), and that of the rat followed within 18 months (RGSPC 2004). Like the human, these genome sequences are constantly updated on their respective NCBI, UCSC, and Ensembl browsers.

Although mutant strains of mice have been a productive source of material for genetic research for the better part of a century, functional genomic analysis of this model organism has been stimulated by three major advances achieved in the 1990s:

- First, the technology for targeted mutagenesis by homologous recombination of the wild-type locus with a disrupted copy has become routine. Time and expense are the only major obstacles to reverse genetics (moving from gene to phenotype), a strategy that is only likely to increase in popularity as gene expression profiling and comparative mapping define candidate genes for numerous traits.

- Second, several groups around the world have embarked on saturation random mutagenesis programs—that is, screens conducted on such a large scale that a point is reached where most new mutations occur in loci already defined by an existing mutation. Dominant mutations are much easier to identify than recessive ones, but even recessive mutations can be recovered in F_3 designs, in which the researcher looks for one-fourth of the grand progeny to show an aberrant phenotype in a sibship, as will be discussed in Chapter 5.

- Third, the expense associated with colony maintenance has led to emergence of "phenomic" analysis, in which mutagenized lines are subject to batteries of biochemical, physiological, immunological, morphological, and behavioral tests in parallel by large research consortia. This approach offers the best prospect for large-scale identification of those genes required for non-lethal phenotypes, supplementing traditional analysis of embryogenesis, skeletogenesis, and skin and coat defects.

Reflecting the centrality of laboratory-bred strains in mouse genetics, most of the genomic resources for mice are organized through a Web site at the Jackson Laboratory in Bar Harbor, Maine (http://www.informatics.jax.org). This Mouse Genome Informatics site (MGI; Figure 1.11) includes a central role for physical and genetic maps, as well as search engines that allow searches by key word, accession numbers, expression patterns, and genomic

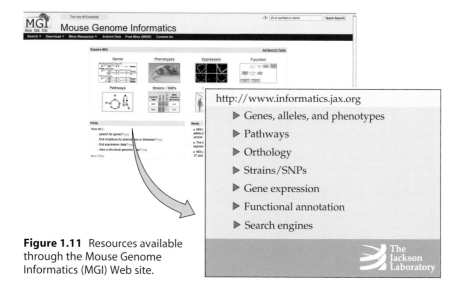

Figure 1.11 Resources available through the Mouse Genome Informatics (MGI) Web site.

location. Tumor biology is also prominently displayed, reflecting the high incidence of tumor classes specific to particular strains of mice. Online comparison of the mouse and human genome sequences facilitates identification of likely regulatory sites and supports gene annotation, as demonstrated by Loots et al.'s (2000) characterization of a coordinate regulator of interleukin expression (Figure 1.12).

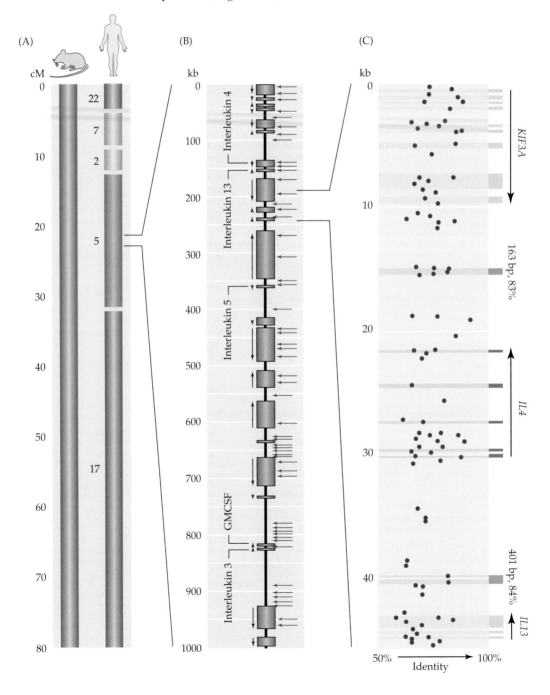

EXERCISE 1.3 *Compare the structure of a gene in a mouse and a human*

Using either the NCBI or Ensmbl Browser, explore the structure of the gene you used in Exercise 1.2 in a mouse and a human (and if possible, in other vertebrates).

SAMPLE ANSWER: *From the Ensembl homepage (http://www.ensembl.org) find the Gene Report for Human IL13 by typing ENSG00000169194 into the search box. The top hit should link to the GeneView page for Human IL13, which allows you to choose alignments with various species or combinations of species. Click on "View Genomic Alignment with Mus musculus" which will bring up graphics of chromosome 5, an overview of a 1-Mb window around IL13, and a "Detailed View" of 3 kb encompassing the gene. You can use the "Zoom" feature to explore the alignment in lesser or greater detail, or navigate to other multiple alignments and sequence feature annotations.*

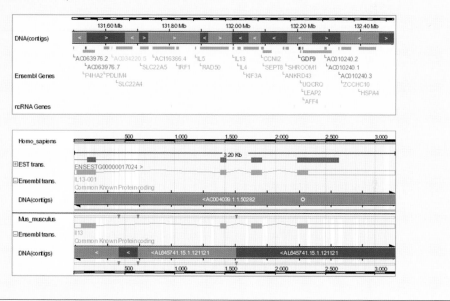

◀ **Figure 1.12 Mouse-human synteny and sequence conservation.** Conservation of gene order and DNA sequence between the human and mouse genomes is observed at three levels. (A) Blocks of synteny between mouse chromosome 11 and parts of five different human chromosomes are indicated (B) Enlarged view of a small region corresponding to the human 5q31 interval. In this approximately 1-Mb region, there is almost perfect correspondence in the order, orientation, and spacing of 23 putative genes, including four interleukins. Within the region, 245 conserved sequences of more than 100 bp with 70% identity were detected, many of which fall in noncoding regions (red arrows). (C) Enlargement of the alignment of 50 kb that includes the genes *KIF-3A*, *IL-4*, and *IL-13*. Blue dots show the distribution of conserved sequences (sequences more than 100 bp long with from 50%–100% identity) between mouse and human. Two of the conserved blocks (red bars, with the indicated levels of identity) fall between genes, whereas most of the others (blue bars) are in the introns and exons of the genes. Such alignments are readily prepared online using PipMaker (**http://pipmaker.bx.psu.edu/pipmaker/**; Schwartz et al. 2000). (B and C after Loots et al. 2000, Figures 1 and 2.)

Three unique features provided by the MGI site are the Pathways, Strains/SNPs, and Orthology pages. Strains/SNPs provides access to the standard laboratory strains, as well as a collection of recombinant inbred lines derived from several of them, that have already been typed for a high density of molecular markers. This information facilitates both quantitative trait dissection and the mapping of new Mendelian mutations as they are generated. Pathways is a browsable database of biochemical pathways known as MouseCyc; and Orthology is a rodent-centric tool for finding related genes in other species.

Other Vertebrate Biomedical Models

Two vertebrate genomes fully sequenced in 2004 included the red jungle fowl (chicken; *Gallus gallus*) and the dog (*Canis familiaris*). The major motivations for sequencing both of these species were biomedical: chickens are an important model for oncogenesis and virology (which also has enormous agronomic implications), while dogs are emerging as an essential model for a wide range of complex diseases such as dermatitis, parasite infection, cancer, arthritis, diabetes, and behavioral disorders. From a biological standpoint, the availability of these genomes was also expected to boost research into avian evolution and poultry science, as well as studies of the impact of artificial selection on canine breed diversity (ICGSC 2004).

More Mendelian disorders have been identified in dogs than in any other vertebrate, and many of these are breed-specific. Interestingly, much effort was put into identifying the most highly inbred dog before sequencing began so as to minimize the impact of polymorphism on assembly of the genome sequence (the lucky canine turned out to be a boxer named Tasha). It is possible to distinguish breeds on the basis of polymorphic microsatellite sequences (Parker et al. 2004) and the degree of sharing and differentiation of the genetic material facilitates dissection of the obvious morphological, physiological, and behavioral differences between chihuahuas and greyhounds, or between beagles and sheepdogs (Ostrander and Kruglyak 2000). For example, a single mutation in the *Igf1* insulin-like growth factor gene contributes to small size in most small breeds (Sutter et al. 2007).

For the purposes of genetic analysis of vertebrate development, the zebrafish *Danio rerio* is an excellent complement to the mouse. Rapid and transparent embryogenesis, ease of culture, the existence of a dense genetic map, and a preponderance of available cell biological tools have established it as a model for the study of embryogenesis, neurogenesis, and organogenesis in particular. Saturation mutagenesis screens have uncovered thousands of genes that are required for the proper development of organs such as the heart and eye; of the musculature and appendages; for axon guidance; and for body symmetry and simple behaviors. In addition, screens for maternal-effect loci are expected to help dissect fundamental aspects of pattern formation in the embryo. The zebrafish genome project can be accessed through the Zebrafish Information Network (http://zfin.org) which provides tools for functional genomic analysis, including extensive anatomical and gene expression atlases, high-density polymorphism databases for fine-structure recombination mapping, and online strain resources. Extensive sequence analysis of two evolutionarily divergent pufferfish with unusually small genomes, *Tetraodon nigroviridis* and *Fugu rubripes*, is also well advanced, while genetic maps are under development for a variety of ecologically and commercially important fish species, including tilapia and other cichlids, sticklebacks, and salmonids.

Animal Breeding Projects

Genome projects to enhance animal and plant breeding efforts are being pursued, building on decades of classical genetic analysis. The Online Mendelian Inheritance in Animals database (http://www.ncbi.nlm.nih.gov/sites/entrez?db=omia) brings together linkage data and genetic maps for over a dozen species of agricultural importance. The site provides search engines that allow researchers to access data on inheritance patterns, molecular linkage, and molecular biology. Information can be accessed by disease or trait, as well as by species.

Although whole-genome sequencing of the cow is complete and that of the horse and pig are well advanced, sequence analysis of farm animals has concentrated on the development of high-density genetic maps and EST databases that aim to identify tags homologous to genes that have been identified in model organisms. Characterization of polymorphism within and among breeds will just as importantly support advancement of quantitative genetic analysis, a discipline that has its origins in animal and plant breeding, and is of ever increasing economic importance. Although Mendelian loci that lead to disease and mortality are a great burden on agriculture, the future benefits of breeding programs lie in improvements in yield, infectious disease resistance, adaptation to climatic conditions, and improved food quality, not to mention maximizing the benefits of transgenic technology. These goals will be met both through enhanced genetic map development and through association studies using SNP technology, as discussed in Chapter 3.

Each of the major farm animal genome projects has its own site or sites. In the United States these are supported by the Agricultural Research Service of the U.S. Department of Agriculture and by the initiatives of individual research groups, many of which are associated with veterinary schools. The Roslin Institute in Edinburgh has overseen the establishment of a series of ArkDBs (http://www.thearkdb.org; Hu et al. 2001), that now provide genome resources for 12 species: cat, chicken, cow, deer, duck, horse, pig, quail, salmon, sea bass, sheep, and turkey. Most of these have nodes, or alternate sites, organized in Europe, Japan, and Australia. Each site provides continuously updated chromosome maps, data from radiation hybrid mapping panels, and marker databases that enable researchers anywhere in the world to carry out linkage studies with the most current marker densities. Organism-specific resources include information on common breeds, meeting and workshop announcements, and links that facilitate comparative genome analyses.

Invertebrate Model Organisms

The first genomes of multicellular eukaryotes to be sequenced in their entirety were those of the nematode worm *Caenorhabditis elegans* (*C. elegans* Sequencing Consortium 1998) and the fruit fly *Drosophila melanogaster* (Adams et al. 2000). These remarkable technological achievements were undertaken in part as proof-of-principle for the sequencing of considerably larger vertebrate genomes and in part for the intellectual excitement of learning what makes a complex organism tick, but primarily as support for traditional molecular genetic research. The subsequent sequencing of a dozen different *Drosophila* species was undertaken largely as a model for how to use comparative genome data to functionally annotate genomes (*Drosophila* 12 Genomes Consortium 2007; Stark et al. 2007), while the sequencing of several other nematode species has more direct biomedical and agronomic importance from the perspective of parasite biology. Most of the tens of thousands of investigators studying these organisms spend several hours each day mining genome data through the FlyBase (http://www.flybase.org) and WormBase (http://www.wormbase.org) Web sites.

A concerted effort has been made to standardize the look and functionality of plant and animal databases. This effort is led by the Generic Model Organism Database (GMOD) project (Figure 1.13). This resource provides web tools and open source software for database construction and management. The Apollo and GBrowse visualization tools provide a familiar environment for examining gene structure (illustrated for *Drosophila* in Figure 1.14), while other modules facilitate exploration of gene ontology and gene expression. Some of the computational issues involved in relational and object-oriented database schemas are discussed in Box 1.3.

Comparison of the content of the fruit fly and nematode genomes confirmed some long-held suspicions, but also revealed a few surprises. Unexpectedly, there are perhaps as many as 50% more genes in the nematode genome than in the fly genome (19,000 versus 13,500), despite the fact that the fly is much more complex at several levels, including number of cells, number of cell types, and nervous system organization. Until recently, it was thought that *C. elegans* represented an ancient, evolutionarily conserved mode of development, but phylogenetic revision suggests that the nematode is most likely to be a highly derived molting protostome, or ecdysozoan. It is possible that an increase in gene number accompanied the evolution of its largely invariant mode of development. Some of the difference in gene number can be attributed to the expansion and contraction of particular gene families, such as the surprising surplus of steroid-hormone receptors in the nematode and the expansion of the olfactory receptor family in *Drosophila*. In any case, it is clear that there is no simple relationship between gene number and tissue complexity—or for that matter, between gene number and DNA content. Possibly the most profound revelation is the high degree of conservation in both species of all of the major regulatory and biochemical pathways, most if not all of which also are identifiable in the unicellular eukaryote *Saccharomyces cerevisiae* and in the various vertebrate genomes (Carroll et al. 2004). However, the expansion and contrac-

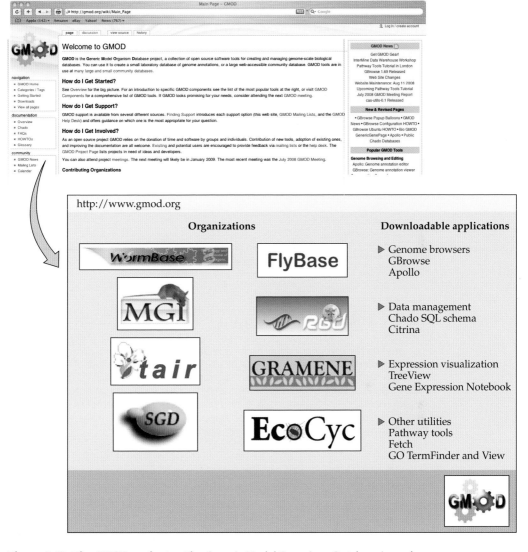

Figure 1.13 The GMOD project. The Generic Model Organisms Database is a collaboration between numerous organizations (indicated in the lower panel) with the aim of providing open source software for genome database construction. Some of the initial applications available for download are indicated on the right.

tion of gene families has played a crucial role in the evolution of development, as discussed by Cañestro et al. (2007) and others.

From the point of view of functional genomics, a major impact of the invertebrate genome projects is the prospect of obtaining identifiable mutations for every single gene of the genomes. In flies, this is being achieved by a combination of saturation mutagenesis and construction of a library of

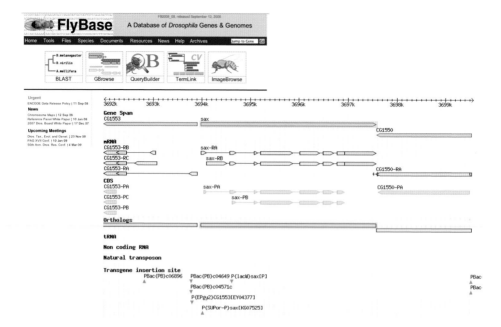

Figure 1.14 *Drosophila* gene annotation.   A typical GBrowse view shows 3 annotated genes in a 7.5-kb region of cytological band 43E of *Drosophila melanogaster*, centered on the *saxophone* gene. Each gene either has a number beginning with CG, or is identified by its standard name (e.g., *sax*). The predicted structure of each set of transcripts is shown, along with the orientation on the chromosome. Further annotation tracks indicating the sites of transposable element insertions (at bottom of image shown here), repeat elements, tRNAs, and predicted genes in *Drosophila simulans*, can be called up at will through a list of "tracks" below the image on the Web page (not shown) http://www.flybase.org/cgi-bin/gbrowse/dmel.

overlapping deficiencies that remove every segment of each chromosome. Some elegant genetic trickery has been employed to enable targeted mutagenesis as well. In the nematode, saturation mutagenesis has been supplemented with RNAi technology, in which double-stranded RNA can literally be fed to the worms in their diet of *E. coli*, with the result that function of the corresponding gene is more often than not reproducibly reduced, if not eliminated. As described in Chapter 5, over 85% of the genes on the five *C. elegans* chromosomes have been knocked out—a forerunner of ambitious functional genomic analyses in other model organisms. These resources are backed up by stock centers at the University of Minnesota (nematodes), and at Indiana University and in Kyoto (flies) that are the lifeblood of invertebrate genetic research. Molecular resources such as SNP databases, probe and genomic clone collections, monoclonal antibody collections, and resources for microarray construction are supported by the respective genome projects.

The publication of genome sequences has bolstered the proposition that invertebrates are not just models for the study of development and physi-

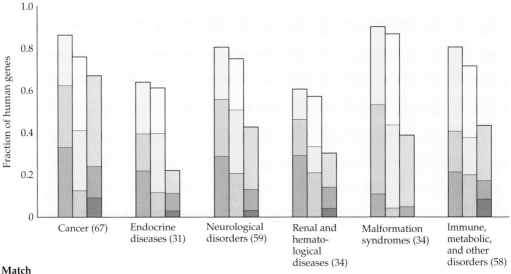

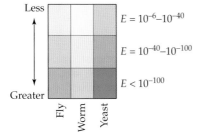

Figure 1.15 Human disease genes in model organisms. Histograms showing the fraction of genes implicated in several classes of human diseases with orthologous genes in the fly, worm, and yeast genomes. Match significance is indicated by depth of color, with the darkest shade indicating the highest significance. Numbers in parentheses indicate the total number of genes in each disease class. (After Rubin et al. 2000.)

ology, but can shed direct light on human disease (Rubin et al. 2000). More than 60% of a representative sample of 289 human genes that are mutated, amplified, or deleted in human diseases have an ortholog in the genome of *D. melanogaster*. The corresponding number is only slightly lower for *C. elegans* and, despite the fact that the yeast *S. cerevisiae* is unicellular, some 20% of human disease genes have orthologs even in that organism. Figure 1.15 shows the fraction of human disease genes in each of six categories that have orthologs in the fly, nematode, and yeast genomes, as detected by sequence similarity at three levels of significance within protein domains. Extrapolating from observations of the conservation of genetic interactions across the animal kingdom, we can confidently expect the genetic analysis of flies and worms to help uncover genes that interact with known disease-promoting loci in humans. It is important to appreciate that there is no presumption that the invertebrate trait is the same as a mammalian trait; rather, the fact that molecular interactions between gene products can be conserved even when they affect distinct processes allows the functional comparison of genes across species (Bier 2005).

BOX 1.3 Managing and Distributing Genome Data

As with many areas of science and technology, genome science has benefited greatly from advances in computing capabilities. Improved computational speed has been important, but a strong argument can be made that the growth of the internet has been even more crucial for genome scientists. In conjunction with the maturation of modern database technology, the World Wide Web has become the natural medium for managing and distributing genomic data.

The emergence of the internet allowed the creation of centralized data warehouses. Just as important, it led to the creation of shared public resources for searching and analyzing the contents of genomic databases. Full-featured Web sites such as those at NCBI (http://www.ncbi.nlm.nih.gov) and EBI (http://www.ensembl.org) provide immediate access to enormous amounts of data and analysis tools, free of charge, from anywhere on the globe. This is a dramatic change from the situation just over a decade ago, when the GenBank database was distributed by paid subscription in a small notebook full of 5.25″ floppy disks!

Networking advances have also been important for within-laboratory data management. Automated capture devices allow data to be added to a lab's database immediately, and with little or no human intervention. Centralized **laboratory information management systems**, or **LIMS**, then allow users at multiple workstations (or even multiple geographic locations) to browse, edit, analyze, and annotate the data. These integrated systems, while they often do not offer cutting-edge analysis tools, provide a relatively seamless work environment and keep the user from reformatting data to satisfy the needs of individual analysis programs. New tools for internet-based data management are available through http://www.geocities.com/SiliconValley/Vista/2207/sql1.html.

Although LIMS systems are invaluable time-savers in the lab, data format remains one of the annoyances of daily work in genome analysis. Sequence editors, databases, and statistical tools often require their own data format. Efforts to create standards exist, but have largely been unsuccessful because of the effort needed to retrofit existing software. Programs such as READSEQ, designed to convert among numerous popular formats, are essential items in the toolkit of genome scientists.

The core item for management and distribution of genomic data is a database system. Most databases can be classified as either **relational databases (RDB)** or **object-oriented databases (OODB).** These systems should be contrasted with familiar spreadsheet and flat file databases, such as text files and Microsoft Excel™, which store data and facilitate searches but are not designed for communication among files. Descriptions of the basic features of the different types of databases can be accessed at numerous Web sites, such as those provided by Microsoft and Oracle, or by the Object Management Group (http://www.omg.org).

Experts debate the relative merits of the two database philosophies. For our purposes, it is well established that RDBs are very effective for storing, searching, and distributing data that fits nicely into tabular format. OODBs are particularly good at

Plant Genome Projects

Arabidopsis thaliana

The first plant genome to be sequenced in its entirety was that of the model organism *Arabidopsis thaliana* (Arabidopsis Genome Initiative 2000), an

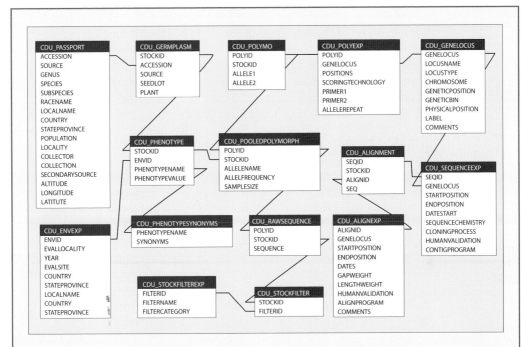

An example of a relational database (RDB) layout from the PanZea maize diversity project, indicating how items in one file are related to fields stored in different files.

handling complex data structures, and are especially useful for performing analyses on sequence "objects." Descriptions of objects include information about the stored data, along with functions for operating on the data objects—a very efficient programming approach.

In contrast to simple techniques for storing data, such as spreadsheets, both RDBs and OODBs allow large amounts of data to be quickly retabulated, sorted, displayed, and queried. Query languages such as SQL (Structured Query Language) have been developed for fast and general searches of databases (explained in a tutorial available at http://www.geocities.com/SiliconValley/Vista/2207/sg17.html). Once the results of a database search are saved in files, scripting languages such as PERL eliminate the tedium of extracting and processing the relevant information returned by the search. PERL scripts can also be written to create data analysis pipelines for repetitive sequence analysis and management tasks. Sample scripts for biologists can be found at http://bioperl.org.

achievement that was the result of a truly international consortium using a top-down approach of tiling across a physical map of clones. The *Arabidopsis* sequence was published in five installments, by chromosome, in *Nature* in 1999 and 2000.

The sequenced portion of 115 Mb is approximately the same size as that of the euchromatic portion of *Drosophila melanogaster*, but contains almost twice as many predicted genes (25,500). This gene complement seems to have evolved via two rounds of whole-genome duplication followed by extensive shuffling of chromosomal regions and considerable gene loss. However, over 1,500 tandem arrays (generally two or three copies) of repeated genes have also been identified, with the result that the majority of genes are duplicated at some level and can be assigned to just 11,000 families. Some geneticists now regard this number as representative of the minimal complexity required to support complex multicellularity and believe it likely that all plant and animal genomes represent modifications of a basic "toolkit" of gene families that evolved more than a billion years ago.

In addition to the high level of segmental duplication (Figure 1.16), the complete sequence of the *Arabidopsis* genome illustrates several features that distinguish plant from animal genomes. Over 800 nuclear genes may be of plastid descent, indicating that transfer of organellar DNA to the nucleus is probably an ongoing process. Many of these genes have lost the protein plastid-targeting signal, however, and so can now function in the cytoplasm. Transposable elements, including novel miniature elements such as MITEs and MULEs, account for at least 10% of the *Arabidopsis* genome; as in animals, these elements are concentrated in repetitive centromeric heterochromatin. Centromeric DNA is enriched for at least 40 families of repeat, and the genomic sequence opens up possibilities for functional characterization of centromeric as well as telomeric structural features.

Plant genomes contain several classes of genes that are either absent from or underrepresented in animal genomes. The products of these "plant-specific" genes include:

- Enzymes required for cell wall biosynthesis.

- Transport proteins that move organic nutrients, inorganic ions, toxic compounds, metabolites, and even proteins and nucleic acids between cells.

- Certain enzymes and other macromolecules required for photosynthesis, such as Rubisco and electron transport proteins.

- Products involved in plant turgor and the responses peculiar to a sessile lifestyle, including phototrophic and gravitropic responses.

- Numerous enzymes and cytochromes involved in the production of the hundreds of thousands of secondary metabolites found in flowering plants.

- A large number of pathogen resistance *R* genes and associated factors. These are generally highly polymorphic, as are mammalian defense-related genes; but unlike the genes of the mammalian immune system, *R* genes are dispersed throughout the genome rather than being localized in a single complex.

Plants share with animals many of the gene families involved in intercellular communication, transcriptional regulation, and signal transduction during development, but there are some notable exceptions. For example, *A. thaliana* lacks homologs of the Ras G-protein family and tyrosine kinase receptors (although there are many serine-threonine kinases), and there are

(A)

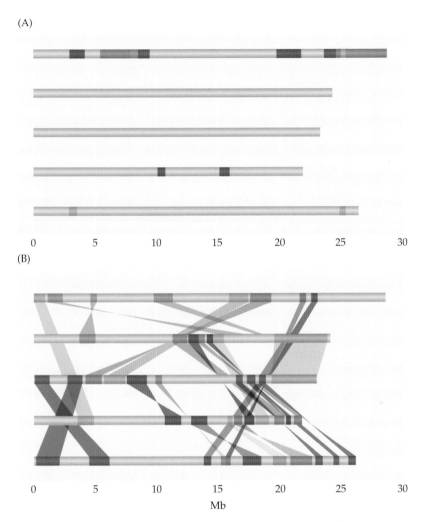

(B)

Mb

Figure 1.16 Chromosome duplications in the *Arabidopsis thaliana* genome.
Whole-genome alignments of this model plant indicate that there have been at least 30 segmental duplications within its genome. (A) Seven intrachromosomal duplications are shown as duplicated blocks of color within three of the five chromosomes; five duplications occur in the first chromosome, and the fourth and fifth chromosomes display one duplication apiece. (B) A schematic depiction of another two dozen interchromosomal segmental duplications. A twist in the band indicates that inversion accompanied the duplication event. (After AGI 2000.)

no obvious Rel, Forkhead, or nuclear steroid receptor class transcription factors. On the other hand, plant genomes encode variant forms of chromatin remodeling proteins, including histone deacetylases and SNF2 ATPases; several unique families of transcription factors; and novel components that mediate plant hormone function.

As with other model organisms, there is much more to the *Arabidopsis* genome project than the complete genome sequence. The Web site for the The *Arabidopsis* Information Resource (TAIR; http://www.arabidopsis.org) allows researchers to integrate the genome sequence with an extensive EST database and with the genetic and physical maps. The site provides links to functional and molecular genetic data and the literature for specific genes, and it shows an ever expanding list of mutant stocks. An alternative resource for *Arabidopsis* and many other plants, UK CropNet (http://ukcrop.net; Dicks et al. 2000), uses common AceDB or WebAce platforms to coordinate genetic and molecular data (see Box 1.3 for a general discussion of genome database management).

One resource that has been particularly useful for quantitative genetic analysis is the availability of several sets of recombinant inbred (RI) lines. Molecular maps of these lines are constantly being updated by individual users, which presents a challenge for the ongoing task of aligning physical and genetic map data and means that the quality of new data must constantly be assessed and monitored. Gene expression data, structural information, primers for genotyping microsatellites and SNPs, and information about mutant stocks are all immediately accessible.

Grasses and Legumes

Genome initiatives are under way for more than 50 different plant species. From an economic standpoint, the most important of these projects are those for the major feed crops—the grasses maize, rice, wheat, sorghum and barley; the forage legumes soybean and alfalfa, the forage rye grasses and fescues, and potential sources of bioethanol such as switchgrass and sugarcane. Several of these genomes are so large (as a result of autopolyploidization and the dramatic expansion of repetitive DNA) that whole-genome sequencing is impractical, and efforts have instead focused on comparative genome methods. Both rice (*Oryza sativa*) and maize (*Zea mays*), however, have relatively small genomes and are such key elements of the agricultural economies of the developed world that complete genome sequences have been prioritized. Draft genomes of the two major rice genome cultivars, *japonica* and *indica*, were published by Syngenta (Goff et al. 2002) and the Beijing Genomics Institute (Yu et al. 2002), respectively, and a complete *Oryza* sequence has since been published (IRGSP 2005). Gene identification in maize has been advanced by techniques such as methylation filtering and high C_0t selection that remove intergenic and highly repetitive sequences, effectively reducing the size of the genome more than fivefold for sequencing purposes (Whitelaw et al. 2003). Comparison of genome sequences of rice and *Arabidopsis* suggests that extensive but complex patterns of synteny will be a useful feature of plant genomics (Figure 1.17).

Medicago (alfalfa) is a true diploid legume that, along with its crucial role in fixing soil nitrogen, constitutes a major part of forage diets. Thus *Medicago* has been selected as the model legume for whole-genome sequencing. The remaining grasses and legumes are the subjects of extensive EST

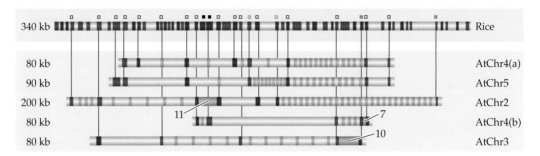

Figure 1.17 Rice-*Arabidopsis* synteny. Twenty of 54 genes in a 340-kb stretch of the rice genome (top) retain the same order in five different 80- to 200-kb regions of the *Arabidopsis* genome (below; not drawn to scale). The conserved genes (red and green boxes) are found on both the rice and *Arabidopsis* strands, but are interspersed by a variable number of different genes (yellow boxes) in *Arabidopsis*. Shaded boxes above the rice chromosome indicate that the conserved gene is in the opposite relative orientation on the *Arabidopsis* chromosome(s).

sequencing and high-resolution genetic map construction in hopes of taking advantage of the expected pervasive synteny within these families. Web sites established by individual research groups integrate research efforts from around the globe. Some useful Web sites include the U.S. Department of Agriculture's Gramene resource (http://www.gramene.org), which is designed to facilitate comparative genome mapping; and organism-specific resources such as MaizeGDB (http://www.maizegdb.org). A common objective of these sites is to link seed stock and real genetic resources to virtual data on linkage and mapping, QTL localization, EST and genome sequences, and biochemical pathway information, among other resources.

The maize, rice, and alfalfa initiatives all focus strongly on quantitative genetic resources, in contrast with the Mendelian focus of many model animal genome projects. Economically important traits include resistance to a broad range of pathogens; flowering time, seed set, grain morphology, and related yield traits; tolerance to drought, salt, heavy metals, and other extreme environmental circumstances; and measures of feed quality such as protein and sugar content. Each of these characters can be improved through combinations of genetic engineering and specialized plant breeding techniques, including introgression of germ line from wild ecotypes. More traditional focus on the genetic basis of heterosis, hybrid vigor, unusual phenotypic segregation, as well as efforts to limit inbreeding depression, will also benefit from improved maps and a more general appreciation of the extent and distribution of polymorphism in the genome.

Given the role of artificial selection in the recent derivation of grasses in particular, these genome projects promise to reveal much information regarding the evolution of domesticated species. Several genes have been identified in maize that were clearly selected during the transition from wild teosinte to cultivated maize, leading to modification of traits such as glume architecture, ear size, and a change from multiple branches to a single stalk

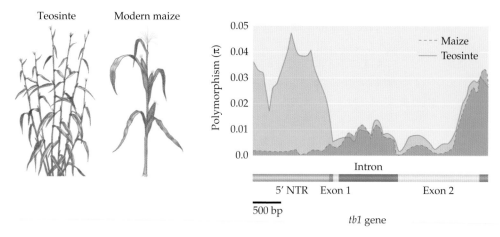

Figure 1.18 *Teosinte branched 1* **and the evolution of maize.**   Modern maize is a derivative of the wild progenitor teosinte, which had multiple tillers. Throughout the coding region of *tb1*, the level of polymorphism (the number of nucleotides that differ between any two alleles in a sliding window along the gene) is substantially the same in a sample of maize and teosinte. However, in the 5′ non-translated region, there is a dramatic reduction in the level of polymorphism in maize relative to that seen in teosinte. (After Wang et al. 1999.)

(tiller). Although folklore suggests that cultivated crops are highly inbred, molecular population genetic data contradicts this notion. In fact, domesticated maize is highly polymorphic relative to most animal species. As in other crop grasses, diversity in domesticated maize is reduced by a maximum of a mere 30% relative to its presumed wild progenitors (Buckler et al. 2001). However, strongly selected portions of loci, such as the upstream regulatory region of *teosinte branched 1*, show a marked reduction in diversity that is indicative of a selective sweep that brought one haplotype associated with the selected site close to fixation, purging linked polymorphism in the process (Figure 1.18). The selected site is thought to upregulate expression of the *tb1* gene, which encodes a repressor of lateral branch elongation, resulting in the evolution of single-tillered plants (Wang et al. 1999). Crop plants are thus established as excellent model systems for studying the molecular basis of morphological divergence, as well as other types of population genetic analysis.

Other Flowering Plants

Over 90 different angiosperm genome projects are active around the world. These include African projects on beans, corn, and fungal pathogens; Australian projects on cotton, wheat, pine, and sugarcane; at least two dozen European projects that include vegetables such as cabbage, cucumber, and pea, and fruits such as apple, peach and plum; and over 50 North American projects as diverse as turf grass, chrysanthemum, almond, papaya, and

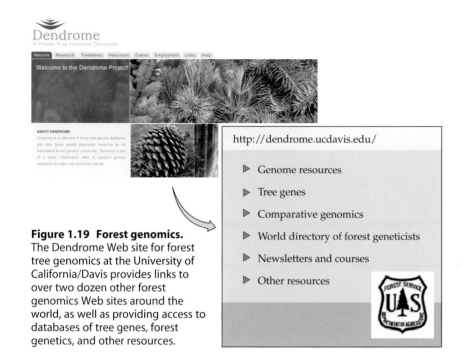

Figure 1.19 Forest genomics.
The Dendrome Web site for forest tree genomics at the University of California/Davis provides links to over two dozen other forest genomics Web sites around the world, as well as providing access to databases of tree genes, forest genetics, and other resources.

poplar. The common denominator among all of these projects is the assembly of genetic maps (and in some cases physical maps) and the placement of a common set of plant genes on them. For some species, large EST sequencing projects are also in place, with the twin objectives of enabling comparative genomic analysis (particularly in regions of synteny) and QTL mapping.

Several model organisms in addition to *Arabidopsis* and the grasses are receiving particular attention as a result of a long history of genetic analysis and/or the potential light these organisms may shed on plant evolution. These include the snapdragon (*Antirrhinum majus*), in which numerous classical flowering mutants were initially isolated as a result of transposable element movement; sunflowers (*Helianthus* spp.) and monkey flowers (*Mimulus* spp.), which are of particular interest in studies of hybrid speciation and adaptation in the wild; and the variants of *Brassica oleracea* (cabbage, kale, Brussels sprouts, broccoli, cauliflower, and kohlrabi), which are in the same family (Brassicaceae) as *Arabidopsis* and are a fascinating model for domestication because they all derive from the spontaneous mutation of genes involved in meristem growth (Purugganan et al. 2000).

Forest trees are an example of an area where genomic analysis has the potential for economic impact where classical genetics has been problematic (Figure 1.19). The first fully sequenced forest tree genome is that of the black cottonwood (*Populus trichocarpa*; Tuskan et al. 2007). High-density genetic maps of spruce, loblolly and several other pines, as well as a few species of *Eucalyptus*, have been established using a combination of AFLP and microsatellite markers and applied to the mapping of Mendelian and quantitative trait loci affecting wood quality, growth, and flowering parameters. These maps can be accessed through the Dendrome Web site (http://dendrome.ucdavis.edu). Marker-assisted selection has the promise to improve desired traits dramatically, at least to the extent that phenotypes measurable in seedlings predict mature qualities. Comparative analyses and transcription profiling of genes involved in wood properties including lignins and enzymes that regulate cell wall biosynthesis (many of which can be identified by large EST sequencing efforts) will also have an impact on forest biotechnology throughout the world.

Several staple crop plants are obviously poised to benefit from genomics, including potato and other tubers, tomato, tobacco, beans, and cotton. Analyzing the genome diversity of ecotypes endemic to the original source of these crops—many of which come from tropical regions—has the potential to affect productivity in developing countries as well as to support yield and quality improvements in countries where monocultures are employed.

The political, legal, and sociological implications of proprietary rights to a plant's germ plasm and the polymorphic DNA sequences identified within it are being addressed as much in courts of law as by international agencies. No plant equivalent of the Human Genome Project's ELSI initiative has yet been established.

Microbial Genome Projects

The Minimal Genome

The first complete cellular genomes to be sequenced were of prokaryotes, starting with *Haemophilus influenzae* (Fleischmann et al. 1995) and quickly followed by *Mycoplasma genitalium* (Fraser et al. 1995), three other bacteria, and, in September of 1997, that of *Escherichia coli* (Blattner et al. 1997). The primary sequences provided immediate information about genome structure (organization of replication, GC content, transposable elements, recombination) and genome content (total number of genes, representation of conserved gene families).

Gene annotation is initially more straightforward for prokaryotes than for eukaryotes, since open reading frames tend to be uninterrupted and genes tend to be closely spaced; however, the assignment of adjacent genes to operons is not trivial. Typically, more than three-quarters of the ORFs in a microbial genome can be assigned a function based on their similarity to genes in other organisms and/or by identifying protein domains. The remaining genes may encode yet-to-be-described functions, they may represent taxon-specific functions, or they may be evolving so rapidly that their function is conserved despite sequence divergence. Bioinformatic approaches are being employed to complete the functional annotation of microbial genomes.

Similarities in gene content between the three bacterial genomes mentioned above, as well as the first sequenced genome of an Archaea (*Methanococcus jan-*

TABLE 1.4 *Number of Genes Involved in Defined Cellular Processes*

	M. genitalium	H. influenzae	E. coli	M. jannaschii
Central intermediary metabolism	6 (1.3)*	30 (1.7)	188 (4.4)	18 (1.0)
Energy metabolism	31 (6.6)	112 (6.4)	243 (5.7)	158 (9.1)
Lipid and fatty acid metabolism	6 (1.3)	25 (1.4)	48 (1.1)	9 (0.5)
Cofactor biosynthesis	5 (1.1)	54 (3.1)	103 (2.4)	49 (2.8)
Amino acid biosynthesis	1 (0.2)	68 (3.9)	131 (3.1)	64 (3.7)
Nucleotide metabolism	19 (4.0)	53 (3.0)	58 (1.4)	37 (2.1)
DNA replication and repair	32 (6.8)	87 (5.0)	115 (2.7)	53 (3.0)
Transcription	12 (2.5)	27 (1.5)	55 (1.3)	21 (1.2)
Translation	101 (21.4)	141 (8.1)	182 (4.2)	117 (6.7)
Regulatory functions	7 (1.5)	64 (3.7)	178 (4.2)	18 (1.0)
Transport and binding proteins metabolism	34 (7.2)	123 (7.0)	427 (10.0)	56 (3.2)
Cell structure	17 (3.6)	84 (4.8)	237 (5.5)	25 (1.4)
Cellular processes	21 (4.5)	53 (3.0)	327 (7.6)	26 (1.5)
Other categories	27 (5.7)	93 (5.3)	364 (8.5)	38 (2.2)
Unclassified	152 (32.3)	736 (42.1)	1,632 (38.0)	1,049 (60.4)
Total	**471**	**1,750**	**4,288**	**1,738**

*Numbers in parenthesis represent percentage of all genes.
Sources: Categories adapted according to scheme of Riley (1997).
Data for *M. genitalium* and *H. influenzae* from Fraser et al. (1995). Data for *E. coli* from Blattner et al. (1997).
Data for *M. jannaschii* from http://www.tigr.org/tigrscripts/CMR2/gene_table.spl?db=arg (1/20/01).

naschii; Bult et al. 1996), are presented in Table 1.4. There are 471 predicted genes in the 0.6 Mb *M. genitalium* genome, 1,750 in the *H. influenzae* genome (1.8 Mb), and 4,288 in *E. coli* strain K12 (4.6 Mb). The average gene length in each species is thus close to 1.1 kb, indicating that differences in genome size are based on changes in gene number, which in turn reflect duplication and divergence in larger genomes, as well as gene loss in small genomes.

It is usually possible to identify specialized metabolic functions that reflect a microbe's adaptation to a particular ecological niche (such as mammalian genital, respiratory, or enteric tracts) by surveying the predicted enzyme and transporter complement. Indeed, one aim of microbial genomics is to be able to predict metabolic phenotypes on the basis of gene content alone, as discussed in Chapter 6. Comparative analysis of sequences in pathogenic strains of *E. coli* and in various pathogenic species in the *Bacillus* (including anthrax) and *Mycoplasma* genera hint at the genetic basis of virulence and pathogenicity and may suggest novel approaches to antibiotic design.

The concept of the **minimal genome** refers to attempts to define the minimum complement of genes that are necessary and sufficient to maintain a free-living organism—in a sense, to define genetically "What is life"? Two general strategies have been pursued to achieve such a definition. A survey of multiple complete bacterial genome sequences to identify a common "core" genome suggests that at least 206 genes are always present in microbial genomes, and hence essential (Gil et al. 2004). By contrast, random insertional mutagenesis of genes in *M. genitalium* suggests that just 100 of its genes can be disrupted without affecting the bacterium's capacity to grow in rich medium, implying that at least 370 genes are required for growth in this species (Glass et al. 2006). Similarly, comprehensive mutagenesis of the 4,100 genes of the *B. subtilis* genome revealed that just 271 of them are indispensible under favorable growth conditions, and most of these are involved in a relatively small number of functions relating to metabolism, cell divi-

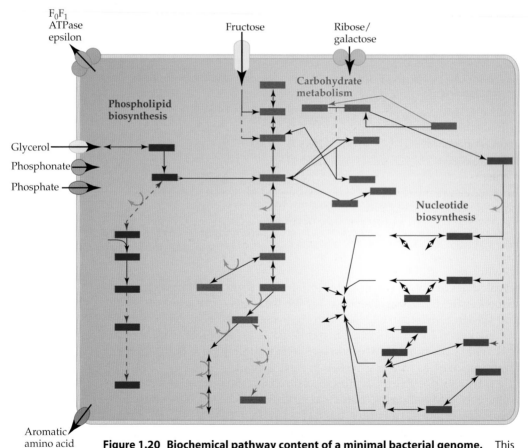

Figure 1.20 Biochemical pathway content of a minimal bacterial genome.   This schematic shows key aspects of the predicted metabolic capacity of *Mycoplasma genitalium*. The locations of biochemical steps performed by genes that have been shown to be nonessential genes (i.e., insertional mutagenesis of these genes results in viable bacteria) are indicated as broken arrows in the three pathways shown. The indicated transporters on the cell surface are also dispensable. (After Glass et al. 2006, Figure 3.)

sion and shape, and synthesis of the cellular envelope (Kobayashi et al. 2003). We can thus infer that life can be supported by a genome of between 250 and 400 genes performing those functions shown for the schematic cell in Figure 1.20. However, it is hard to imagine such a minimal organism possessing any competitive advantage outside of a petri dish.

Efforts are also under way to build a viable organism from scratch by stitching together artificially synthesized genes, building on the assembly of a poliovirus from published sequence data (Cello et al. 2002). In addition to this re-synthesis of a poliovirus, a complete *M. genitalium* genome has been built stepwise by combining overlapping 7 kb cassettes into successively larger fragments, first in vitro and then in *E. coli* and yeast vectors (Gibson et al. 2008). Mycoplasmas are ideal for this strategy since they use the UGA codon to encode tryptophan, whereas UGA is read as a standard stop codon in other microbes; this crucial difference reduces the potential toxicity of the gene fragments. Since it has also proved possible to transplant the genome of *M. mycoides* into *M. capricolum* (Lartigue et al. 2007), it is only a matter of time before synthetic genomes containing completely artificial mixtures of genes are assembled and devised for a wide range of purposes.

One of the first insights provided by microbial genome analysis was the realization that bacterial genomes are much more modular and evolutionarily labile than hitherto appreciated. An intriguing feature of the *E. coli* genome is the evidence for genome plasticity in the form of repetitive sequences and insertions of several families of IS class transposable elements, as well as the presence of cryptic prophages and prophage remnants. Another element of bacterial plasticity, uncovered by combining comparisons of phylogenetic distributions with studies of the distribution of GC content and codon biases, is **horizontal gene transfer**, or gene exchange across species (Woese 2000). Adaptation and divergence in microbes is now seen to be associated with a dynamic process of loss and gain of genes, notwithstanding the conservation of a common core of othologs that can be revealed by careful phylogenetic analysis of quartets of completed genome sequences (Daubin et al. 2003).

Sequenced Microbial Genomes

The **Comprehensive Microbial Resource (CMR)** of the J. Craig Venter Institute (http://cmr.jcvi.org/tigr-scripts/CMR/CmrHomePage.cgi) provides online access to complete sequences and associated resources for over 400 microbial genome sequences. The CMR provides tools for searching, summarizing, or analyzing each genome, as well as comparative tools that facilitate alignment, annotation, and pathway analyses. Standardized compact or circular chromosomal displays (Figure 1.21) show each gene color-coded according to its orientation and the predicted molecular function of its product, and include mouse-over links to background information, publications, and data on each individual sequence annotation.

(A)

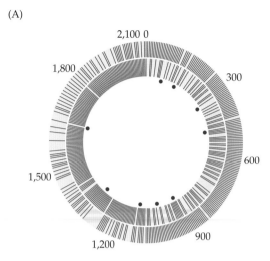

Figure 1.21 Representation of a typical microbial genome.   Circular (A) and linear (B) CMR representations of the 2.1 Mb genome of *Streptococcus pneumoniae*. The outer and inner circles in (A) represent genes encoded on the two strands of the chromosome, with a clear asymmetry that also corresponds with GC content bias. The red dots indicate the locations of nine clusters of genes that are missing in two other strains of the same species and demarcate potential virulence factors. (B) A close-up view of a 25-kb portion of the chromosome provides evidence for genes on each strand from sequence matches to a gene from an HMM model (blue), BLAST (yellow), or genes in the omniome database of microbial genes (pink). (A after Tettelin et al. 2001.)

(B)

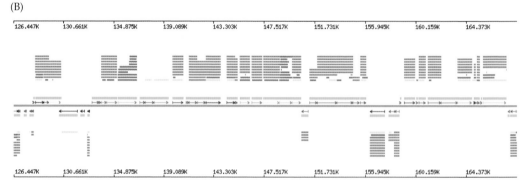

The Microbial Genomes page of the Entrez Genome Project site (Figure 1.22) lists over 1,600 microbial sequences at various stages of completion; almost half of these are already complete. Over 80 of the genomes sequenced are from the kingdom Archaea. The species genomes are listed according to their size, GC content, phenotype, growth requirements, and any pathogenic characteristics. Each genome can be displayed simply by clicking on the taxon name. For complete genomes, a series of tools are available, including taxonomic maps of closely related genes in other species (T), a table listing all predicted proteins (P), clusters of orthologs (COGs C), proteins predicted to have similar structures (D), a BLAST tool for the microbe (L), a list of conserved domains in the proteins (S), a pairwise whole-genome sequence alignment plot (G), a taxonomic plot of related genes in any two other species (X), and FTP site for sequence and annotation download (F), and a link to publications (R).

Considerable microbial genome sequencing is also performed at the Joint Genome Institute of the U.S. Department of Energy (http://microbialgenomics. energy.gov). This Web site lists approaching 500 microbial genomes of inter-

---

EXERCISE 1.4 *Compare two microbial genomes using the CMR*

Use the Comprehensive Microbial Resource to contrast the annotated cell adhesion genes in the genomes of two species of Pseudomonas. *Compare these genes with the annotated cell adhesion genes in* E. coli *and in* B. anthracis. *Also compare the gene content of the latter two species.*

Answer: *In the menu bar on the CMR homepage, select "Tools," choose the "Lists" link, then the "Gene Lists" section and the "Genes by Role" category. This page links to lists of all of the annotated genes in each family of functions. Open the "Cellular Processes" heading, select the species you are interested in, and search for the term "Cellular Adhesion" on the page that comes up. Most species have a handful of cell adhesion genes, but* Pseudomonas putida *has 37 such loci, and* P. syringae D3000 *has 21, 15 of which have known roles in pilus biogenesis (pili are threadlike structures that are important in bacterial cell adhesion). By contrast, there are just four genes in the cell adhesion list in* E. coli 536, *and none in* E. coli K12. B. anthracis *Ames only has two adhesion proteins.*

For the second part of the exercise, return to the CMR homepage, choose the "Genome Tools" folder and then the "Role Category Pie Chart" tool. Select representative genomes for the two species and examine the tables and pie graphs that are produced. These graphics show that between one-third and one-half of the genes of B. anthracis *and* E. coli *encode "hypothetical" proteins of unknown function.*

---

est for a wide range of applications, from environmental waste clean-up to energy production and biotechnology enhancement. The Integrated Microbial Genomes resource (Markowitz et al. 2008) allows for comparative analysis and annotation of most sequenced microbes, while MicrobesOnline (Alm et al. 2005) provides additional tools for high-end genomic analyses.

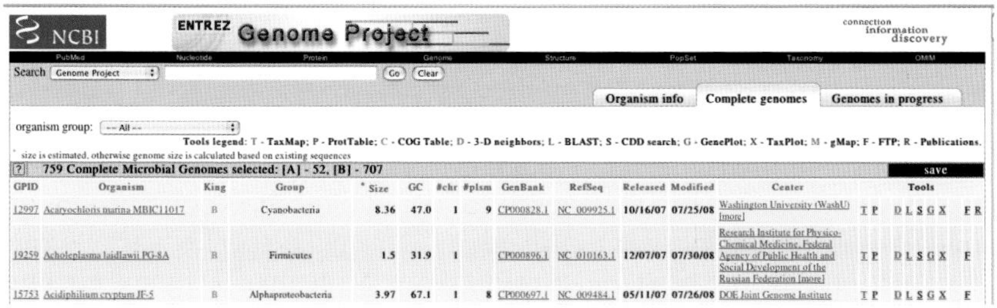

Figure 1.22 The Entrez Genome Project's main page for microbial genomes.
Abbreviations at right are the links to tools mentioned in the text for this site, found at http://www.ncbi.nlm.nih.gov/genomes/lproks.cgi.

Funding for microbial genome research is largely driven by medical research and economic motives, but there is a strong evolutionary component to the sampling strategy. Broad sampling increases the power of phylogenetic analysis, which is a valuable tool for the functional classification of members of gene families. Sampling from a wide variety of habitats, including extreme environments such as thermal vents, is also likely to lead to the identification of novel genes, or gene products adapted for functions that can be utilized by biotechnologists (such as the *Taq* thermally stable DNA polymerase). Table 1.5 provides a survey of the range of sequenced microbial genomes.

Yeast

The complete genome sequence of the budding yeast *Saccharomyces cerevisiae* was published in a supplement to the journal *Nature* in May of 1997 (Mewes et al. 1997). The 12 Mb sequence was assembled from roughly 300,000 sequence reads generated in over 100 laboratories and had an initial estimated error rate of 3 in 10,000 bases. Approximately 6,000 predicted genes are spread over 16 chromosomes, which include 53 identified regions of clustered gene duplications. Slightly less than half of these genes were functionally annotated (experimentally characterized and/or identified by homology to sequences in other species) prior to completion of the genome sequence.

Chromosome displays and individual gene listings can be accessed through two main sites: the *Saccharomyces* page of the Munich Information Center for Protein Sequences (MIPS; http://mips.gsf.de/genre/proj/yeast/index.jsp), and the *Saccharomyces* Genome Database (SGD) at Stanford University (http://www.yeastgenome.org). An enormous amount of information is available on the SGD locus pages for all *Saccharomyces* genes, including mutational, structural, biochemical, and transcription data. The SGD site provides links to basic information, as well as "additional information" links that provide extensive detail, including historical notes on the annotation of the gene (the "Locus History" link, which also alerts readers to contradictions in the literature, changes in nomenclature, and other sources of potential confusion); several types of functional data ("Function Junction") and expression data ("Expression Connection"); mapping data; and advanced protein and sequence resources.

Sequences of two other well studied yeasts—the fission yeast *Schizosaccharomyces pombe* and *Candida albicans*, a yeast commonly associated with opportunistic human infections—are now complete as well. Genomic analysis of numerous other fungi of agricultural and biotechnological interest is underway, including those of the allergenic *Aspergillus* species, various fruit white rots, and the rice blast *Magnaporthe grisea*. More than 60 fungal genome sequences are compiled at the http://fungalgenomes.org Web site.

TABLE 1.5 *Sequenced Microbial Genomes (October 2008)*

Archaea (31)	Crenarchaeota	Hyperthermophiles
	Euryarcheota	Sulphur metabolizers Halophiles Methanophiles Thermophiles
	Nanoarcheota	Nanoarchaeum
Bacteria (421)	Actinobacteria	*Corynebacterium* (bioengineering) *Mycobacterium* (animal pathogens) *Tropheryma whipplei* (reduced genome) *Streptomyces* (bioengineering)
	Aquificae	*Aquifex* (chemolithoautotroph)
	Bacteroides	Human gut/oral bacteria
	Chlamydiae	*Chlamydia pneumoniae* Other *Chlamydia* (intracellular pathogens)
	Cyanobacteria	*Synechococcus, Plochlorococcus*, etc.
	Deinococcus	*D. radiodurans* (radiation resistance)
	Firmicutes	*Bacillus* (pathogens, incl. anthrax) *Clostridium* (tetanus, other toxins) *Listeria* (food-borne pathogen) *Staphylococcus* (incl. *S. aureus*) Lactobacillae, streptococci (gut bacteria) Mycoplasmas (small animal pathogens)
	Proteobacteria	Alpha (*Rhizobium, Rickettsia, Caulobacter*) Beta (*Bordetella, Neisseria*) Delta (*Bdellovibrio* , a predatory bacterium) Epsilon (*Helicobacter* group) Gamma (diverse *E. coli* group)
	Spirochaetes	*Borrelia, Treponema* (human parasites)
	Other bacteria	Fusobacteria, Chlorobacteria, Planctomycetes, Thermotogae
Viruses (3)		dsDNA, ssRNA viruses

Source: JCVI Comprehensive Microbial Resource at http://cmr.jcvi.org/tigr-scripts/CMR/CmrHomePage.cgi. Numbers in parentheses refer to complete genome sequences as of October 2008. Compare with 16 Archaea and 123 Bacteria in July of 2004.

EXERCISE 1.5 *Examining a gene in the Saccharomyces Genome Database*

Answer the following questions about the yeast GAL4 gene (or a gene of your choice):

a. Write down two aliases (alternate names) for GAL4

b. List the three genes on either side of GAL4 in the S. cerevisiae genome

c. What are the three GO annotations for the gene?

d. How many amino acids are there in the GAL4 protein?

e. What is the phenotype of a GAL4 mutant?

ANSWER: *Each of these questions can be answered directly from the GAL4 page found by a quick search from the SGD home page. (a) One alias is GAL8Y while the systematic name is listed as YPL248C. (b) Clicking on the small gene image at the top right brings up a local chromosomal view, from which you can see that the three genes to the left are ICY2, RPL36B, and GYP5, while those to the right are YPL247C, RBD2, and YPL245W (the "W" and "C" represent Watson and Crick strands for unannotated genes). (c) The three GO annotations (see Box 2.5) are listed as transcriptional activator (Molecular Function), galactose metabolism and DNA-dependent regulation of transcription (Biological Process), and nucleus (Cellular Component). (d) You can download the protein sequence of 881 amino acids by linking out to the Entrez Protein page (and examine the protein structure through the PDB site). (e) Null mutants cannot use galactose as a carbon source.*

Several complementary approaches to functional characterization of the complete set of yeast genes have been taken. **Systematic mutagenesis** uses homologous recombination to knock out open reading frames in heterozygous diploid cells, thus allowing complete loss of function to be monitored in haploid cells after induction of sporulation. Winzeler et al. (1999) deleted almost one-third of the yeast ORFs and found that 17% of these were essential for viability in rich medium, while 40% showed quantitative growth defects in rich or minimal medium. Subsequently, over 96% of the yeast genome has been tagged with molecular barcodes associated with gene deletions. A second, complementary approach is a **genetic fingerprinting** strategy, in which natural selection is allowed to sort through a large population of massively mutagenized cells growing in a chemostat under a variety of conditions. These experiments similarly found that about half (157 of 268) of the genes tagged by a transposable element insertion resulted in a detectable reduction in fitness, with many of the genes that showed weak effects not having been characterized previously (Smith et al. 1996). In order to characterize pairwise **genetic interactions** in a systematic manner, Tong et al. (2004) crossed a set of 132 query mutants to a panel of 4,700 viable yeast deletion mutants. This approach detected over 4,000 functional interactions involving more than 1,000 genes.

A fourth major experimental approach to functional annotation of the yeast genome has been **microarray analysis**, which is discussed in detail in Chapter 4. Using microarrays, changes in the transcriptome have been characterized throughout the cell cycle; during meiosis; in response to environmental shifts, including anoxia and nutrient source; and in a variety of mutant backgrounds. The so-called "compendium" strategy tracks coordinated changes in transcription under a variety of experimental circumstances, assigning genes to functional groups on the basis of co-regulation. Transcriptional data is also being supplemented with protein expression data and proteomic data, as discussed in Chapter 5.

Parasite Genomics

The World Health Organization (WHO) has singled out 10 tropical diseases that affect billions of people worldwide for intensive research, in the hope of eventually eradicating the pathogenic agents responsible (Table 1.6). Mosquito-transmitted viral infection is responsible for one of the target diseases, Dengue hemorrhagic fever. Seven diseases are carried by water- or insect-borne eukaryotic parasites: malaria, leishmaniasis, American and African trypanosomiasis (Chagas' disease and sleeping sickness, respectively), lymphatic filariasis (elephantiasis), river blindness (onchocerciasis), and schis-

TABLE 1.6 *Tropical Diseases Caused and/or Transmitted by Eukaryotic Organisms*

	Vector	Parasite	Number of humans at risk
Dengue hemorrhagic fever	Mosquito	Virus	2.5 billion
Leishmaniasis	Sandfly	Protozoan	350 million
Malaria	Mosquito	Protozoan	2.4 billion
African sleeping sickness	Tsetse fly	Protozoan (*Trypanosoma brucei*)	60 million
Chagas disease (American)	Insect	Protozoan (*Trypanosoma cruzi*)	100 million
Elephantiasis	Mosquito	Filarial nematode	1 billion
River blindness	Blackfly	Nematode	120 million
Schistosomiasis	Water	Platyhelminth (fluke)	500 million
Tubeculosis	Human	Bacterium	2 billion
Leprosy	Human	Bacterium	Near eradication

tosomiasis (bilharzia). The final two infections, tuberculosis and leprosy, are caused by airborne bacteria and are transmitted from human to human.

The Microbial Sequencing Centers of the J. Craig Venter Institute (JCVI) and Broad Institute provide access to 30 disease-related microbial genome projects including viruses, pathogenic protozoa, and several insect vectors. A related site, the Pathogenic Functional Genomics Resource Center, is dedicated to the provision of experimental tools for research into these organisms. Six of the most pathogenic microbes identified by the National Institute of Asthma and Infectious Disease are brought together at a single site, the Pathema Microbial Resource Center, which is sponsored by the JCVI (http://pathema.jcvi.org). These include the agents responsible for anthrax, botulism, melioidosis, glanders, gas gangrene, and amebiasis, some of which are regarded as potential bioterrorism agents.

Numerous parasites carried by invertebrate and animal hosts cause widespread diseases in North America and Europe, including *Cryptosporidium*-induced diarrhea, zoonotic toxoplasmosis, and Lyme disease. Several infectious diseases affecting immunocompromised patients are also emerging. Nematode genomics in particular has direct potential for practical benefits, building on the extensive functional analysis of the *C. elegans* model system. Crop damage caused by parasitic plant nematodes costs billions of dollars each year and causes untold human suffering around the world. In addition, globally as many as 1 billion people may be infected by intestinal and other nematodes that cause diseases including elephantiasis and numerous intestinal disorders such as ascariasis and tricuriasis. These parasites also infect farm animals and pets in the form of hookworm and heartworm.

Current approaches to disease control focus on public health issues such as improving basic living conditions and educating the public about disease transmission, but have traditionally included chemical interventions (e.g., application of pesticides, administration of antibiotics) that may have deleterious environmental consequences and often lead to increased resistance among the pathogens they are intended to destroy. The complex lifestyle of many of these parasites also limits the effectiveness of chemical intervention, and insect hosts have proven extremely difficult to control. Hence genomic approaches are being supported by WHO, several pharmaceutical giants, and the major genome sequencing centers.

The aims of parasite genomics include:

- Identification of species-specific genes that may be used to generate antigens for vaccination.

- Better understanding of the developmental genetics associated with key transitions in the life cycle, suggesting targets for drug design.

- Polymorphism surveys that address the population biology of the parasites, informing epidemiological studies.

- In the case of malaria in particular, considerable effort has been devoted to the genomics of the mosquito vector (*Anopheles* spp.), including mapping of loci that affect parasite (*Plasmodium*) transmission.

Web resources providing access to genomic data for these and other projects can be accessed through TIGR, NCBI, and the European Bioinformatics Institute parasite genomics server (http://www.ebi.ac.uk/parasites/parasite-genome.html).

Metagenomics

An emerging area of microbial genomics has been dubbed **metagenomics** or **environmental sequencing**. Both terms are used to refer to the sequencing of literally thousands of genomes' worth of DNA extracted from an environment such as ocean water, soil, or intestinal flora. The main reason for taking such an approach is that the vast majority of bacteria cannot be cultured in vitro, which implies that our knowledge of microflora is both limited and biased by sampling of a relatively small fraction of the biota. The idea behind metagenomics is to circumvent this problem by directly cloning DNA fragments isolated from an appropriate size fraction of biological material that largely excludes viruses and multicellular organisms, without any attempt to grow the microbes outside of their environment. The difficulty with this approach is that there is so much microbial diversity that it will take hundreds of human genome equivalents to assemble more than a handful of complete genomes, and most fragments are just gene pieces that cannot easily be assigned to separate species. Nevertheless, the approach has greatly increased the number of known protein families (Yooseph et al. 2007) and new tools are being developed that allow assemblages of related genomes to be clustered together, facilitating studies of the distribution of microbial and viral genetic diversity across the globe (Rusch et al. 2007).

Much insight into both microbial and genic diversity has been gained from the identification of novel genes in seawater. Such knowledge has the potential to change oceanographers' understanding of the mechanisms of photosynthesis and of global carbon and nitrogen cycling, as suggested by the discovery of a previously unsuspected type of rhodopsin generated in uncultured proteobacteria from a sample taken from Monterey Bay (Béjà et al. 2000). Such proteorhodopsin genes have since been detected in hundreds more samples, suggesting that light harvesting need not be coupled to chlorophyll in cyanobacteria, as hitherto had been thought. By using degenerate gene-specific primers to survey samples from deep sea hydrothermal vents, Campbell et al. (2003) were able to detect a key enzyme involved in carbon fixation in bacteria associated with a polychaete worm, providing insight into how life is sustained in such an extreme environment. Venter et al. (2004) industrialized environmental sequencing by generating more than a Gigabase of sequence derived from about 1,500 liters of seawater from the nutrient-poor Sargasso Sea. In the process, they identified over one million new genes (approximately equivalent to the number of genes in existing databases) and almost 150 new types of bacteria in an estimated sample of 1,800 total species! The still-untapped oceanic microbial diversity is more than an order of magnitude greater than this, and protist diversity is similarly unexplored.

Similar strategies are being employed to understand phenomena such as the bacterial contribution to digestion in termites and beetles in the soil, the function of probiotic bacteria in the human gut, and microbial community structure in the rhizosphere, expanding the reach of genomics from organismal biology to functional ecology. Metagenomics data is brought together at the CAMERA (Community Cyberinfrastructure for Advanced Marine Microbial Ecology Research and Analysis) Web site at http://camera.calit2.net.

Summary

1. Genome science is a new way of looking at biology that unites genetics, molecular biology, computational biology, and bioinformatics.

2. The core aspects of genome science include the utilization of organismal databases; assembly of physical and genetic maps; genome sequencing; gene annotation; characterization of sequence diversity; expression profiling; proteomics; functional genomics; and comparative genomics.

3. Genetic and physical maps are the foundation upon which most genome projects are built. The two types of maps can be unified because of the direct relationship between recombination rate and chromosomal distance, as well as by way of cytogenetic maps.

4. "Synteny" refers to the conservation of gene order among groups of several to tens or even hundreds of genes in divergent species. Synteny is the basis for comparative gene mapping.

5. The Human Genome Project (HGP) has assembled a complete sequence of the human genome, and is now working to promote use of the sequence information in the domains of general biology, public health, and social policy. Additional research has proceeded apace along lines fostered by a rapidly expanding commercial sector.

6. Mouse, rat, dog, and chimpanzee genome sequences have been completed, and offer the most immediately accessible vertebrate model organisms for genetic analysis of physiology, development, cell biology, and behavior.

7. A wide range of animal genome projects include cow, horse, pig, sheep, dog, numerous fish species, and the two major invertebrate models, *Drosophila*, and *Caenorhabditis*.

8. The first complete plant genome sequence was that of the mustard weed *Arabidopsis*, and significant progress has been made toward completion of the rice and maize genomes. Additional plant genomes being studied include legumes, grasses, fruits, vegetables, forest trees, and ornamentals.

9. Over 400 bacterial genome sequences are available, providing a vast resource for modeling of biochemistry, metabolism, and microbial ecology, as well as opening up new avenues of biotechnological research.

10. In addition to the budding yeast *Saccharomyces cerevisiae*, several other fungi of industrial, biomedical, and agricultural importance have genome projects.

11. Genome science promises new strategies for dealing with several of the world's most devastating parasitic diseases using avenues of genetic research that were hitherto unimaginable with classical methods.

12. Metagenomics is an emerging approach that sequences DNA extracted from an environmental medium such as soil, ocean water, or intestinal flora. Although its methods do not typically result in complete gene sequences, metagenomics has increased the number of known protein families and has led to the discovery of novel genes and increased insight into genic and bacterial diversity, as well as adding to our knowledge of the mechanisms of certain biogeochemical cycles.

Discussion Questions

1. Now that the human genome sequence is nearly complete, how should the excess sequencing capacity at major genome centers be used? Is it good use of public funds to keep sequencing more and more genomes? If so, which genomes should be prioritized?

2. The volume and expense of genome data raises new issues in relation to peer review of research and publication of results. How can editors and reviewers oversee data quality? Should journals demand that companies make complete genome sequences freely available?

3. How can the commercial, medical, and agricultural benefits of genome science be extended to developing countries? Are there intellectual property issues raised by genomic analysis of species and populations in one country, and should ownership of the data be retained by that country or transferred to the company that pays for the research?

4. Propose a genome project for an animal or plant species of particular interest to you. What will be your key aims, both in terms of generation of genomic resources and posing of scientific questions?

5. Contrast the impact of genomics on the fabric of basic and applied biological research with the impact that molecular biology had in the 1970s. What will be the next revolution?

Literature Cited

Adams, M. D. et al. 1995. Initial assessment of human gene diversity and expression patterns based upon 83 million nucleotides of cDNA sequence. *Nature* 377: 3–174.

Adams, M. D. et al. 2000. The genome sequence of *Drosophila melanogaster*. *Science* 287: 2185–2195.

Alm E. J. et al. 2005. The MicrobesOnline Web site for comparative genomics. *Genome Res.* 15: 1015–1022.

AGI (Arabidopsis Genome Initiative). 2000. Analysis of the genome sequence of the flowering plant *Arabidopsis thaliana*. *Nature* 408: 796–815.

Aparicio, S. et al. 2002. Whole-genome shotgun assembly and analysis of the genome of *Fugu rubripes*. *Science* 297: 1301–1310.

Bailey, J. A., R. Baertsch, W. Kent, D. Haussler and E. E. Eichler. 2004. Hotspots of mammalian chromosomal evolution. *Genome Biol.* 5: R23.

Béjà, O. et al. 2000. Bacterial rhodopsin: Evidence for a new type of phototrophy in the sea. *Science* 289: 1902–1906.

Benson, D. A., I. Karsch-Mizrachi, D. Lipman, J. Ostell and D. L. Wheeler. 2007. GenBank. *Nucl. Acids Res.* 35(DB issue): D21–D25.

Bier, E. 2005. *Drosophila*, the golden bug, emerges as a tool for human genetics. *Nat. Rev. Genet.* 6: 9–23.

Birney, E. et al. 2004. An overview of Ensembl. *Genome Res.* 14: 925–928.

Blattner, F. R. et al. 1997. The complete genome sequence of *Escherichia coli* K-12. *Science* 277: 1453–1462.

Buckler, E. S., J. M. Thornsberry and S. Kresovich. 2001. Molecular diversity, structure, and domestication of grasses. *Genet. Res.* 77: 213–218.

Bult, C. J. et al. 1996. Complete genome sequence of the methanogenic archaeon *Methanococcus jannaschii*. *Science* 273: 1058–1073.

Campbell, B. J., J. L. Stein and S. C. Cary. 2003. Evidence of chemolithoautotrophy in the bacterial community associated with *Alvinella pompejana*, a hydrothermal vent polychaete. *Appl. Environ. Microbiol.* 69: 5070–5078.

Cañestro, C., H. Yokoi and J. H. Postlethwaite. 2007. Evolutionary developmental biology and genomics. *Nat. Rev. Genet.* 8: 932–942.

Carroll, S. B., J. K. Greiner and S. D. Weatherbee. 2004. *From DNA to Diversity: The Evolution of Animal Body Plans*. Blackwell Science, Malden, UK.

C. elegans Sequencing Consortium. 1998. Genome sequence of the nematode *C. elegans*: A platform for investigating biology. *Science* 282: 2012–2018.

Cello, J., A. V. Paul and E. Wimmer. 2002. Chemical synthesis of poliovirus cDNA: Generation of infectious virus in the absence of natural template. *Science* 297: 1016–1018.

Collins, F. S. and D. Galas. 1993. A new 5-year plan for the U.S. Human Genome Project. *Science* 262: 43–50.

Collins, F. S. et al. 1998. New goals for the U.S. Human Genome Project: 1998–2003. *Science* 282: 682–689.

Collins, F. S. et al. 2003. A vision for the future of genomics research. *Nature* 422: 835–847.

CSAC (Chimpanzee Sequencing and Analysis Consortium). 2005. Initial sequence of the chimpanzee genome and comparison with the human genome. *Nature* 437: 69–87.

Daubin, V, N.A. Moran and H. Ochman. 2003. Phylogenetics and the cohesion of bacterial genomes. *Science* 301: 829–832.

Dehal, P. et al. 2001. Human chromosome 19 and related regions in mouse: Conservative and lineage-specific evolution. *Science* 293: 104–111.

Deloukas, P. et al. 1998. A physical map of 30,000 human genes. *Science* 282: 744–746.

Delseny, M. 2004. Re-evaluating the relevance of ancestral shared synteny as a tool for crop improvement. *Curr. Opin. Plant Biol.* 7: 126–131.

Dicks, J. et al. 2000. UK CropNet: A collection of databases and bioinformatics resources for crop plant genomics. *Nucl. Acids Res.* 28: 104–107

Drosophila 12 Genomes Consortium. 2007. Evolution of genes and genomes on the *Drosophila* phylogeny. *Nature* 450: 203–218.

Ferguson-Smith, M.A. and V. Trifonov. 2007. Mammalian karyotype evolution. *Nat. Rev. Genet.* 8: 950–962.

Fleischmann, R. D. et al. 1995. Whole-genome random sequencing and assembly of *Haemophilus influenzae* Rd. *Science* 269: 496–512.

Fraser, C. M. et al. 1995. The minimal gene complement of *Mycoplasma genitalium*. *Science* 270: 397–403.

Gil, R., F.J. Silva, J. Peretó, and A. Moya. 2004. Determination of the core of a minimal bacterial gene set. *Micro. Molec. Biol. Rev.* 68: 518–537.

Gibson, D. G. et al. 2008. Complete chemical synthesis, assembly, and cloning of a *Mycoplasma genitalium* genome. *Science* 319: 1215–1220.

Glass, J. I. et al. 2006. Essential genes of a minimal bacterium. *Proc. Natl Acad. Sci. USA* 103: 425–430.

Goff, S. A. et al. 2002. A draft sequence of the rice genome (*Oryza sativa* L. ssp. *japonica*). *Science* 296: 92–100.

Green, R. E. et al 2006. Analysis of one million base pairs of Neanderthal DNA. *Nature* 444: 330–336.

Gyapay, G. et al. 1994. The 1993–1994 Genethon human genetic linkage map. *Nat. Genetics* 7: 246–339.

Hamosh, A., J. Scott, D. Amberger, D. Valle and V. A. McKusick. 2000. Online Mendelian Inheritance in Man (OMIM). *Hum. Mutat.* 15: 57–61.

Hattori, M. et al. 2000. The DNA sequence of human chromosome 21. *Nature* 405: 311–319.

Hu, J. et al. 2001. The ARKdb: Genome databases for farmed and other animals. *Nucl. Acids Res.* 29: 106–110.

Hudson, T. J. et al. 1995. An STS-based map of the human genome. *Science* 270: 1945–1954.

ICGSC (International Chicken Genome Sequencing Consortium). 2004. Sequence and comparative analysis of the chicken genome provide unique perspectives on vertebrate evolution *Nature* 432: 695–716.

IHGSC (International Human Genome Sequencing Consortium). 2001. Initial sequencing and analysis of the human genome. *Nature* 409: 860–921.

IHGSC (International Human Genome Sequencing Consortium). 2004. Finishing the euchromatic sequence of the human genome. *Nature* 431: 931–945.

IHMC (International HapMap Consortium). 2007. A second generation human haplotype map of over 3.1 million SNPs. *Nature* 449: 851–861.

IRGSP (International Rice Genome Sequencing Project). 2005. The map-based sequence of the rice genome. *Nature* 436: 793–800.

Jeffreys, A. J. et al. 2004. Meiotic recombination hot spots and human DNA diversity. *Phil. Trans. R. Soc. Lond. B Biol. Sci.* 359: 141–152.

Kobayashi, K. et al. 2003. Essential *Bacillus subtilis* genes. *Proc. Natl. Acad. Sci. USA* 100: 4678–4683.

Kuhn, R. M. et al. 2007. The UCSC Genome Browser database: Update 2007. *Nucleic Acids Res.* 35(DB issue): D668–D673.

Lander, E. S. et al. 1987. MAPMAKER: An interactive computer package for constructing primary genetic linkage maps of experimental and natural populations. *Genomics* 1: 174–181.

Lartigue, C. et al. 2007. Genome transplantation in bacteria: Changing one species to another. *Science* 317: 632–638

Levy, S. et al. 2007. The diploid genome sequence of an individual human. *PLoS Biol.* 5: e254.

Loots, G. G. et al. 2000. Identification of a coordinate regulator of interleukins 4, 13, and 5 by cross-species sequence comparisons. *Science* 288: 136–140.

Markowitz V. M. et al. 2008. The integrated microbial genomes (IMG) system in 2007: Data content and analysis tool extensions. *Nucleic Acids Res.* 36 (DB issue): D528–D533.

Mayer, K. et al. 2001. Conservation of microstructure between a sequenced region of the genome of rice and multiple segments of the genome of *Arabidopsis thaliana*. *Genome Res.* 11: 1167–1174.

McKusick, V. A. 1998. *Mendelian Inheritance in Man: Catalogs of Human Genes and Genetic Disorders*, 12th Ed. Johns Hopkins University Press, Baltimore.

McVean, G.A., S. Myers, S. Hunt, P. Deloukas, D. Bentley and P. Donnelly. 2004. The fine-scale structure of recombination rate variation in the human genome. *Science* 304: 581–584.

Mewes, H. W. et al. 1997. Overview of the yeast genome. *Nature* 387 (Suppl.): 7–9.

MGSC (Mouse Genome Sequencing Consortium). 2002. Initial sequencing and comparative analysis of the mouse genome. *Nature* 420: 520–562.

Murphy, W. et al. 2000. A radiation hybrid map of the cat genome: Implications for comparative mapping. *Genome Res.* 10: 691–702.

Noonan, J. P. et al. 2006. Sequencing and analysis of Neanderthal genomic DNA. *Science* 314: 1113–1118.

Ostrander, E. A. and L. Kruglyak. 2000. Unleashing the canine genome. *Genome Res.* 10: 1271–1274.

Parker, H. G. et al. 2004. Genetic structure of the purebred domestic dog. *Science* 304: 1160–1164.

Purugganan, M., A. Boyles and J. Suddith. 2000. Variation and selection at the CAULIFLOWER floral homeotic gene accompanying the evolution of domesticated *Brassica oleracea. Genetics* 155: 855–862.

RGSPC (Rat Genome Sequencing Project Consortium. 2004). Genome sequence of the Brown Norway rat yields insights into mammalian evolution. *Nature* 428: 493–521

Riley, M. 1997. Functions of the gene products of *Escherichia coli. Microbiol. Rev.* 57: 862–952.

RMGSAC (Rhesus Macaque Genome Sequencing and Analysis Consortium). 2007. Evolutionary and biomedical insights from the rhesus macaque genome. *Science* 316: 222–234.

Rubin, G. M. et al. 2000. Comparative genomics of the eukaryotes. *Science* 287: 2204–2215.

Rusch, D. B. et al. 2007. The Sorcerer II Global Ocean Sampling expedition: Northwest Atlantic through eastern tropical Pacific. *PLoS Biol.* 5: e77.

Schwartz, S. et al. 2000. PipMaker: A Web server for aligning two genomic DNA sequences. *Genome Res.* 10: 577–586.

Smith, V., K. Chou, D. Lashkari, D. Botstein and P. O. Brown. 1996. Functional analysis of the genes of yeast chromosome V by genetic footprinting. *Science* 274: 2069–2074.

Stark, A. et al. 2007. Discovery of functional elements in 12 *Drosophila* genomes using evolutionary signatures. *Nature* 450: 219–232.

Sutter, N. B. et al. 2007. A single *IGF1* allele is a major determinant of small size in dogs. *Science* 316: 112–115.

Tettelin, H. et al. 2001. Complete genome sequence of a virulent isolate of *Streptococcus pneumoniae. Science* 293: 498–506.

Tong, A. H. et al. 2004. Global mapping of the yeast genetic interaction network. *Science* 303: 808–813.

Tuskan, G. A. et al. 2006. The genome of black cottonwood, *Populus trichocarpa* (Torr. & Gray). *Science* 313: 1596–1604.

Venter, J. C. 2007. *A Life Decoded: My Genome, My Life.* Viking, New York.

Venter, J. C. et al. 2001. The sequence of the human genome. *Science* 291: 1304–1351.

Venter, J. C. et al. 2004. Environmental genome shotgun sequencing of the Sargasso Sea. *Science* 304: 66–74.

Wang, D. G. et al. 1998. Large-scale identification, mapping, and genotyping of single-nucleotide polymorphisms in the human genome. *Science* 280: 1077–1082.

Wang, R., A. Stec, J. Hey, L. Lukens and J. Doebley. 1999. The limits of selection during maize domestication. *Nature* 398: 236–239.

Whitelaw, C. A. et al. 2003. Enrichment of gene-coding sequences in maize by genome filtration. *Science* 302: 2118–2120.

Winzeler, E. A. et al. 1999. Functional characterization of the *S. cerevisiae* genome by gene deletion and parallel analysis. *Science* 285: 901–906.

Woese, C. 2000. Interpreting the universal phylogenetic tree. *Proc. Natl. Acad. Sci. USA* 97: 8392–8396.

Yooseph, S. et al. 2007. The Sorcerer II Global Ocean Sampling expedition: Expanding the universe of protein families. *PLoS Biol.* 5: e16.

Yu, J. et al. 2002. A draft sequence of the rice genome (*Oryza sativa* L. ssp. *indica*). *Science* 296: 79–92.

2 Genome Sequencing and Annotation

The first objective of most genome projects is to determine the DNA sequence either of the genome or of a large number of transcripts. This endeavor leads both to the identification of all or most genes and to the characterization of various structural features of the genome. This chapter explains the basic principles of how blocks of DNA sequence are obtained, and how these blocks are serially assembled first into contiguous stretches of sequence (contigs) and ultimately into a whole-genome sequence. Emerging sequencing methods that are paving the way toward "genomes-in-a-day for $1,000" are also described. Subsequently, the essence of the bioinformatic strategies for sequence alignment (since alignment is the basis of sequence assembly), comparison of cDNA/EST and genomic sequences, and annotation of open reading frames is described. In addition to identifying individual genes, DNA sequences reveal information about other features of the genome, including repetitive elements; centromeres and telomeres; variable distribution of GC content; and evolutionarily conserved elements of yet-to-be-determined function. This chapter concludes with a discussion of how genes are annotated by comparison with, and evolutionary analysis of, similar predicted protein sequences from other organisms.

Automated DNA Sequencing

The Principle of Sanger Sequencing

Almost all of the genome-scale sequencing performed, up to and including the Human Genome Project, has made use of the basic chain termination method developed in 1974 by Frederick Sanger. The idea behind Sanger's method is to generate all possible single-stranded DNA molecules differing in length by one nucleotide that are complementary to a template. The template starts at a common 5′ base and extends anywhere up to 1 kilobase or more in the 3′ direction. These single strands of DNA are labeled in such a way as to allow us to infer the identity of the 3′-most base in each molecule. Separation of the molecules according to size by electrophoresis results in a ladder of bands, with each adjacent band corresponding to a class of molecule differing by the addition of one particular base. The sequence is then "read" from this ladder, as shown in Figure 2.1.

The sequencing reaction is primed by annealing an oligonucleotide of 20 or so bases to a denatured template. The template is usually a plasmid that contains a cloned piece of DNA. The same "universal" primer can be used in all reactions, since it is complementary to a short sequence in the plasmid adjacent to the inserted DNA fragment that is being sequenced. For some applications, genomic DNA is sequenced directly, in which case the primer is specific to a sequence within the gene. A form of the enzyme DNA polymerase is then used to catalyze synthesis of the complementary strand in the presence of all four dNTPs.

The trick used to generate molecules that differ in length by one nucleotide is to randomly terminate polymerization of the growing single strand sequence by the incorporation of a dideoxynucleotide (ddNTP). Whenever a ddNTP incorporates, the absence of a hydroxyl group on the sugar-phosphate backbone means that there is nothing for the next dNTP to attach to, and polymerization is effectively terminated. Once the sequencing reaction terminates, the DNA is again denatured and the fragments are separated by virtue of the differential retardation of the migration of molecules of different length and electric charge through a semiporous matrix such as that produced by an acrylamide polymer.

Throughout the 1980s, most sequencing was done manually. The process was labor-intensive: each reaction was set up individually, polyacrylamide gels for electrophoretic separation were prepared by hand, and the sequence bands were read by the human eye. A radioactive label, typically ^{33}P or ^{35}S, was incorporated into the sequencing product as part of one of the dNTPs, and four separate reactions had to be set up for each sequence—one for each dideoxy terminator. The four reactions were run side by side on a large slab gel, and were stopped after an appropriate time so that all products within the desired size range could be visualized by exposing the gel to X-ray film. With a separation of 0.5 mm between bands corresponding to each molecule differing in length by one nucleotide, generally only up to 500 bases could be read from a single set of four lanes on a 30-cm gel. At least three runs of different durations had to be performed to sequence any fragment longer than 1,000 base pairs. Manual sequencing was also limited by problems associated with poor resolution of short stretches of sequence, notably GC-rich regions that tend to compress, leading to ambiguities that could not always be resolved by sequencing the opposite strand.

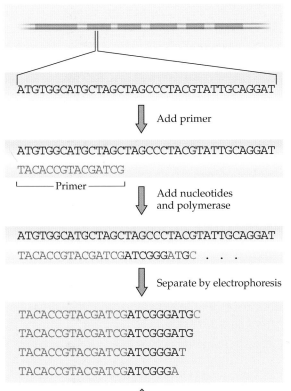

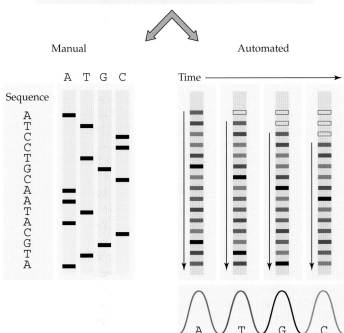

Figure 2.1 The principle of dideoxy (Sanger) sequencing. Single-strand molecules are synthesized from a template and randomly terminated by the addition of a labeled dideoxynucleotide (ddA, ddT, ddC, ddG), then separated by electrophoresis. The sequence is visualized either by radioactivity (manual sequencing; one lane per base on a fixed gel) or fluorescence (automated sequencing; each base is identified by computer as it emerges from a single lane).

High-Throughput Sequencing

In the 1990s, several advances were made that enabled the automation of Sanger sequencing. Genome-scale sequencing would be impossible without these new techniques and equipment:

- Four-color fluorescent dyes replaced the radioactive label. Attachment of these dyes to the ddNTPs results in a fluorescent tag directly marking just the terminated DNA molecule. Consequently, a single sequencing reaction spiked with all four ddNTPs is sufficient to sequence any template. (For some applications, the dyes are attached to the primer, in which case four reactions are still performed but may be pooled in a single lane before electrophoresis.)

- Rather than stopping electrophoresis at a particular time, the products are scanned for laser-induced fluorescence just before they run off the end of the electrophoresis medium. The sequence is collected as a set of four "trace files" that indicate the intensity of the four colors; a peak in the trace distribution implies that the particular base was the last one incorporated at the position. Such traces can be read automatically, as described in the next section, resulting in enormous savings in time and reducing the scoring errors that inevitably creep into manual readings.

- Improvements in the chemistry of template purification and the sequencing reaction, including use of bioengineered thermostable polymerases that can read through secondary structure with high fidelity, has extended the length of high quality sequence. Reads greater than 1,200 bp are possible with current technology, though the 500–900 bp range is more common.

- Slab gel electrophoresis gave way to capillary electrophoresis with the introduction in 1999 of Applied Biosystem's ABI Prism 3700 automated sequencers, which in 2003 were updated with ABI Prism 3730 DNA analyzers. These sequencers give extremely high quality, long reads; save time and money by abolishing the laborious and often frustrating step of gel pouring; and add a new level of automation in that the capillaries are loaded by robot from 96-well plates rather than by hand. Each machine can handle twenty 96-well plates per day, or approximately 2 Mb of sequence—which is two orders of magnitude greater than the output of a single investigator just a decade earlier. Other automated DNA sequencing systems are also available, including GE Healthcare's MegaBACE 4000 and Li-Cor's 4300 DNA Analysis systems.

Reading Sequence Traces

The reading of raw sequence traces, or **base calling**, is now routinely performed using automated software that reads bases, aligns similar sequences, and provides an intuitive platform for editing. Genome-scale sequencing requires only minimal human input—which is a good thing, because if it took just a quarter of an hour for a single person to properly edit any given read, it would take seven people a very boring week to process the output from one day's operation of a single ABI Prism 3700. Automated software such as the freely available **phred** program developed at the University of Washington (http://www.phrap.org/phredphrapconsed.html; Ewing et al. 1998; Ewing and Green 1998), or commercial equivalents, convert traces into sequences that can be deposited in a database within seconds after the completion of a sequencing run. These programs assign probability scores to the accuracy of each base call as the trace is read, and this information is utilized in subsequent alignment steps.

Whereas human observers integrate multiple pieces of information in calling a base from a sequence trace, software uses an algorithm in which the process is broken down into a series of steps, as outlined in Figure 2.2. First, the four traces corresponding to the four fluorescence spectra are merged into a single file, maintaining the register of the peaks. Next the computer calculates where it expects to find a peak, using an averaging process based on the mean distance between peaks over some stretch of the sequence. This ensures that an N is called in place of an A, C, G, or T where no base is seen, and that only a single base is called where two appear to be present. The most efficient algorithms can adjust for the increase in spacing between peaks that occurs as the run proceeds, as well as for variation due to changes in local GC content.

Subsequently, the algorithm detects local maxima for each of the four channels and checks to see that each peak occurs with the predicted spacing relative to the adjacent peaks. Since not all peaks are the same height, and runs of the same base can show reduced peak resolution, algorithms typically employ a threshold that computes the relative magnitude of a local minimum and maximum to decide whether a peak is real. Similarly, on occa-

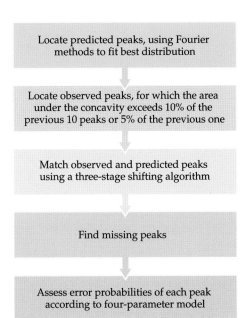

Locate predicted peaks, using Fourier methods to fit best distribution

Locate observed peaks, for which the area under the concavity exceeds 10% of the previous 10 peaks or 5% of the previous one

Match observed and predicted peaks using a three-stage shifting algorithm

Find missing peaks

Assess error probabilities of each peak according to four-parameter model

Figure 2.2 The phred base-calling algorithm. Programs such as phred convert computer-generated traces into base sequences and assess the probable accuracy of each base call.

sion the tail of one peak can be higher than the maximum of the next peak of a different nucleotide, so the program must be able to call the correct peak rather than simply detecting the highest signal. Where two peaks are called but there is only space for one, or vice-versa, peaks can be either omitted or split based on probability measures. Uncalled peaks are inserted if there is a local maximum of one dye that was not initially called as a peak. Automated base calling usually takes less than a half a second per 1-kb trace and results in a string of letters corresponding to the 5′-to-3′ order of nucleotides.

Good software should also be able to identify instances where two different bases are present at the same site. This can occur if an individual is heterozygous, or if two or more samples have been pooled together. Alignment with other sequences greatly assists in the identification of such single nucleotide polymorphisms (SNPs), but in the first stage of trace reading, these may simply be designated as unassigned or ambiguous bases and labeled N. Probability scores can also be used to flag sites as potential SNPs.

Several common problems can lead to errors in sequencing, some of which are shown in Figure 2.3. For example, the first 50 or so bases of a read are typically "noisy" due to the anomalous migration of short DNA fragments that contain bulky dyes (Figure 2.3A). Similarly, traces become progressively less uniform as a run proceeds and the effects of diffusion are amplified while the relative mass differences between successive fragments

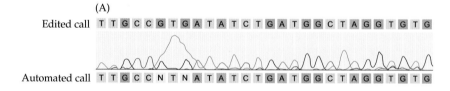

(A)

Edited call   T T G C C G T G A T A T C T G A T G G C T A G G T G T G

Automated call   T T G C C N T N A T A T C T G A T G G C T A G G T G T G

Figure 2.3 Automated sequence chromatograms. Each trace is accompanied by two lines of sequence: the automatic call below the trace, and a manually edited call above it. (A) This sequence shows "noisiness" typical of the first 30 or so bases of a run. (B) A decline in sequence quality typically occurs after about 800 bp. (C) The middle two rows show a segment of two sequences that are polymorphic for both SNPs and an indel.

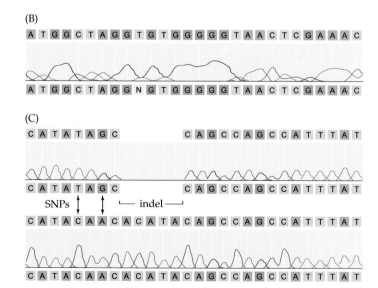

(B)

A T G G C T A G G T G T G G G G G G T A A C T C G A A A C

A T G G C T A G G N G T G G G G G G T A A C T C G A A A C

(C)

C A T A T A G C   C A G C C A G C C A T T T A T

C A T A T A G C   C A G C C A G C C A T T T A T

SNPs ↑   ↑   └─ indel ─┘

C A T A C A A C A C A T A C A G C C A G C C A T T T A T

C A T A C A A C A C A T A C A G C C A G C C A T T T A T

decreases (Figure 2.3B). In Figure 2.3C, alignment of the sequence traces of two different alleles reveals SNPs and insertion-deletion (indel) polymorphisms. Dye-terminator chemistry deals quite well with compression problems, but anomalies such as reduced signals of G following A must be accounted for. Not all reactions work as well as they should, resulting in low signal-to-noise ratios and effects such as a large peak of one dye that overwhelms the true signal; these must also be recognized either by the software or manual editors.

EXERCISE 2.1 *Reading a sequence trace*

Read the following DNA sequence obtained by direct sequencing of a single individual organism, assuming that on the trace green is A, red is T, blue is C, and black is G. Remark on any ambiguities in the sequence.

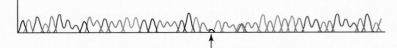

ANSWER: *The sequence reads:*

GCTCATTTGAATAACCTGAAATGCA**N**CAA(C/T)AACACACATTCATTTATC

Note that the base labeled N (arrow) is impossible to call due to poor sequence quality. Four nucleotides further downstream, two peaks of the same height are observed at the same location, suggesting that the individual is heterozygous at this site for a C/T single nucleotide polymorphism (SNP).

All automated base-calling algorithms make errors, and some deal with particular types of ambiguity better than others. For this reason, it has become standard to assign error probability estimates to each base based on measures of the consistency of peak heights. The error probability, P, is then converted to a phred score, q, which is 10 times the negative logarithm of P. Thus, a phred score of less than 13 means that there is a greater than 0.05 probability that the base is incorrectly called, while a score of 30 or more means that the associated error probability is 0.001. In general, scores above 20 are regarded with high confidence.

Contig Assembly

The "finishing" stage in sequencing a stretch of DNA longer than a single clone entails alignment, editing, and error correction. All of these steps are customarily done with sequence editing software, an example of which is the combination of the **phrap** assembler and **consed** graphic editor (Gordon et al. 1998) that complement the phred base-caller and are widely used in academic settings. High-level commercial editors also aim to link assembled sequences immediately to relational databases and annotation applications. The key features of successful editors are:

- The use of color to illustrate key features such as the different bases, quality scores, regions of sequence conservation, and contrasts between automated and manual base calls.

- The ability to view and navigate along the actual traces of the sequences being compared, and to tag ambiguities and features of interest with notes.

- Easy display of the complementary strand.

- Tools for manual sequence editing, including inserting and deleting bases, without disrupting the original trace files yet propagating the edits throughout the assembly as appropriate (including adjustments to linked output files as requested by the editor).

- A flexible alignment algorithm that implements user-defined alignment parameters.

- Computation of probability scores associated with a calculated consensus sequence.

- The ability to identify potentially polymorphic sites.

- Provision of tools to guide error correction.

Editing starts with the chromatogram files (for example, of the types .abi or .scf standard chromatogram format) that contain the fluorescence trace profiles for each of the four channels of a read. These files are generally left untouched by editors. They are converted to base call files by programs such as phred (which creates .phd files) that include the base calls, quality scores, peak positions, and any tagged information provided by human annotators. If a separate assembly program such as phrap is used, it creates a third input file (.ace) that includes information on the alignment, the consensus contig sequence, and quality scores for each base in the consensus. After manual editing, multiple versions of .phd and .ace files may be stored, each of which can be called up later for subsequent re-editing, perhaps following addition of newly obtained sequences.

Consed provides a graphic interface that allows the user to call up and interactively edit individual reads and/or contigs assembled by phrap. The aligned-reads window shown in Figure 2.4 presents the consensus sequence as well as each individual sequence that contributes to an alignment. Gray shading indicates the quality of each base call, with white indicating high quality. The user can manipulate features of the display, including the use of color to indicate quality or match, and the inclusion of tags. Navigation along a sequence is facilitated either by scroll bars or, for contigs that exceed several kilobases in length, by manual entry of numerical positions, and can be performed using criteria that automatically identify features such as ambiguous bases, regions of poor alignment, high-quality reads that dis-

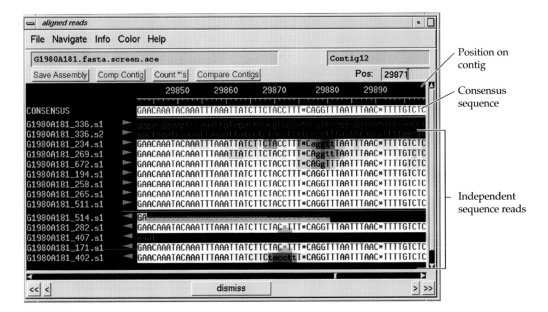

Figure 2.4 An aligned-reads window in consed.   The consensus sequence is shown across the top, as calculated by phrap from the individual files included in the alignment below. Gray shading and lowercase lettering indicate reduced quality of a particular base call. Asterisks show the location of a potential insertion or deletion, one example of which may be a true polymorphism, indicated by the yellow bar highlighting a region of discrepancy between two high-quality regions. The pink arrowheads indicate which strand was sequenced, in this case suggesting that the discrepancy may be an artifact of which strand is represented. The light blue bar marks a region of misalignment. Scroll bars and clickable options allow the user to navigate and edit the entire assembly. (From Gordon et al. 1998.)

agree with the consensus, and homopolymeric stretches. Editing of individual bases is facilitated by calling up the trace file, which allows visual comparison of two or more traces, again using a series of interactive tools.

The alignment algorithms used by various assemblers are generally modifications of methods introduced by Needleman and Wunsch in 1970 and by Smith and Waterman in 1981; these methods are described in Box 2.1. Not all alignments of the same set of sequences are identical, primarily because they assign different default parameters—and use them differently—to weight the effects of insertions and deletions relative to single-base mismatches. Algorithms that also use quality scores from the contributing sequence reads add further complexity, as does weighting attached to the frequency of mismatches. Although default settings are the easiest to use, they are not necessarily the settings most likely to yield the correct alignment. Manual inspection remains the most efficient way to resolve ambiguities, but is too time-consuming to be performed on a genome scale.

BOX 2.1 Pairwise Sequence Alignment

The single most important class of bioinformatics tools are those dealing with pairwise alignment of DNA and amino acid sequences. We shall see that pairwise sequence alignment and database searching are essentially the same computational problem, and that the underlying algorithms for pairwise alignment are found in a variety of other bioinformatics applications.

Consider the following two simple sequence alignments:

	Alignment 1	Alignment 2
Sequence 1	ACGCTGA	ACGCTGA
Sequence 2	A--CTGT	ACTGT--

Intuitively, we recognize that the first alignment is "superior" to the second. Why? Because the residues in the same column are mostly identical (four exact matches), and there is no obvious way to improve the alignment quality. In contrast, the second alignment has less sequence identity (three exact matches), and we can recognize the potential improvement offered by "sliding" the CTG subsequence of sequence 2 to match the CTG in sequence 1.

The goal of pairwise sequence alignment is to formalize this intuitive procedure. We seek alignments that contain high levels of sequence identity, few mismatches and gaps, and little opportunity for improvement. We will work under the assumption that the observed similarity in sequences to be aligned is the result of either random chance (in which case the similarity is likely of no interest) or the result of a shared evolutionary origin.

By placing residues from two sequences in the same column of an alignment, we are implying that they are **homologous**, meaning that the two residues are descendants of a common ancestral residue. Homology is not to be confused with sequence identity; two residues may be identical without being homologous. Two positions in an alignment column can have any of the relationships shown in the figure. Parts C and D of Figure A point out that it is possible to have the same true alignment arise through different evolutionary events; when we are dealing with pairwise alignments, the data do not allow us to distinguish between insertion and deletion events, thus explaining the origin of the term **indel**.

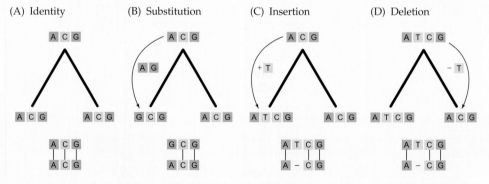

Figure A Common evolutionary events and their effects on alignment. The alignments are shown beneath diagrams of how they might occur. Two positions in an alignment may contain identical residues (A); they may hold different residues as the result of one or more substitution events (B); or one of the sequences can hold a gap or indel, the result of insertion (C) or deletion (D) events.

Of course, sequences do not come to us in aligned form. We must infer the location of past indel and substitution events. Several computational approaches for performing this task exist, but the most widely used are score-based methods in which inferred events in an alignment are penalized by amounts proportional to their rarity. For example, we may assign substitutions a penalty of –1 unit, while less common indels receive a penalty of –5 units. A match (identity) might receive a score of +3 units. The score for a candidate alignment is then the sum of the penalties at each of the positions in the alignment. Using this scoring scheme, the four alignments in Figure A have scores of 9, 5, 4, and 4, respectively. If we search through all possible alignments and compute the score for each, we can identify the alignment (or alignments) with the optimal score and use it as our best guess for the true alignment.

It is worthwhile to understand the computational approach for finding the optimal pairwise alignment of two sequences. Understanding the algorithms enhances the ability to work with database searching tools, and the computational approach is widely used in many other bioinformatics tools. The problem is defined as taking as inputs two sequences and inserting gaps into one or both of them in such a way to maximize the total alignment score. Needleman and Wunsch (1970) described the first computationally tractable algorithm for finding optimal pairwise alignments. The algorithm is an example of *dynamic programming*, and it guarantees that we find all optimal alignments in a manner that avoids exhaustive searching of all possible alignments of two sequences of lengths m and n.

Consider the example given at the beginning of this box, where sequence 1 has 7 residues (ACGCTGA) and sequence 2 has 5 residues (ACTGT). One way to find the optimum alignment would be to compute every possible alignment score, which is not so difficult in this example but would be computationally intensive if the two sequences were ten times longer, or if one sequence was being aligned with a database of thousands of sequences. Instead, the Needleman-Wunsch approach uses a recursive algorithm based on the addition of one position at a time to the best alignment of residues up to that point. That is, if we already know the best alignment of the first $(i-1)$ residues of sequence 1 with the first $(j-1)$ residues of sequence 2, then we can quickly compute the best alignment of residues i against j. This is because thre are only three ways to terminate the longer alignment, illustrated as follows for alignment of the first four residues ($i = j = 4$):

1. Having already aligned residues 1 through $(i-1)$ of sequence 1 and 1 through $(j-1)$ of sequence 2, when we add both residues i and j, we add the match score or penalty that they generate (in this case, a penalty due to a C-G mismatch):

    ACG
    | | .   (Score = 3 + 3 – 1 = 5) → | | . .   (score = 5 – 1 = 4)
    ACT   ACGC
                                               ACTG

2. Having already aligned residues 1 through $(i-1)$ of sequence 1 and 1 through j of sequence 2, when we add residue i we incur the penalty for aligning residue i with a gap:

    AC–G
    | | . | (Score = 3 + 3 – 5 + 3 = 4) → | | . | (score = 4 – 5 = –1)
    ACTG   AC–GC
                                                   ACTG–

3. Having already aligned residues 1 through i of sequence 1 and 1 through $(j–1)$ of sequence 2, when we add residue j, we incur the penalty for aligning a gap with residue j:

    ACGC
    | | .   (Score = 3 + 3 – 1 – 5 = 0) → | | .   (score = 0 – 5 = –5)
    ACT–   ACGC–
                                                   ACT–G

(*Continued on next page*)

BOX 2.1 *(continued)*

Clearly, the best option was adding both residues without a gap. Formally, the score for alignment of *i* residues of sequence 1 against *j* residues of sequence 2 is given by:

$$S(i,j) = \max \begin{array}{l} S(i\text{-}1, j\text{-}1) + c(i,j) \\ S(i\text{-}1,j) + c(i,\text{-}) \\ S(i,j\text{-}1) + c(\text{-},j) \end{array}$$

where $c(i,j)$ is the score for alignment of residues *i* and *j* and takes the value 3 for a match or –1 for a mismatch, and $c(\text{-},j)$ is the penalty for aligning a residue with a gap, which takes the value –5 given the default match and penalty scores. (Note that assigning different scores may give different optimal alignments.)

We can display this algorithm using a two-dimensional matrix. The (i,j) entry in the matrix is the best score for aligning the first *i* residues of sequence 1 (on the *y*-axis) with the first *j* residues of sequence 2 (the *x*-axis). The optimal alignment score of two sequences of length *m* and *n* can then be found in the upper right of the matrix, the (m,n) element.

We begin by filling in the first row and first column of the matrix, which corresponds to initial gaps (e.g., aligning the A in sequence 2 against a gap in sequence 1 costs –5; aligning the first three positions of sequence 1 against initial gaps in sequence 2 costs –15).

$i = 7$	A	−35						
$i = 6$	G	−30						
$i = 5$	T	−25						
$i = 4$	C	−20						
$i = 3$	G	−15						
$i = 2$	C	−10						
$i = 1$	A	−5	3					
$i = 0$	−	0	−5	−10	−15	−20	−25	
	−		A	C	T	G	T	
$j =$		0	1	2	3	4	5	

From this starting point, we can fill in each cell in the table using the algorithm. The entry for S(1,1), is the maximum of the following three events:

$$S(0,0) + c(A,A) = 0 + 3 \qquad = 3$$
$$S(0,1) + c(A,\text{-}) = (\text{-}5) + (\text{-}5) = \text{-}10$$
$$S(1,0) + c(\text{-},A) = (\text{-}5) + (\text{-}5) = \text{-}10$$

In a similar fashion, we find S(2,1) as the maximum of three values: (–5)–1 = –6; (3)–5 = –2; and (–10)–5 = –15. The best entry is the addition of the C indel to the A–A match, for a score of –2. If we fill in the entire matrix, we get the following:

A	−35	−27	−19	−9	−3	1
G	−30	−22	−14	−4	2	2
T	−25	−17	−9	−1	−1	7
C	−20	−12	−4	0	4	3
G	−15	−7	1	5	4	−1
C	−10	−2	6	1	−4	−9

A	−5	3	−2	−7	−12	−17
−	0	−5	−10	−15	−20	−25
	−	A	C	T	G	T

If you are having trouble following the algorithm, just imagine filling in the cells of the matrix by exhaustively computing the best score for each alignment of i by j residues. The first column shows what would happen if you aligned each of the i residues of sequence 1 with gaps; the second column if you aligned them against A and a series of gaps, and so on. The entry for row i and column j indicates the optimal score for aligning i residues of sequence 1 against j residues of sequence 2. These scores were computed by iterative algorithm rather than exhaustive computations for each possibility. The score for the optimal alignment of the whole sequences is indicated in the top right hand corner, the (m,n) cell, in this example alignment of 7 against 5 residues having a score of 1.

In order to see what the actual alignment associated with this score looks like, we perform a **traceback** step. Beginning at the upper right (m,n) element, we determine which of the three possible "directions" was taken to reach that cell. For example, the 1 in the (7,5) cell could only be reached by the addition of the mismatch AT, the value of 1 being obtained by adding a mismatch penalty of –1 to the 2 found in the (6,4) cell. Had the (7,5) cell been reached from the (7,4) cell, the score would have been (–3) + (–5) = –8; had it been reached via adding an indel from the (6,5) cell the entry would have been 2 + (–5) = –3. We will indicate the taken directions with arrows, and step our way back through the entire matrix to the (0,0) cell.

A	−35	−27	−19	−11	−3	1
G	−30	−22	−14	−6	2	2
T	−25	−17	−9	−1	−1	7
C	−20	−12	−4	0	4	3
G	−15	−7	1	5	4	−1
C	−10	−2	6	1	−4	−9
A	−5	3	−2	−7	−12	−17
−	0	−5	−10	−15	−20	−25
	−	A	C	T	G	T

A cell with two arrows indicates two equally good alignments, and any continuous path from the upper right cell to the lower left cell represents an alignment with the optimal score. For example, the value of –4 in the (4,2) position could be obtained from either of two directions—either by adding a penalty of –5 to the 1 in cell (3,2), or by adding a match of 3 to the –7 in cell (3,1). Every such split doubles the number of best-scoring alignments. In this example, this means that there is a pair of optimal alignments, each having score 1, for these two sequences, and these are:

```
ACGCTGA        ACGCTGA
A--CTGT        AC--TGT
```

Note that each alignment includes four matches, one mismatch, and two indels. The ambiguity has to do with which C in sequence 1 aligns with the C in sequence 2.

EXERCISE 2.2 *Computing an optimal sequence alignment*

Compute the best possible alignment for the following two sequences, assuming a gap penalty of –5, a mismatch penalty of –1, and a match score of +3. Would your answer be any different if the gap penalty was also –1 (rather than –5)?

                    AGCGTAT   *and*   ACGGTAT

ANSWER: *Three possible high-quality alignments are:*

```
    (1) AGCGTAT        (2) AGC- GTAT        (3) AGCG- TAT
        | ··||||           |-|-||||             |-||-|||
        ACGGTAT            A-CGGTAT            A- CGGTAT
```

The second and third alignments will produce the same score with these penalties, namely (6 × 3) – (2 × 5) = 8 with a gap penalty of –5; or (6 × 3) – (2 × 1) = 16 with a gap penalty of –1. By contrast, the first alignment gives a score of (5 × 3) – (2 × 1) = 13. Thus, the first alignment is best with a large gap penalty, but either of the other two alignments would be better with the smaller gap penalty.

The final task of a sequence editor is to help resolve gaps and ambiguities. To this end, the "Autofinish" function within Consed (Gordon et al. 2001) designs finishing reads by suggesting primers and identifying templates that will help to bridge gaps between contigs. Consed also facilitates the comparison of two or more contigs for consistency—for example, by identifying regions that may be erroneously inverted in one assembly, or contain an insertion or site of recombination that may have generated a chimeric clone. The program will also identify local regions of poor sequence and design primers that can be used for a new round of PCR amplification and sequencing across the ambiguous region. This is particularly important in a genome project where a pre-set, genome-wide error acceptance rate has been established, as design of further experiments to improve the sequence quality can be automated.

Once a multiple sequence alignment has been produced, a consensus sequence is derived on the basis of the quality of the reads at each position. If one or two of the reads in a contig disagree with the others, and if these reads are of low quality, the consensus will disregard them and instead reflect the high-quality sequences. Furthermore, by combining the probability scores of multiple reads, confidence in the actual consensus is greatly increased. *It is this averaging of multiple reads that provides high confidence in automated DNA sequence determination and circumvents the impossibility of manually verifying every sequence in a genome.* That is to say, in high-throughput sequencing applications, it is much more efficient to sequence every stretch of DNA multiple times and derive a consensus than to manually curate a single trace for each sequence. Nevertheless, most genome sequencing projects make the original trace files available for individual investigators to examine online.

Emerging Sequencing Methods: The Next Generation

Although dideoxy-based sequencing has supported the vast majority of genome research to date, a number of novel sequencing technologies are on the horizon (Bentley 2006). Sequencing a mammalian genome still costs $30 to $50 million, so it is not surprising that considerable investment has been made in the development of technologies that will dramatically reduce sequencing costs; the target is $1,000 per human genome by the year 2010.

Three commercial platforms for genome-scale sequencing were introduced in 2007, each with the capacity to sequence whole bacterial genomes for just several thousand dollars. These can also be used to re-sequence strains of model organisms such as *Drosophila* or *Arabidopsis*, to generate profiles of mRNA abundance, or to target specific regions of mammalian chromosomes for re-sequencing to detect polymorphisms and mutations, among other applications. All three of these platforms generate hundreds of megabases of sequence in a couple of days, but require high-performance computing hardware and extreme data storage capacity. Their basic principles are schematicized and compared with traditional Sanger sequencing in Figure 2.5. Flash presentations demonstrating these technologies are available from the respective manufacturers' Web sites.

The **454 Life Sciences** platform relies on massively parallel pyrosequencing to generate millions of 100-bp reads (Margulies et al. 2005). The target DNA is fragmented and coupled to beads, which are then emulsified with a mixture of nucleotides and DNA polymerase that facilitates amplification and the deposition of millions of copies of a single fragment on a bead into a picoliter-sized chamber. The sequence on each bead is then determined by pyrosequencing (see Figure 3.21) in which nucleotide incorporation is associated with a flash of light. The four nucleotides are added sequentially in a cycle that is repeated several hundred times. Since the light intensity is proportional to the number of nucleotides incorporated at each step, the sequence is inferred from the successive intensities of light flashes associated with the incorporation of each nucleotide. Millions of individuals reads are then aligned with the reference genome to which they are being compared, or are assembled de novo into an original genome sequence.

The **Illumina Genome Analyzer** uses technology originally developed by Solexa. It is conceptually similar to polony sequencing (see below), except that the cell-free clones are arrayed on an optically transparent surface. Single-stranded, 200-bp fragments of DNA are ligated to 5′and 3′adapters that are complementary to a sea of primers on the array surface. These support conversion into bridges of double-stranded DNA that are amplified on the array, resulting in "microislands" of paired fragments sticking up into solution, anchored at either end. These microislands are sequenced one base at a time using fluorescently labeled dideoxynucleotides, which are read and then deprotected so that they will accept a new nucleotide in the next cycle. The four

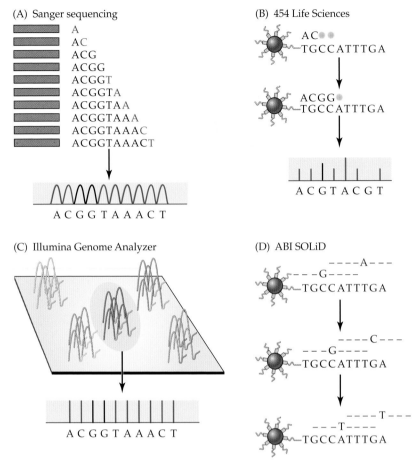

Figure 2.5 Principles of massively parallel genome resequencing. (A) Traditional Sanger sequencing reads each nucleotide as labeled fragments of increasing length electrophorese past a laser detector. (B) Pyrosequencing (454 Life Sciences) captures the intensity of signal emitted as each successive nucleotide is added, with peak height proportional to the number of nucleotides in succession. (C) Reversible terminator technology (Illumina Genome Analyzer) reads one base at a time from locally amplified cell-free clones on an optically transparent surface. (D) Ligation sequencing (ABI SOLiD) depends on inference of the sequence from incorporation of chains of 9-mer oligonucleotides. (After Bentley 2006, Figure 1.)

nucleotides are added sequentially in 50 or more cycles, starting with a primer complementary to one of the adapters. After the first reaction is completed, sequencing from the other primer supports the generation of paired 50-mer sequences. Together these sequences provide better alignment to the reference genome than a single read would.

The **ABI SOLiD** (**S**upported **O**ligonucleotide + **Li**gation **D**etection) system is a hybrid between sequencing by hybridization and primer extension on a template. Beads with amplified fragments attached to them are analyzed in massively parallel fashion, but the sequence is read according to a unique approach, every fifth base at a time. All possible 8-mer oligonucleotides are hybridized to the template simultaneously, but only those that have a specific two-base-pair combination in the middle will bind strongly enough to the template to allow the oligonucleotide to ligate to the growing strand. The identity of the fifth base is then decoded according to the color of the fluorescent dye at the unligated end of the 8-mer. The last three nucleotides are cleaved off, and a new cycle of oligonucleotide hybridization begins. After five to seven cycles of ligation, the newly synthesized strand is melted off and a second round of ligation cycles is initiated with a new primer that is offset from the first primer by one base. After five such rounds, the reads are merged to infer a sequence. The requirement that each oligonucleotide provides a readout of two adjacent bases ensures high accuracy of the sequence read. Assembly of the genome sequence proceeds by computational alignment with a reference genome.

An intriguing method for small-laboratory scale genome analysis is known as **polony sequencing** (Figure 2.6). Polonies are PCR colonies: rather than cloning a DNA molecule into a plasmid and growing it in cells, or

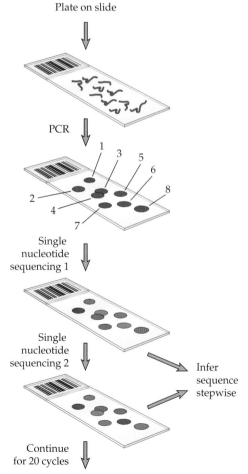

Plate on slide

PCR

Single nucleotide sequencing 1

Single nucleotide sequencing 2

Infer sequence stepwise

Continue for 20 cycles

Figure 2.6 Single-molecule polony sequencing.   A dilute solution of DNA molecules is prepared in an acrylamide matrix and plated onto a glass microscope slide. In situ PCR produces thousands of tiny colonies of DNA ("polonies"), which become templates for the incorporation of single dye-labeled dNTPs. The slide is read after each cycle of incorporation of a new base. Each polony produces a short (20–25 nucleotides) sequence. These short sequences are assembled computationally into a contiguous sequence.

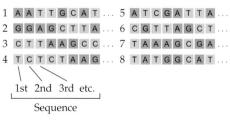

1 A A T T G C A T ... 5 A T C G A T T A ...
2 G G A G C T T A ... 6 C G T T A G C T ...
3 C T T A A G C C ... 7 T A A A G C G A ...
4 T C T C T A A G ... 8 T A T G G C A T ...

1st 2nd 3rd etc.

Sequence

amplifying it in a plastic tube, the PCR reaction is performed in the acrylamide matrix on a glass microscope slide. The DNA to be sequenced is greatly diluted so that single molecules form PCR templates a few microns apart from one another. This results in tens of thousands of tiny colonies of identical DNA molecules, and cycles of fluorescent nucleotide incorporation followed by scanning of the slide in a microarray reader leads to the sequencing of several kilobases of DNA per slide.

Several other approaches are under development, or are being used in specific resequencing applications. These include:

- *Sequencing by hybridization (SBH)* makes use of the complementarity of the two strands of DNA molecules to detect whether an exact match to an oligonucleotide is present in a sample of DNA. As described in Chapter 3, this principle underlies variant detector arrays (VDA) that contain all possible oligonucleotides in a molecule several kilobases in length, as well as mismatches at each site. The company Perlegen has used this technology to resequence entire primate chromosomes and to scan individual human genomes for 1.5 million SNPs.

- *Mass spectrophotometric (MS) techniques* have been demonstrated for the sequencing of fragmented DNA molecules up to 50 bases in length. The identity of fragmented oligonucleotides can be determined by reference to the time of flight of a set of standards through a vacuum chamber. Making use of advances in information theory, in principle it should be possible to determine the full sequence of a molecule that is divided into all possible oligonucleotides, thereby also facilitating de novo sequencing. MS methods are rapid and likely to become cost-effective. The Sequenom corporation uses this approach in massively parallel genotyping based on the differential mass of polymorphic nucleotides.

- Ultrafast and relatively inexpensive sequencing of long DNA fragments might be achieved using *nanopore sequencing strategies* (Deamer and Akeson 2000). One approach monitors changes in electrical current as the nucleotide base blocks a tiny membrane pore as a single strand of DNA or RNA passes through it. Another, being pursued by US Genomics, channels the nucleic acid through a single-molecule fluorescence reader. Such methods promise to allow the sequencing of molecules hundreds of kilobases long in just a few minutes.

- Several other single-molecule approaches still employ the basic Sanger approach, but avoid the reagent and liquid-handling costs associated with capillary-based automated sequencing. Pacific Biosciences use *single-molecule real time (SMRT) technology* to visualize the incorporation of each base into the growing complementary strand of an isolated strand of DNA. The fluorescent dyes are located on the phosphate group rather than the base of the nucleotide, and each of 10 base incorporations per second is detected in a parallel array of miniscule chambers. In theory, molecules kilobases long may be read, and it is possible this approach may lead to the generation of a human genome sequence in 15 minutes for around $100.

A plausible extension of all these emerging techniques will be the ability to obtain a thorough read-out of gene expression at the cellular level by lyzing cells and directly reading the sequences of thousands of complete mRNA molecules.

Genome Sequencing

Whole-chromosome sequences are reassembled from the sequences of hundreds of thousands of fragments, each typically between 500 and 1,000 bp in length. Two general strategies for fragmentation and reassembly are used, known as **hierarchical sequencing** and **shotgun sequencing**.* The distinction between them is that in the hierarchical approach, the first step is to develop a low-resolution physical alignment that is used to ensure the sequence is obtained in large ordered pieces, whereas the shotgun approach simply breaks the genome into small, sequenceable units and relies on computer algorithms to assemble them. The difference is summarized in Figure 2.7. Most new genome sequencing projects adopt the whole-genome shotgun approach, bypassing the use of hierarchical strategies to guide assembly.

---

*For an historical overview of genome sequencing, starting with the first recombinant DNA molecule and moving through early sequencing developments, the first genome sequence (SV40), map-making, and proposals for the human genome project, see http://www.sciencemag.org/feature/plus/sfg/human/timeline1.dtl at Science Online.

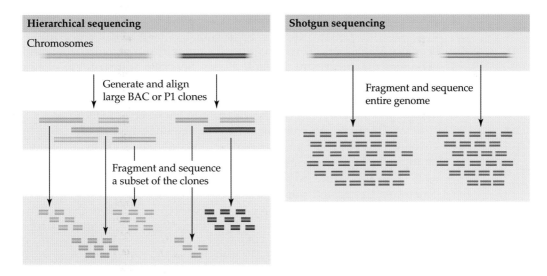

Figure 2.7 Hierarchical versus shotgun sequencing.   In hierarchical sequencing, the chromosomes are first cloned as large BAC or P1 fragments up to 200 kb. These are physically ordered, and a subset that gives a minimal overlap for complete genome coverage is chosen for shotgun sequencing. In the whole-genome shotgun approach, no attempt is made to order the clones in advance. Instead, the whole genome is assembled using computer algorithms that order contigs based on their overlapping sequences.

Hierarchical Sequencing

Hierarchical strategies, which are also known as *top-down*, *map-based*, or *clone-by-clone* strategies, were developed in the late 1980s, at a time when the cost of chemicals was exorbitant, high-volume automated protocols had not yet been developed, and computers were not yet powerful enough to handle whole-genome shotgun sequencing. The general approach of breaking a genome into smaller and smaller units, the relative locations of which are known before sequencing commences, had two further advantages: it fostered assembly of high-resolution physical and genetic maps; and it allowed groups working around the globe to form consortia and work together without risking repetition, since each group could concentrate on a particular chromosome.

The technology for cloning large fragments of genomes and maintaining them in a stable state progressed rapidly throughout the 1990s, with the result that the *E. coli*, *S. cerevisiae*, *C. elegans*, and *A. thaliana* genome projects were the beneficiaries of ever more efficient techniques of hierarchical sequencing. The basic strategy of hierarchical sequencing remains applicable to any organism, and in fact was the used by the publicly funded human genome project.

The first step in top-down sequencing is to clone the genome as manageable units of some 50–200 kilobases in length. Large insert cloning vectors include bacterial artificial chromosomes (BACs, which carry up to 300 kb of DNA), P1 phage-derived PAC clones (~100 kb), and smaller phage-derived cosmids or fosmids (~50 kb). As shown in Figure 2.8, BAC and PAC clones are similar in organization to ordinary plasmid cloning vectors, with the major difference that vector-encoded sequences help keep the copy number close to one per cell, which limits the potential for recombination and maintains the integrity of the large clones. Special features of PAC clones allow the initiation of phage packaging during the cloning step, excision of the high-copy number plasmid sequences after bacterial infection, and induction of replication immediately prior to harvesting of DNA.

DNA libraries are constructed by partial digestion or shearing of genomic DNA by sonication, after which the fragments are ligated into a multiple cloning site (mcs) in the vector of choice using standard recombinant DNA procedures. The source of genomic DNA can be the whole genome, or it can come from a chromosome isolated by pulsed-field gel electrophoresis or some other method. Researchers typically aim for five- to tenfold redundancy, meaning that every portion of the genome should be represented at least five times in the library of clones. Since each clone will have different ends, it should in principle be possible to select a scaffold of clones that form a contiguous sequence covering a whole chromosome—a **tiling path**—by aligning the regions of overlap (Figure 2.9). The tiling path can be assembled using a combination of three methods: hybridization, fingerprinting, and end-sequencing.

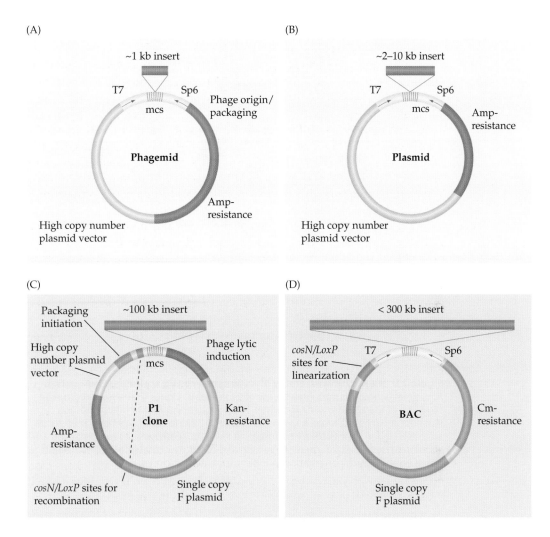

Figure 2.8 Cloning vectors used in genome sequencing. Different size inserts are cloned in different types of vectors. Phagemids (A) are derivatives of M13 that can shuttle between double-stranded plasmid and single-stranded bacteriophage forms. Like standard cloning plasmids (B), the multiple cloning site (mcs) into which the foreign DNA is inserted is flanked by primer binding sites for sequencing. P1 clones (C) are assembled as plasmids, but are converted into phage using the *cosN* or *LoxP* recognition sites for Ter and Cre recombinases. As with BAC clones (D), copy number within *E. coli* cells is kept low by use of the F plasmid origin of replication.

Hybridization. All of the clones in a library that carry a particular sequence can be identified rapidly by hybridizing a small radioactively or chemically labeled probe containing the sequence to a filter on which is printed an array

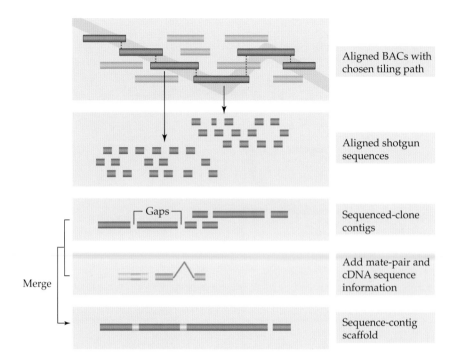

Figure 2.9 Hierarchical assembly of a sequence-contig scaffold (supercontig).
A minimal tiling path through a library of aligned BAC clones that ensures complete
coverage of the chromosome is chosen. After sequencing independent shotgun
libraries for each BAC, small gaps in the sequenced clone contigs remain. These gaps
are closed as far as possible by merging the two BAC sequences, as well as by the
addition of mate-pair information (yellow) and cDNA structural information (red),
which establishes the orientation and distance between cloned segments.

of tens of thousands of clones (Figure 2.10A). Robots are used to pick
colonies and replica plate them on nitrocellulose membranes in high-density grids. If the sequence is unique in the genome, it should "light up" only
a small number of clones, which can be compared with one another. Subsequently, the end of one of these clones can be used as a probe in a fresh screen
for adjacent clones, in a procedure known as chromosome walking (see Figure 1.2). Alternatively, the library can be interrogated with thousands of random probes from throughout the genome, leading to the generation of
islands of overlapping clones that eventually will be linked by some other
method.

Fingerprinting. The most cost-effective way to assemble contigs of large
insert clones is to compare and align them according to restriction digest
profiles (Figure 2.10B,C). Restriction enzymes that cleave DNA at 6-bp recognition sites cut on average just under once every 4 kb, and thus liberate
between 20 and 30 fragments when used to cleave a 100-kb BAC clone.

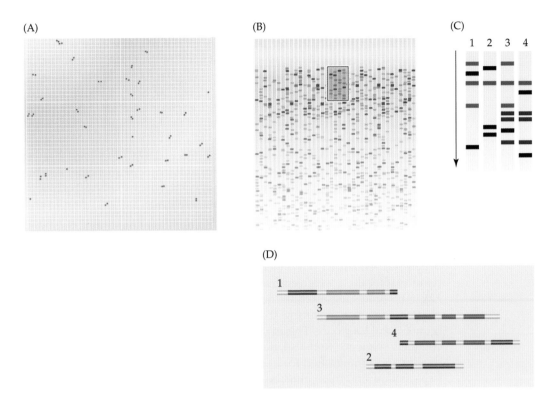

Figure 2.10 Aligning BAC clones by hybridization and fingerprinting. (A) A macroarray of BAC clones is probed with a short, radioactive fragment to identify all BACs that carry a specific fragment. (B) These clones are digested with a restriction enzyme, end-labeled, and separated by gel electrophoresis. (C) Software converts the bands to a virtual profile, shown hypothetically for a small portion of four clones (highlighted box in part B). Shared bands (red or blue) imply that the two clones share the same sequence. Green indicates the vector band common to all clones. (D) The fingerprint profile is then converted into a BAC alignment. In this example, clone 2 does not share any bands with the others and so is placed into a separate BAC contig, while the other three clones form a tiling path.

These fragments can be separated either by agarose or acrylamide gel electrophoresis; they are then placed into size bins (following appropriate statistical normalization that accounts for variability among gels). Software has been developed that contrasts fragment profiles and aligns the clones according to regions of overlap (Figure 2.10D). Given sufficient redundancy, fingerprinting procedures regularly result in the assembly of Mb-length contigs, and suggest a minimum set of clones that are then individually subject to shotgun sequencing.

End-sequencing. A common way to identify clones spanning the gaps that remain after fingerprinting is to sequence both ends of the collection of BAC clones. Once a critical threshold of assembled sequence has been achieved, there is a high probability that at least one end of a BAC will lie within an assembled region, implying that the other end of the clone either extends into the gap, or even links to an adjacent contig. For example, end-sequencing 10,000 BAC clones will provide a sequence tag every 5 kb along a 10-Mb genome; this density will close most gaps of less than 50 kb. End-sequencing is also an important component of the techniques used to verify sequence assemblies.

Once a tiling path has been chosen, the individual BAC clones are sheared into small fragments that are subcloned for automated sequencing. A common cloning vector is an M13-derived phagemid, which exists as both double-stranded plasmid DNA and single stranded phage, facilitating rapid and efficient template preparation. With this vector, only short fragments package efficiently, and reads are most readily obtained in a single direction, up to 1 kb. An alternative method is to clone 2- to 3-kb fragments into a plasmid vector and sequence from both ends. Shearing the BAC into small pieces is done by sonication, which ensures that the ends of each fragment are unique. Enough sequencing reactions are performed to approach tenfold coverage of each sequence, ensuring that the majority of any given sequence is sequenced at least once. This in turn allows computational reassembly of the complete sequence, and also increases sequence accuracy as regions of low quality are compensated for in other reactions. Inevitably, however, small gaps remain, and these must be filled in the sequence finishing phase.

Shotgun Sequencing

In shotgun sequencing, computer algorithms are used to assemble contigs derived from hundreds of thousands of overlapping sequences. As in hierarchical sequencing, the aim is to achieve five- to tenfold redundancy for each fragment of the sequence, but the contigs are generated from a plasmid library that has been constructed from a single whole genome rather than from individual BAC clones. Multiple sequences are aligned using algorithms that successively screen out repetitive sequences, overlap reads of the same sequence, generate contigs and scaffolds, and resolve repeats. At that point, a genome sequence may be more than 90% assembled, but the finishing phase—closing gaps, cleaning up ambiguities, and improving the overall quality—can take as much time and effort as the initial shotgun phase. In general, the details of sequence assembly are only superficially described in publications and users are asked to trust the assemblies produced by the genome sequencing centers without actually examining the raw data themselves.*

---

*The initial proposal for shotgun sequencing of large genomes, including calculations of the necessary clone and sequence coverage, can be found in Weber and Meyers (1997) along with a rebuttal by Green (1997). See also Venter et al. (1998) for the announcement of Celera's intention to proceed with sequencing of the human genome.

There are a number of different genome assembly programs that are often used in parallel to generate the most consistent final alignment. Several of these are available as open source software from http://sourceforge.net, notably the **JCVI Celera** and **AMOS** (**A M**odular **O**pen **S**ource) assemblers. Three others that see wide usage are the Broad Institute's **Arachne**, the Sanger Institute's **Phusion**, and Baylor College of Medicine's **Atlas** assemblers. All utilize variations on the following screening, overlapping, and contig ordering steps.

The **Screener** is employed to mask (that is, to mark and hide) sequences that contain repetitive DNA (including microsatellites with a less than 6-bp repeat) and known families of interspersed repeats such as LINE elements, Alu repeats, retrotransposons, and ribosomal DNA. These sequences are not removed, but they are screened out of the alignment algorithms so that they do not contribute to determining overlap. Once short chromosome fragments have been assembled, however, the length and location of repetitive sequences is taken into account in subsequent stages of genome assembly. Cloning vector sequences are also trimmed, and attempts are made to recognize contaminating bacterial, viral, or other extraneous sequences.

The **Overlapper** compares every unscreened read against every other unscreened read, searching for overlaps of a predetermined length and identity. This overlapping process is essentially the same as performing a BLAST search (Box 2.2) of each sequence read against each other sequence read. In the case of the human genome sequence assembly, overlaps of at least 40 bp with no more than 6 differences were required. The differences allow for misidentification and other sequencing errors, as well as for the presence of polymorphisms present in heterozygotes or in libraries assembled from multiple individuals. Overlaps of this specificity have a probability of appearing once in every 10^{17} comparisons, and so are exceedingly unlikely to occur twice in a single genome (unless the sequence is recently duplicated). Parallel processing on 40 supercomputers, each with 4 gigabytes of RAM, allowed the 27 million screened human sequence reads to be overlapped in less than 5 days.

At the end of the overlapping process, a genome is usually ordered into thousands of DNA fragments that unambiguously lie adjacent to one another and usually form one long stretch of sequence from a piece of chromosome. These are called **contigs** (short for **contig**uous DNA segments). As the fold coverage of the genome increases, the average length of contigs increases (Figure 2.11A). The contig N50 is a commonly used measure of genome sequence quality: it refers to the length of the smallest contig required to ensure that at least half of the genome is included in a contig. This quantity is regarded as a more robust indicator of quality than the average contig size, because contaminating sequences and fragments in highly repetitive regions of the genome tend not to assemble and so drag down the average size of contigs.

BOX 2.2 Searching Sequence Databases Using BLAST

Without a doubt, the single most extensively utilized bioinformatics tool is the Basic Local Alignment Search Tool, or BLAST. The BLAST program by Altschul et al. (1990) is used to search large databases of molecular sequences, returning sequences that have regions of similarity to a **query sequence** provided by the user.

In Box 2.1, we described the Needleman-Wunsch algorithm for global alignment of a pair of sequences, in which all residues from both sequences are included. In order to understand the workings of BLAST and similar search tools (notably FASTA), it is necessary to introduce the concept of **local alignment.** Local alignment algorithms attempt to find *isolated regions* in sequence pairs that have high levels of similarity. It is this property that makes local alignment ideal for database searches. The user provides a query sequence, which is then compared to the entire database. One can envision the concatenation of all sequences in the database as the **target sequence.** BLAST then searches for regions of the target sequence with similarity to the user's query sequence. Thus, we see the fundamental importance of pairwise alignment methods in bioinformatics.

Interpreting BLAST Output

Algorithms guaranteed to find the best local alignment of two sequences have been developed (Smith and Waterman 1981). However, as is the case with many bioinformatics applications, in practice these methods are often too slow to use. BLAST is an effective heuristic search method that is not guaranteed to find the best local alignment, but has been especially effective in practice. In this box we will use an example that pays particular attention to interpreting the output of the widely used NCBI BLAST program.

DeSalle et al. (1992) extracted 16S ribosomal DNA from an insect preserved in 25-million-year-old fossilized amber. BLAST can be used to verify that the amplified DNA was not a laboratory contaminant, and to identify the closest living relative of the ancient organism. Figure A shows some results from BLAST after probing GenBank using a 92 bp fragment from the fossilized insect (GenBank accession number S45649).

The BLAST report (graphical portion in Figure A, textual portion in Figure B) ranks **hits** in order of a measure of statistical significance called the **E-value,** with the most significant hits listed first. Formally, the E-value is the number of hits with the same level of similarity that you would expect by chance if there were no true matches in the

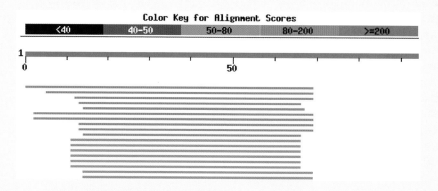

FIGURE A A BLAST output.   The graphical overview of BLAST clearly indicates which local region of our query sequence has similarity to a sequence in the database. The NCBI implementation of BLAST also has many features available through use of your mouse, including the actual alignment, along with a variety of useful summary statistics.

database. Thus, a hit with an E-value of 0.01 would be expected to occur once every 100 searches even when there is no true match in the database; a hit with E-value 1.0 is expected every time you search.

The E-value is similar in spirit to the traditional *p*-value of statistical hypothesis tests. The definition of a *p*-value is the probability of obtaining a result as extreme as the observed one, if in fact there is truly no effect. In the context of database searching, the desired *p*-value would be the probability of finding a sequence similarity as similar as the observed match if there were really no true matches in the database. Clearly, the E-value doesn't satisfy this definition, since it can take any positive value. Conveniently, though, the E-value approximates the *p*-value when it is small, say less than 0.1. Since we are primarily interested in "unusual" hits, it is typically safe to interchange E-values and *p*-values.

For the results of the example search, the E-values for the top hits are all very small—even the largest is a mere 3×10^{-13}, strongly suggesting that the similarities are not the result of random chance.

If we click on the value in the "Score" column, a display of the actual local alignment appears.

Scoring Matrices

Although BLAST has many settings that can be adjusted by the user, it is wise to leave most adjustments to experts. The exception to that statement is the choice of a scoring matrix. In Box 2.1, we saw a very simple scheme for scoring sequence matches and mismatches: all mismatches received the same penalty. A scoring matrix extends this idea to allow some mismatches to be penalized less than others. For instance, a leucine-isoleucine mismatch might be penalized less than a more biochemically radical leucine-tryptophan substitution.

Some of the best scoring matrices derive their scores from empirical collections of sequence alignments. The PAM family of scoring matrices is based on the pioneering work of Margaret Dayhoff and her colleagues (Dayhoff et al. 1972). The penalties were derived from closely related species to avoid the complications of unobserved multiple substitutions at a single position. More recently, it was recognized that the goal of most database searches was to identify very ancient homologies. A scoring matrix based on data from distantly related sequences was the logical step, resulting in the BLOSUM family of scoring matrices (Henikoff and Henikoff 1992).

When possible, the selection of a scoring matrix should depend on the level of similarity that one hopes or expects to find when performing a database search. Practical experience suggests that the BLOSUM62 matrix is quite suitable for general use.

```
                                                               Score    E
Sequences producing significant alignments:                   (bits)  Value

gi|256517|gb|S45649.1|S45649    16S rRNA [Mastotermes electrod...  139   5e-33
gi|12005612|gb|AF246514.1|AF246514   Drosphila ornatipennis ...    86   7e-17
gi|11119031|gb|AF304735.1|AF304735   Sphyracephala bipunctipe...   84   3e-16
gi|3552018|gb|AF086859.1|AF086859    Mystacinobia zealandica l...   84   3e-16
gi|256518|gb|S45650.1|S45650    16S rRNA [Mastotermes darwinie...   84   3e-16
gi|15341487|gb|AF403473.1|AF403473   This canus 16S ribosomal...   82   1e-15
gi|15341483|gb|AF403469.1|AF403469   Icaridion debile 16S rib...   82   1e-15
gi|13435200|ref|NC_002697.1|   Chrysomya chloropyga mitochond...   82   1e-15
gi|13384216|gb|AF352790.1|AF352790   Chrysomya chloropyga mit...   82   1e-15
gi|3552016|gb|AF086857.1|AF086857    Calliphora quadrimaculata...   82   1e-15
gi|15341485|gb|AF403471.1|AF403471   Malacomyia sciomyzina 16...   80   4e-15
gi|15341463|gb|AF403449.1|AF403449   Helcomyza mirabilis 16S ...   80   4e-15
```

FIGURE B A local alignment.   Here we see the representation of several of the reported matches. The primary section of the BLAST output is a list of the best matches to our query sequence. Following the exact match to our query sequence itself, we see that the top matches are to known insects, such as *Drosphila ornatipennis*, providing confirming evidence that the extracted DNA was, in fact, from an insect.

(A)

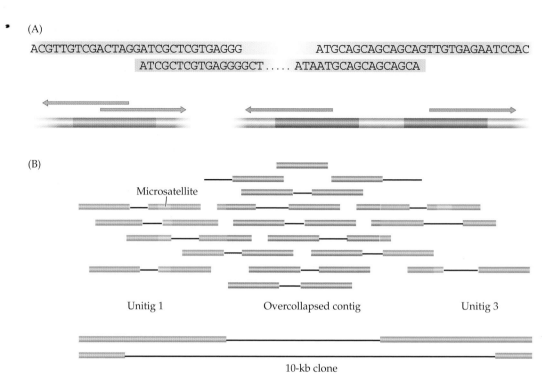

ACGTTGTCGACTAGGATCGCTCGTGAGGG   ATGCAGCAGCAGCAGTTGTGAGAATCCAC

ATCGCTCGTGAGGGGCT . . . . . ATAATGCAGCAGCAGCA

(B)

Microsatellite

Unitig 1   Overcollapsed contig   Unitig 3

10-kb clone

Figure 2.11 Unitigs and repeat resolution. (A) Sequence alignment between two or more shotgun clones can arise between unique sequences (left) or repetitive sequences (right). (B) The Overlapper aligns contigs, which are identified as unique contiguous sequence alignments (also called unitigs) or overcollapsed repeats (blue). Two unitigs can be aligned and oriented by using mate-pair sequence information from the ends of longer (10- or 50-kb) clones, as shown at the bottom, while mate-pairs from 2-kb fragments allow assembly of scaffolds despite the presence of simple repeats such as microsatellites (light blue) that are masked before performing alignments.

Contigs can be further assembled into sequence **scaffolds**, which are a series of ordered contigs in the correct position and alignment but that are separated by gaps that remain to be filled (Figure 2.11B). The scaffold N50 is another measure of sequence quality. In a finished genome sequence, the number of scaffolds should be the same as the number of linkage groups and chromosomes, with centromeric heterochromatin usually lying within a large gap in scaffolds that correspond to whole chromosomes. In practice, though, draft genome sequences fall far short of this quality. For example, the 5′ rhesus macaque genome sequence (RMGSAC 2007) was assembled from 15 Gb of raw sequence and covered 98% of the 2.87 Gb of euchromatin in 122,580 scaffolds with an N50 of approximately 25 Mb, and in 301,039 contigs with an N50 of just less than 25 kb. Half the genome was covered by 36 scaffolds and 32,000 contigs.

Most of the remaining gaps are due to repeats, and are resolved by computationally aggressive and error-prone steps, recognizing that it is better

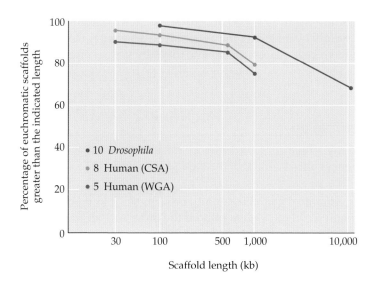

Figure 2.12 Proportion of fly and human genomes in large scaffolds. The plot shows the percentage of scaffolds that have a length greater than that indicated for the *Drosophila* 10×, human 8× (compartmentalized shotgun assembly, CSA) and human 5× (whole-genome assembly, WGA) sequences generated by Celera. The fly and CSA assemblies include shredded sequences generated from BAC clones by public genome sequencing efforts.

to produce large scaffolds that can be verified than to leave the genome in thousands of disconnected pieces. Figure 2.12 shows the estimated coverage of the fly and human whole genomes after initial assembly: in both cases, 84% or more of the genome was covered by scaffolds at least 100 kb in length, while most scaffolds were in the Mb range. Increasing sequence coverage from 5× to 10× results in a 10% increase in the proportion of scaffolds of lengths up to 1 Mb. The choice of level of coverage must balance genome size, availability of funds, and the needs of the project, but obviously the more sequences, the more complete the assembly will be. Where a closely related species has already been sequenced, it may be possible to hang a 3× coverage assembly on the scaffold of that species, but this risks failure to observe genome rearrangements.

At least three different approaches can be taken to closing any remaining gaps. Where it is clear that a gap is solely due to a repetitive sequence that was screened out, this repeat sequence can be reinserted into the assembly, with the proviso that it may be a generic repeat sequence rather than the actual sequence (including unique polymorphisms) at that position. Where it is not clear what the gap sequence may be, it will often be possible to identify a clone that spans the gap, because the mate-paired ends of the clone derive from two different scaffolds, and then to carefully sequence just that clone. For this reason, shotgun sequencing projects include some fraction of 50-kb clones in addition to 2-kb and 10-kb clones. Alternatively, gaps may be bridged by sequences from different projects, including cDNA sequences (two exons may be separated by a large intron that falls in the gap) and comparative alignment with other species in which two adjacent scaffolds always map to the same locus. This information at least links the scaffolds together, and directs finishing efforts to design PCR primers to bridge the gaps.

The scaffolds are then assigned to chromosomal locations. This is achieved most simply by matching the assembly to previously cloned gene sequences that have been placed on genetic maps by linkage. Representative clones can also be placed on cytological maps by fluorescent in situ hybridization (FISH) to chromosome spreads.

Sequence Verification

The veracity of any whole genome sequence must be assessed at three levels: its completeness, the accuracy of the base sequence, and the validity of its assembly.

- *Completeness.* Microbial genomes tend to be sequenced in their entirety, but may contain small gaps (generally of the order of 1 kb or less) that prove difficult to close. It should be recognized, however, that *any* sequence is that of a single isolate and that some length polymorphism, including insertion and deletion of whole genes, is to be expected. Most higher eukaryotic genomes contain large stretches of heterochromatin that are simply excluded from analysis and may never be sequenced except for small islands. These can include essential genes, such as the *Raf* serine kinase gene that is dispersed in heterochromatin in the vicinity of the centromere of the second chromosome of *Drosophila*.

- *Accuracy* is assessed by probability scores, as described earlier in this chapter (see the section "Reading Sequence Traces," pp. 68–71), and can always be increased simply by sequencing more clones to cover a specific region, possibly using novel chemistry to resolve ambiguities.

- *Validity of assembly* is not trivial to assess, and can be approached either by measuring internal consistency or by contrasting the assembly with genetic or pre-existing physical maps. Two excellent measures of internal consistency are (1) the error rate in the alignment of predicted restriction profiles with observed fingerprints, and (2) the correct spacing of paired end-sequences from clones of different sizes. The spacing between mate-pair reads from 50-kb clones is particularly useful for confirming the orientation and extent of repeat content. Disagreements are called breakpoints.

In the case of the human genome, at the time of publication of the draft assemblies in February 2001, hundreds of inconsistencies and gaps were reported for each chromosome. As shown in Figure 2.13, some of these differentiated the public and private drafts by more than half the length of a chromosome. Since the two "finished" chromosomes, 21 and 22, showed only a handful of breakpoints, it is likely that most of these represent discrepancies that arose during the assembly process. Finishing is a very expensive process, and for some species the draft genome is sufficient because individual investigators with an interest in a particular region will generally resolve ambiguities in the course of their research. However, proper ordering of clones is essential where the genome sequence is used to guide

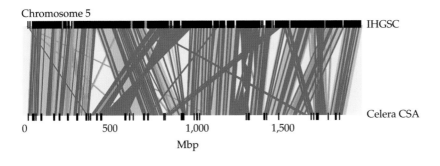

Chromosome 5

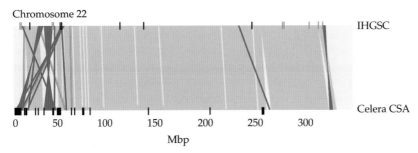

Chromosome 22

Figure 2.13 Alignment of two draft human genome assemblies.   First-draft genome assemblies are far from complete, as indicated by the discrepancies between the public (IHGSC) and Celera compartmentalized shotgun assembly alignments for chromosomes 5 and 22. Green links indicate matched alignment; orange lines indicate sequence blocks greater than 50 kb that are out of order, and blue lines blocks that are also in the opposite orientation. Small black notches indicate breakpoints, and light blue notches are runs of 10 kb of Ns. (Redrawn from data available at Science Online: www.sciencemag.org/cgi/content/full/291/5507/1304/DC1.)

linkage and association studies as described in the next chapter, and draft genome sequences can be deficient in this regard.

Genome Annotation

EST Sequencing

The most direct way to identify a gene is to demonstrate that a fragment of the genome is transcribed reproducibly. This can be done using microarrays as described in Chapter 4, but elucidation of the structure of a transcript requires sequencing of one or more cDNAs. These are reverse-transcribed (complementary DNA, or cDNA) copies of mRNAs, and a collection of cloned cDNAs is known as a **cDNA library**. Full-length cDNAs are usually too long to be sequenced in a single reaction, so the term **expressed sequence tag** (**EST**) is used to describe a single sequence that represents a partial sequence of a cDNA. Since most genes are represented by multiple transcripts produced by alternative splicing, and since each transcript can be represented by several ESTs covering different stretches of sequence, single genes can be represented by multiple ESTs. Nevertheless, random sequencing of ESTs is an efficient approach to documenting the diversity of transcripts that are present in tissue.

The first step in generation of a cDNA library is isolation of the polyA fraction of RNA from a tissue. After the mRNA strand is converted to DNA, generally using a polyT oligonucleotide to prime reverse transcription, the double-stranded DNA fragments are cloned into a plasmid. The vast majority of expressed sequences will be present at a frequency less than one clone in 10,000; each of several hundred sequences will constitute between 0.01% and 1% of the total transcripts; and there will be a handful of sequences that are present at even higher frequencies. Consequently, sequencing of random clones results in a high level of redundancy: at least half of the transcripts will be represented two or more times well before all the different transcripts are identified. The efficiency with which new genes are identified can be improved using several strategies that help "normalize" the library, for example by the use of subtractive screening prior to sequencing of the cDNA fragments (Carninci et al. 2000).

Because a large proportion of genes are differentially expressed in time and space, as well as under a variety of environmental conditions or between the two sexes, it is necessary to sample a wide array of tissues in order to identify as many genes as possible. Both the public and private sectors have developed atlases of human gene expression, drawing on hundreds of different cDNA libraries, many of which are derived from cancers or transformed cell lines. Clone sequences can be downloaded from the NCBI's dbEST database (Boguski et al. 1993) at http://www.ncbi.nlm.nih.gov/dbEST, or from Ensembl at http://www.ensembl.org. Similarly, the vast majority of genome projects are developing EST collections that are an essential resource in the confirmation of the correspondence between open reading frames and expressed genes, and these are available both through the above URLs and at the home Web sites for each organism.

Prior to the completion of the Human Genome Project, estimates of the number of genes based on the number of observed ESTs exceeded 100,000—four to five times what we now believe to be the true number of human genes. The two main reasons for such overestimates are the prevalence of alternative splicing, and artifacts due to incompleteness of cDNA fragments, as diagrammed in Figure 2.14.

Alternative splicing refers to the phenomenon whereby the same gene gives rise to multiple different transcripts. These differ according to the combination of exons that are incorporated in the mature mRNA transcript. The average human gene contains between 10 and 15 exons and encodes three or more different proteins as a result of alternative splicing. When different exons of a single gene are included along with the 3'-most exon in a collection of sequence tags, they will initially be predicted as multiple genes. Complementary DNA synthesis is a notoriously error-prone process, and internal priming can occur, also giving rise to different ESTs from one gene.

(A) Gene structure

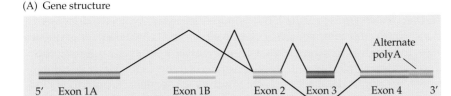

(B) Complete cDNAs

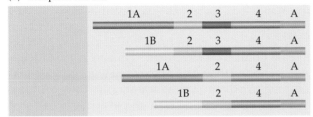

(C) ESTs

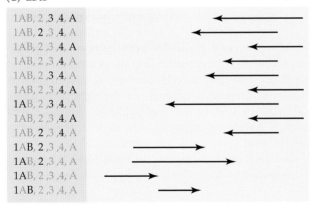

Figure 2.14 Relationship between gene structure, cDNA, and EST sequences. (A) Alternative splicing and the use of alternate 5′ start sites or alternate polyA signals results in (B) the generation of multiple cDNA transcripts from individual genes. (C) EST sequences can be derived from either the 5′ or 3′ end of a cDNA clone and are generally incomplete, resulting in the representation of different exons in different clones from the same gene.

Further, the probability that the 5′ end of a clone corresponds to a true start site of transcription decreases as a function of the length of the transcript. As a result, EST sequences derived from the 5′ end often commence at internal sequences, and will tend to identify different sequences that may be interpreted as belonging to different genes. Errors can also occur close to the 3′ end as a result of internal deletions, or artifacts of mis-splicing, while some genes have alternate natural 3′ ends.

Comparison of EST and genomic DNA sequences readily resolves the issue of single genes that are represented as independent ESTs. Proper definition of gene structure, however, requires isolation and sequencing of multiple full-length cDNAs from high-quality libraries. The position of the 5′ end is usually confirmed with techniques such as RNA extension or RNAse protection. The representative major transcript for a gene is known as a **RefSeq** (reference sequence), but genome browsers typically display evidence for the full complement of alternate transcripts by default. In sequence files, the predicted coding sequence of an mRNA—that is, excluding the 5′ and 3′ untranslated regions—is annotated as the CDS. Many organisms now have public databases of full-length cDNA sequences, notably including the H-InvDB of over 21,000 full-length human cDNA sequences at http://www.h-invitational.jp (Imanishi et al. 2004), the *Drosophila* GOLD collection, and the mouse FANTOM and the *Arabidopsis* RIKEN full-length cDNA projects.

Ab Initio Gene Discovery

Protein-coding sequences within a whole-genome sequence can be identified by a process known as *ab initio* **gene discovery**, using software that recognizes features common to protein-coding transcripts. These features include the existence of long open reading frames, particularly ones for which the codon bias is typical of that observed for the species being studied; proximity of transcriptional and translational initiation motifs and 3′ polyadenylation sites; and splicing consensus sequences at putative intron-exon boundaries.

In bacteria and in eukaryotes with small, compact genomes, programs such as GeneFinder and Grail have been found to accurately predict in excess of 90% of all true genes. The genomes of higher eukaryotes, however, tend to have numerous introns and extensive noncoding intergenic regions—not to mention quirks such as *trans*-splicing and genes located within the introns of other genes—that make *ab initio* gene discovery considerably more difficult. Programs such as Genie, Genscan, HMMgene, and FGENES, incorporate statistical approaches into **hidden Markov models (HMMs)** to help overcome some of these difficulties (Box 2.3; reviewed by Rogic et al. 2001).

Irrespective of which discovery algorithm is used, all computationally identified putative genes must be confirmed by a second line of evidence before being elevated to gene status in the genome annotation. The six standard types of direct evidence are:

1. Identity to a previously annotated cDNA reference sequence.

2. A match to one or more EST sequences from the same organism.

3. Similarity of the nucleotide or conceptually translated protein sequence to such sequences from other organisms in GenBank or other databases.

4. Protein structure prediction that matches a domain in the PFAM database.

5. Association with predicted promoter sequences, including a TATA box consensus sequence, proximity to clusters of recognized transcription factor binding sites, and often (in mammals) proximity to a CpG island.

6. Proof that a mutation in the sequence produces an observable phenotype.

Each of these procedures can be automated to some extent, but human oversight appears to be indispensable for sifting through discrepancies and noise. For most genome projects, users in the scientific community at large are encouraged to submit corrections and updates to annotations as they identify them.

The three biggest deficiencies of computational gene discovery methods are imprecise or incomplete characterization of gene structure; characterization of false positive genes; and failure to identify true genes. Comparison of predicted with previously characterized genes indicates that *ab initio* methods correctly identify all of the intron-exon boundaries for at most only three-quarters of the genes in higher eukaryotic genomes, and in some cases considerably fewer. Given the complexity of alternative splicing in most of these cases, thorough characterization of gene structure by sequencing of complete cDNAs is regarded as a routine component of thorough annotation. Comparison of the genome sequences of related species is also likely to considerably improve the annotation of gene structure, particularly as multiple sequence alignments become available.

Perhaps the more important fact is that probably 80–90% of all true genes *are* identified, with a less than 10% false positive rate. False positives include some putative genes for which there is simply no evidence either of transcription or similarity to other genes in the database, as well as unfiltered pseudogenes and/or genes associated with transposons that are unlikely to have a function in the organism. False negatives can be detected using a different gene finding program, or by adjusting the stringency of the original program's search parameters, or even by shifting the properties of the database that is searched (Gopal et al. 2001).

Confirmation that an annotated gene corresponds to a true gene ultimately depends on functional data. Many genes were first identified following characterization of a mutation or polymorphic variant of the gene associated with an abnormal or extreme phenotype. As described in Chapter 5, reverse genetic **gene knockout** strategies—modifying a gene so that it is nonfunctional and seeing what effect this has on the phenotype—provide perhaps the most direct methods for confirming not just that a gene is expressed, but also that it is functional.

The expression of genes for which no EST or cDNA counterpart has been identified can be established by specifically searching for transcripts from the predicted locus. An efficient way to do this is to identify open-reading-frame sequence tags (OSTs: Reboul et al. 2001), by amplifying predicted genes from a cDNA library using primers designed to anneal to both ends of putative open reading frames. Alternatively, fluorescently labeled cDNA can be hybridized to microarrays constructed with genomic DNA probes from the candidate genes (Penn et al. 2000). Transcription from loci that are not predicted to encode proteins is also commonly observed when cDNA is hybridized to so-called tiling arrays of short oligonucleotides representing all genomic sequences (Shoemaker et al. 2001). Some of this signal may be attributed to nonfunctional priming of transcription, some to production of regulatory RNAs, and some to genes that just are not predicted by bioin-

BOX 2.3 Hidden Markov Models and Gene Finding

Many of the bioinformatics tools used for predicting features or functions of sequences belong to the family of methods that rely on **hidden Markov models,** or **HMMs.** HMMs allow for local characteristics of molecular sequences to be modeled and predicted within a rigorous statistical framework, and also allow the knowledge from prior investigations to be incorporated into analyses.

An Example of the HMM

It is perhaps simplest to introduce HMMs with an example. Let us suppose that every nucleotide in a DNA sequence belongs to either a "normal" region (N) or to a GC-rich region (R). As the name implies, the bases G and C are more prevalent in the GC-rich categories. Furthermore, let us assume that the normal and GC-rich categories are not randomly interspersed with one another, but instead have a patchiness that tends to create GC-rich islands located within larger regions of normal sequence. A short representation of such a sequence might be:

NNNNNNNNNRRRRRNNNNNNNNNNNNNNNNNNNNRRRRRRRRNNNN

The **states** of the HMM are the two categories, either N or R. The two states emit nucleotides with their own characteristic frequencies. The word "hidden" in the name "hidden Markov model" refers to the fact that the true status of the states is unobserved, or hidden. Keeping in mind that sites in GC-rich regions are more likely to be either G or C, a possible DNA sequence having this underlying collection of categories is:

TTACTTGACGCCAGAAATCTATATTTGGTAACCCGACGCTAA

At first glance, this may appear to be a typical random collection of nucleotides; the observed 60% AT and 40% GC is not too far from the expected counts from a "random" sequence. However, if we focus on the red GC-rich regions, we see that the frequency of GC in them is 83% (10/12), compared to a GC frequency of 23% (7/30) in the other sequence regions. The main use of HMMs is to identify these types of features in sequences.

   HMMs have the ability to capture both the patchiness of the two classes and the different compositional frequencies within the categories. They have proven invaluable in applications such as gene finding, motif identification, and prediction of tRNAs. In general, if we have sequence features that we can divide into spatially localized classes, with each class having distinct compositions, HMMs are a good candidate for analyzing or finding new examples of the feature.

   Let us now become more specific. Suppose that studies of other DNA sequences reveal that normal and GC-rich regions have the following base compositions (the compositional biases are exaggerated for the purpose of illustration):

	A	T	G	C	Mean length
Normal (N)	0.3	0.3	0.2	0.2	10
GC-rich (R)	0.1	0.1	0.4	0.4	5

The process of obtaining these frequency estimates from empirical data is called *training* the HMM. An HMM describing these data is represented as in Figure A.

   The workings of an HMM can be broken into two steps: the assignment of the hidden states, and then the emission of the observed nucleotides *conditional* on the hidden states. To illustrate, consider the simple sequence TGCC. One possible way that this sequence could arise is from the set of hidden states NNNN. In the case that we know (or assume)

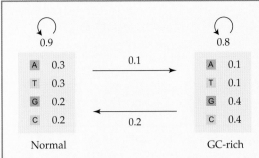

FIGURE A Training the HMM. The *states* of the HMM are the two categories, N or R. *Transition probabilities* govern the assignment of states from one position to the next. In the current example, if the present state is N, the following position will be N with probability 0.9, and R with probability 0.1. The four nucleotides in a sequence will appear in each state in accordance to the corresponding *emission probabilities*.

the hidden states, the probability of the observed sequence is simply a product of the appropriate emission probabilities: $Pr(TGCC \mid NNNN) = 0.3 \times 0.2 \times 0.2 \times 0.2 = 0.0024$, where, for example, the notation $Pr(T \mid N)$ is the *conditional probability* of observing a T at a site given that the hidden state is N.

The HMM provides the probability of this occurrence. In general, this probability is computed as the sum over all hidden states as:

$$Pr(sequence) = \sum Pr(sequence \mid hidden~states)Pr(hidden~states)$$

using a simple application of the rules of conditional probability (see, for example, Moore and McCabe 1999). The description of the hidden state of the first residue in a sequence introduces a technical detail beyond the scope of this discussion, so we simplify by assuming that the first nucleotide is in a Normal region. With that simplification in hand, there are $2 \times 2 \times 2 = 8$ possible sets of hidden states, or **paths** through the Markov model, and the TGCC sequence could have been produced by any of these paths. We apply the law of total probability, in conjunction with conditional probability rules, to obtain:

$$\begin{aligned} Pr(TGCC) = &Pr(TGCC \mid NNNN)Pr(NNNN) + Pr(TGCC \mid NNNR)Pr(NNNR) + \\ &Pr(TGCC \mid NNRN)Pr(NNRN) + Pr(TGCC \mid NRNN)Pr(NRNN) + \\ &Pr(TGCC \mid NNRR)Pr(NNRR) + Pr(TGCC \mid NRNR)Pr(NRNR) + \\ &Pr(TGCC \mid NRRN)Pr(NRRN) + Pr(TGCC \mid NRRR)Pr(NRRR) \end{aligned}$$

To illustrate the calculation, (recall that we assumed the first position was in a normal state):

$$\begin{aligned} &Pr(TGCC \mid NNNN)Pr(NNNN) = \\ &Pr(T \mid N)Pr(G \mid N)Pr(C \mid N)Pr(C \mid N) \times Pr(N-N)Pr(N-N)Pr(N-N) = \\ &(0.3 \times 0.2 \times 0.2 \times 0.2) \times (0.9 \times 0.9 \times 0.9) = 0.00175 \end{aligned}$$

Similarly,

$$\begin{aligned} &Pr(TGCC \mid NNRR)Pr(NNRR) = \\ &Pr(T \mid N)Pr(G \mid N)Pr(C \mid R)Pr(C \mid R) \times Pr(N-N)Pr(N-R)Pr(R-R) = \\ &(0.3 \times 0.2 \times 0.4 \times 0.4) \times (0.9 \times 0.1 \times 0.8) = 0.000691 \end{aligned}$$

Notice that the observed sequence is slightly more likely for the first path, NNNN. If we extend this notion and compute the probability of the sequence for all possible paths, we can use the path that contributes the maximum probability as our best estimate of the unknown hidden states. For the sample sequence, one finds that the most probable path is in fact NNNN, which is slightly higher than the path NRRR (.00123), pointing out the role of spatial clustering. If the fifth nucleotide in the series were also a G or C, the path NRRRR would be more likely than NNNNN, providing evidence for the existence of a GC-rich

(Continued on next page)

BOX 2.3 *(continued)*

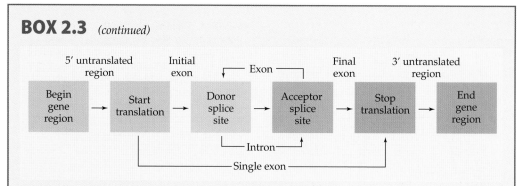

FIGURE B Schematic of the hidden states included in an HMM.   Boxes denote *signal sensors* for regulatory elements, coding region start sites, intron donor and acceptor sites, and translation stop sites; arrows indicate *content sensors* for intergenic regions, exons, and introns. Each of these regions emits nucleotides with frequencies characteristic of that region, with these frequencies being obtained by training the HMM on data sets of many known genes.

island. This approach is one of several for predicting the hidden states (see Sonnhammer et al. 1998 for a thorough discussion).

Gene Finding

We conclude our treatment of HMMs with a more complex example, that of predicting genes from large unannotated DNA sequences. Figure B displays a version of the HMM implemented in the GENSCAN program (Burge and Karlin 1997). Rather than go into extensive details of the intricacies of the model, the figure simply points out the main features.

   We close by pointing out that gene prediction is an inexact science. Different prediction methods will give different predictions. Most of the time the predictions are quite similar, however. We illustrate this fact with a region of gene predictions from the UCSC Human Genome Browser (http://genome.ucsc.edu), shown in Figure C.

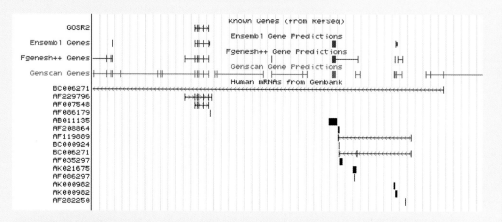

Figure C Predicting genes.   Three different prediction methods (Ensembl, Fgenesh, and Genscan) were used on a region of chromosome 17 that includes the well-annotated GOSR2 gene. The black images below indicate the location of matching cDNA/EST sequences.

formatic algorithms. Since over one-quarter of the genes in most newly sequenced genomes do not show similarity to genes in other organisms, the development of novel tools for the characterization of predicted but not yet described genes is a major goal of functional genomics.

Regulatory Sequences

Comparative genomics can be used to annotate the regulatory sequences of genes, which are often recognized as conserved non-coding sequences. These sequences tend to be located 5′ to the transcription start site, but may also be intronic or located at the 3′ end of the gene, as much as several tens of kilobases from the transcript sequence. The general approach known as **phylogenetic footprinting** was first demonstrated through detection of regulatory elements in a globin gene by comparison of mouse and human genomic DNA, and has since become a standard method for directing molecular biologists to candidate regulatory regions. Multiple sequence alignments provide the most information for detecting conserved elements, particularly when phylogenetic information is used with a statistical model of the rates of evolution of conserved and non-conserved sequences (Blanchette and Tompa 2002). Two popular open-source programs for visualizing pairwise sequence alignments are PipMaker (http://bio.cse.psu.edu; see Figure 1.12) and VISTA (http://www-gsd.lbl.gov/vista).

As it has become apparent that as many as 30% of the functionally defined binding sites for transcription factors in human regulatory regions are not conserved in rodents (Dermitzakis and Clark 2003), novel computational methods for finding regulatory elements are being developed. The approach shown in Figure 2.15, called **phylogenetic shadowing**, is conceptually similar to phylogenetic footprinting with the difference that several closely related species are compared rather than pairs of distantly related ones. (The software eShadow is available online at http://eshadow.dcode.org; Ovcharenko et al. 2004.) Boffelli et al. (2003) showed that by applying a model of sequence evolution to an alignment of five or six primates, they could discern 10 candidate binding sites, each 50–100 bp long, in the *apo(A)* promoter, most of which were confirmed to bind to a nuclear protein extract and to be necessary for normal levels of transcription. The technique has been used to compare fungal and plant sequences as well.

Description of computational methods for detecting short regulatory motifs is deferred to Box 4.4. Many of these methods are based on searching for clusters of motifs that have been shown experimentally either (1) to bind to specific transcription factors, (2) to be required for transcription of reporter genes, or (3) to be overrepresented in potentially regulatory genomic DNA (Rajewsky et al. 2002). Gibbs sampling approaches can be used to find single short motifs, but there is typically a high false-positive rate, both because of the potential for error in finding short sequences and the confounding effects of other positive and negative regulatory elements in promoters.

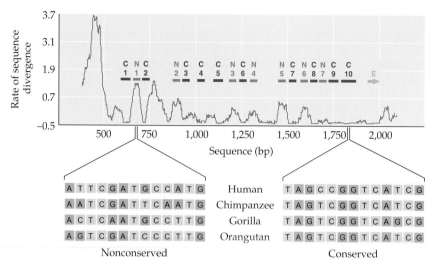

Figure 2.15 Phylogenetic shadowing. A multiple alignment of sequences from at least five closely related species allows fitting of a statistical model that formulates the ratio of likelihoods that each window of 50 nucleotides is evolving at a relatively fast or slow rate. For the apo(A) promoter, this led to the identification of 10 conserved elements (C) and 7 nonconserved elements (N). Oligonucleotides of several of the C elements were found to bind nuclear extracts, while disruption of most of these and several other C elements resulted in reduced transcription from a reporter gene in cell culture. Hypothetical examples of conserved and nonconserved sequences are illustrated.

Non-Protein Coding Genes

The identification of genes whose product is not a protein but a functional RNA is complicated by several features that distinguish these genes. First, for the most part the transcripts are not polyadenylated, and so are not represented in standard cDNA libraries. Second, the constraint on sequence divergence is at the level of secondary structure rather than codon sequence, so sequence divergence between species is often too great to identify the genes purely by sequence similarity. And thirdly, relatively little is known about the function and distribution of non-protein coding RNAs (ncRNAs; Table 2.1) other than those involved in transcriptional processing and translation.

Nevertheless, experience from analysis of the content of mitochondrial genomes, as well as knowledge gained from detailed biochemical analysis of ribosomal complexes, has led to the development of algorithms that are thought to identify a substantial fraction of functional RNAs. As with protein-coding genes, first-pass annotation is unavoidably incomplete and imprecise, but intrinsic interest in the biology of RNA-mediated catalysis ensures the ongoing characterization of these gene classes.

Transfer RNAs (tRNAs) fold into a characteristic cloverleaf structure by virtue of the assembly of short stretches of base-pairing in which the local

TABLE 2.1 *Classes of Non-coding RNAs in the Human Genome*

Class	Function	Number	Localization
tRNA	Protein synthesis	~500	Dispersed large clusters
rRNA	Protein synthesis	~200 each	Tandem arrays
U snRNAs	Splicing	<20 each	Dispersed in clusters
snoRNAs	rRNA modification	~100	Dispersed single copy
Others	Various	~20 ??	Single copy

Source: As reported by IHGSC (2001).

complementarity of sequence is more conserved than the actual sequences involved. This structure is identified by using software such as tRNAscan-SE (Lowe and Eddy 1997), which has an advanced search grammar based on sequence similarity as well as base-pairing potential to sort true tRNAs from likely pseudogenes. Although there are 61 codons to be decoded (excluding the three stop codons), only 45 anticodons are required because "wobble" pairing allows third-position U or C nucleotides to be recognized by the same tRNA species. As shown in Figure 2.16A, there are actually 48 classes of identified tRNA in the human genome but, as in all eukaryotes, most of these are present in multiple copies, yielding a total of 497 canonical human tRNAs. The frequency of each tRNA type correlates with codon usage in mRNAs (Figure 2.16B), and is thus thought to be responsible in part for the evolution of codon biases in genomes. Transfer RNAs are dispersed throughout genomes, but tend to occur in clusters containing multiple different classes.

Ribosomes contain four types of ncRNA: the 28S, 5.8S, and 5S rRNAs found in the large subunit, and the 18S small subunit rRNA. Each of these is represented by between 150 and 300 copies in higher eukaryotic genomes, in which they are found in single arrays of tandemly duplicated genes (although related single copy sequences can also be identified in dispersed locations). The highly repetitive nature of these clusters makes thorough structural characterization of rRNA complexes difficult, due both to biases against sequencing repetitive DNA in genome projects and problems associated with the assembly of such regions. As with the tRNAs, many dispersed copies of the genes are likely to be pseudogenes, relatively recently generated insertions that are accumulating mutations and perform no physiological function.

Eukaryotic rRNA is heavily modified in the nucleolus by protein-RNA complexes, including a family of small nucleolar RNAs (snoRNAs). Two families of these, the C/D box and H/CA families, guide site-specific methylation and pseudouridylation, and 84 single-copy rRNA genes have been found dispersed throughout the human genome.

The splicing of primary transcripts into mature mRNA is mediated by protein-RNA complexes that include spliceosomal RNAs designated as U1 through U12. The genes for these molecules also occur in loosely structured

Figure 2.16 Transfer RNA content in the human genome. (A) The table indicates the percentage of codons for each color-coded amino acid below the diagonal, and the number of tRNAs with the associated anticodon above the diagonal in each box. (B) The plot shows that codon bias (expressed as the percentage of the particular amino acid that is encoded by each codon) is only partially correlated with the number of tRNAs with the complementary anticodon. Thirteen possible codons are not associated with exactly complementary tRNAs, and so must be decoded by "wobble" anticodons.

(A)

	NUN		NGN		NAN		NCN	
UNU	46	0	18	10	44	1	45	0
UNC	54	14	22	0	56	11	55	30
UNA	8	8	15	5				
UNG	13	6	6	4			100	7
CNU	13	13	29	11	41	0	9	9
CNC	19	0	32	0	59	12	19	0
CNA	7	2	28	10	26	11	11	7
CNG	40	6	11	4	74	21	21	5
ANU	36	13	24	8	47	1	15	0
ANC	48	1	36	0	53	33	24	7
ANA	16	5	28	10	43	16	20	5
ANG	100	17	12	7	57	22	20	4
GNU	18	20	26	25	47	0	17	0
GNC	24	0	40	0	53	10	34	11
GNA	11	5	23	10	43	14	25	5
GNG	47	19	11	5	57	8	24	8

(B)

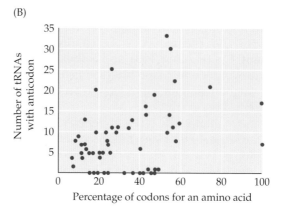

Number of tRNAs with anticodon

Percentage of codons for an amino acid

tandem arrays of up to 20 copies, though some are dispersed as single copies. Identification of these genes requires sequence-matching to previously characterized RNAs, and like the small RNAs that mediate the editing of messages, they are likely to be both underrepresented and improperly characterized in shotgun genome sequences.

Several other classes of ncRNA are known to exist, including RNA components of enzymes such as telomerase and RNase P. These genes have been identified through sequence identity to purified products. In addition, a small number of non-protein coding regulatory genes have been identified by cloning of genetic lesions that affect an RNA that does not include an ORF. For example, the *Xist* gene has no open reading frame, but is required for dosage compensation (X-chromosome inactivation) in mammals. Searches are also under way for transcripts from portions of the genome that are not predicted to encode a protein yet may produce an important product. The possibility exists that eukaryotic genomes encode numerous genes whose functional product is a cryptic RNA.

A large family of such genes came to light only in the early 2000s with the discovery of microRNAs, which are short (~22 nucleotide) hairpin RNAs that bind to 3′ untranlated regions of mRNAs and function in gene regulation either by repressing translation or promoting mRNA degradation. An online miRNA database, miRBase (http://microrna.sanger.ac.uk; Griffiths-Jones 2004) catalogs several thousand known miRNAs from numerous organisms, along with their predicted structure, and genomic locations. The founding members of this class of gene are *LET4* and *LIN7* from the nematode *C. elegans*, which were first identified as regulators of developmental timing. There are over 100 known miRNAs in *C. elegans*, many of which appear to be conserved in other animals and even in plants.

Structural Features of Genome Sequences

There are numerous structural features of genomes that are of biological interest. These include the distribution of repetitive elements; variation in GC content; simple sequence repeats (SSRs, including microsatellites and minisatellites); degree and arrangement of segmental duplications; and the structure of the centromeres and telomeres. Analysis of such sequences is relevant to the basic principles of genome evolution, as well as to understanding the role of chromatin structure in gene regulation and chromosome replication.

Repetitive sequences. Interspersed repetitive sequences account for just a few percent of some eukaryotic genomes, such as yeast and *Drosophila*, but constitute more than 50% of the human genome and over 90% of the genomes of species as diverse as certain crickets, lilies, and amoebae. As a consequence, genome size varies over several orders of magnitude, whereas gene *content* in multicellular organs varies over less than one order of magnitude. This **C-value paradox** is hypothesized to arise as a result of lineage-specific differences in the relative rates of expansion of repetitive sequences, and deletion of nearly neutral sequences. A small shift in the balance of these two processes might lead to a rapid expansion or contraction of genome size, but it is not yet known whether or how natural selection shapes the balance. Analysis of the age and frequency distribution of repetitive sequences, as well as the deletion and substitutional mutation rates in pseudogenes in whole genomes, promises to shed light on the mechanisms and causes of the evolution of genome size (Petrov et al. 2000).

At least five classes of repetitive elements can be recognized:

1. Transposon-derived repetitive elements.
2. Inactive mRNA-derived copies of cellular genes (pseudogenes).

3. Simple sequence repeats (microsatellites and short tandem repeats/VNTRs).

4. Segmental duplications of up to 300 kb.

5. Blocks of noninterspersed repeats, including ribosomal gene clusters and heterochromatin.

All of these are described in detail in most standard genetics textbooks. Transposable elements (transposons) can be grouped into a large number of families by phylogenetic analysis, but are generally grouped into the four categories shown in Figure 2.17. One category transposes by way of a DNA intermediate, using a transposase enzyme to catalyze excision and reinser-

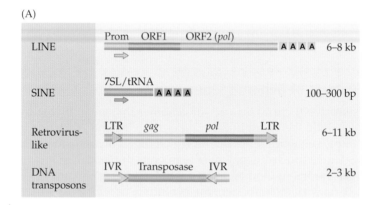

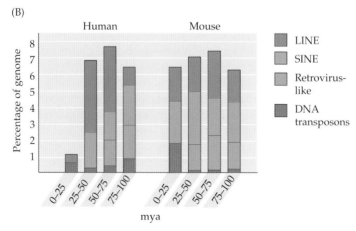

Figure 2.17 Classes of repetitive sequences and their frequency in mammalian genomes. (A) The structure of the four major classes of repetitive elements (LINEs and derivative SINEs, retrovirus-like LTR elements, and DNA transposon families). The average length of each type of sequence is shown at the right. (B) The percentage of the human and mouse genomes each class of element constitutes. Age bins were estimated from the degree of sequence divergence from the consensus for each family of element. The human genome has relatively more LINEs than the mouse, but the rate of transposition slowed around 25 million years ago (mya). (After IHGSC 2001.)

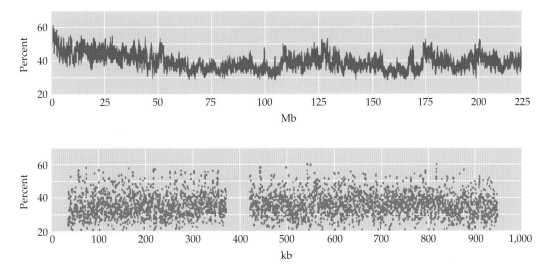

Figure 2.18 Distribution of GC content along human chromosome 1.   GC content varies between 20% and 65% at several different levels of resolution, including for the entire 220 Mb of chromosome 1 averaged over 1-Mb windows (top) and within just 1 Mb for 200-bp windows (bottom). A gap in the IHGSC sequence can be seen at the 400-kb mark on the 1-Mb scale. (After Venter et al. 2001 and IHGSC 2001.)

tion, and includes the famed *Drosophila* P-elements, bacterial IS and yeast Ty elements, and the widespread mariner family of transposons. The other three categories transpose by way of an RNA intermediate that is converted to DNA by a reverse transcriptase enzyme. These include autonomously transposing LINEs of up to 8 kb; nonautonomous derivative SINEs (which use the LINE enzyme to transpose) just a few hundred bases in length; and retrovirus-like long terminal repeat (LTR) containing elements.

GC content. Genomes show wide variation in their overall GC content. Factors contributing to this variation are thought to include environmental temperature of microbial niches, levels of methylation (particularly of cytosine, which mutates to thymidine at an elevated rate when methylated), and recent transposon activity (the nucleotide biases of transposons often differ from those of the remainder of the genome).

Over stretches of hundreds of kilobases, GC content should vary by less than 1% as a result of random sampling, but most genomes show a mosaic of nucleotide bias ranging over as much as 30% (Figure 2.18). Boundaries between regions with different GC content are not sharp, but nevertheless may be identified computationally, and some authors divide the genome into **isochores** that may have structural and biological significance (Bernardi 2000). Karyotypic bands revealed by nuclear dyes such as Giemsa tend to correlate with GC content (dark bands being more AT-rich), possibly reflecting their propensity to coil into superstructure, but clearly other features of the DNA contribute to chromatin assembly. CpG dinucleotides are underrepresented in mammalian genomes overall, but cluster as CpG islands between 0.5 and 2 kb in length that are significantly enriched just upstream of genes (Larsen et al. 1992).

Simple sequence repeats. **Microsatellites** are defined as simple sequence repeats (SSRs) with a repeat length of up to 13 bases, whereas longer repeats give rise to **minisatellites**. The most common classes of SSRs are dinucleotide, trinucleotide, and tetranucleotide repeats, which occur at a rate of at least one every 10 kb in a wide range of eukaryotic genomes. SSRs may arise by a variety of mechanisms, probably most commonly replication slippage, but transposon-associated microduplications have also been documented, and a fraction of SSRs could be conceived by substitutional mutations as well. Once established, unequal recombination between repeats can generate stepwise changes in repeat number greater than the standard single repeat increase or decrease (Calabrese et al. 2001).

Microsatellites have a much higher mutation rate than standard sequences—up to 0.001 gametes per generation—and have a high probability of back mutation. They are extremely useful in estimating evolutionary relationships between populations within a species, but generally evolve too rapidly to be phylogenetically informative between species (for example, in *Drosophila*, many microsatellites found in one species are not even present in sibling species). Human microsatellites average at least 10 alleles with heterozygosity per locus over 80%, and hence they have become a crucial tool in forensic analysis and paternity testing. The U.S. Federal Bureau of Investigation considers a set of 13 SSRs to uniquely describe each American, although issues relating to the inference of individual identification in the courtroom are still under debate. SSRs are also important in human genetic studies, both as markers for pedigree analysis and because a family of diseases (including Huntington's disease and some forms of muscular dystrophy and ataxia) can be traced in part to trinucleotide expansion (Table 2.2; Cummings and Zoghbi 2000).

Segmental duplications and conserved non-genic Sequences. Segmental duplications are remarkably common in vertebrate and plant genomes. Three percent of the human genome, for example, shows a match to another sequence at 90% or more identity over at least a kilobase. Duplications are almost as likely between chromosomes as within them. Some duplications cover more than 50 Mb, and up to one-quarter of the genes in higher eukaryotic genomes may have a paralog. This level of duplication is one of the major surprises to emerge from genome sequences, and the consequent potential for redundancy is just beginning to be explored in relation to homeostasis of developmental and physiological processes, potential for evolutionary divergence, impact on genetic heterogeneity and the etiology of disease, as well as the technical demands it places on gene mapping. In invertebrates, a substantial fraction of genes are tandemly duplicated, and it is well known that this phenomenon constitutes a considerable impediment to functional genetic dissection.

TABLE 2.2 *Microsatellite-Associated Human Diseases*

Disease	Gene	Type	Number of repeats		Location
			Normal	Disease	
Fragile X syndrome	*FMR1*	CGG	6–53	60–230	5′ UTR
	FMR2	GCC	6–35	60–200	5′ UTR
Myotonic dystrophy	*DMPK*	CTG	5–37	50–'000s	3′ UTR
Friedreich ataxia	*X25*	GAA	7–34	34–100	Intron 1
Kennedy disease	*AR*	CAG	9–36	38–62	Coding
Huntington disease	*HD*	CAG	6–35	36–120	Coding
Haw River syndrome	*DRPLA*	CAG	6–35	49–88	Coding
Spinocerebellar ataxia	*SCA1*	CAG	6–44	39–82	Coding
	SCA2	CAG	15–31	36–83	Coding
	SCA3	CAG	12–40	55–84	Coding
	SCA6	CAG	4–18	21–33	Coding
	SCA7	CAG	4–35	37–300	Coding
	SCA8	CTG	16–37	110–250	Coding
	SCA12	CAG	7–28	66–78	5′ UTR

Source: From Cummings and Zoghbi (2000).

As much as 5% of mammalian genome sequence is highly conserved even in marsupials, whereas only 2% (at most) of a typical mammalian genome encodes transcripts. It is estimated that the human genome contains about 60,000 conserved non-genic sequences (CNGs), each generally at least 100 nucleotides in length and over 85% identical across the Mammalia (Dermitzakis et al. 2003). The distribution of nucleotide substitutions in many of these elements are distinct from those typically associated with ncRNAs and regulatory motifs. Observations such as this suggest that some of the so-called "junk" in genomic DNA will eventually be assigned a function.

Structure of centromeres and telomeres. The evolutionary dynamics of centromeric and telomeric regions are markedly different from those of general euchromatin. As a rule, heterochromatin consists of long, highly repetitive stretches of DNA and includes junkyards of transposable elements, interchromosomal duplications, and even insertions of large chunks of mitochondrial DNA. For this reason, centromeric chromatin is not typically sequenced in its entirety. The structural properties of centromeres that mediate mitosis and meiosis remain to be characterized in most species, but are yielding to comparative sequence analysis (Hall et al. 2004). It is also clear that heterochromatin contains interspersed unique genes, a striking example being relatively gene-rich regions of otherwise heterochoromatic Y chromosomes in several species.

Telomeric DNA similarly remains to be thoroughly characterized, in part because it is often refractory to traditional cloning, in part because it consists of long tandem repeats. These repeats are actually generated during replication by a telomerase enzyme that uses an RNA template to prime lagging strand DNA synthesis. Subtelomeric regions in the human genome contain relatively short stretches of interchromosomal duplications that may be a remnant of the process by which new telomeres are created during karyotype evolution.

The ENCODE Project. After the completion of the first human genome sequences, attention turned to the analysis of human genetic variation (see Chapter 3) and the gathering of functional information about the structure and expression of sequences throughout the genome. An **Encyclopedia of DNA Elements** (ENCODE) Project was initiated (ENCODE Project Consortium 2007 and 36 papers in a special issue of *Genome Research* June 2007) that in its pilot phase involved performance of over 200 different genomic assays on a targeted 30-Mb sequence, or 1%, of the human genome. Fourteen distinct regions—each around 1 Mb in length—were chosen because they were already well annotated with respect to biological function (for example, the *HoxA* cluster on chromosome 7), while another 30 smaller regions were picked to ensure as broad a coverage of different types of chromosomal element as possible. Model organism ENCODE projects have also been initiated.

Over 400 million experimental data points were collected in the first few years of the ENCODE project, which is jointly funded by public and private foundations in the United States, Europe, and Japan. The experiments are summarized at the Consortium's browser, http://genome.ucsc.edu/ENCODE, which provides access to data on both the biochemical and biological properties of sequence elements. Most of the assays are designed to define the structure and expression of transcripts; to identify which sequences are chemically modified or bound to proteins that have roles in gene expression and DNA replication; and to facilitate computational comparison of mammalian species. Sophisticated procedures for assembly, alignment, and annotation, not to mention statistical analysis of the data, continue to be developed. Among the major findings of this effort are:

- The majority of all nucleotides are transcribed, whether as part of rare coding transcripts, noncoding RNAs, or random transcripts that may have no biological function.

- Many genes have multiple, previously undetected, transcription start sites, and regulatory sequences are as likely to be upstream as downstream of the major start sites.

- Chromatin is modified by a variety of mechanisms that result in highly ordered and cell-type-specific patterns of histone association and accessibility to regulatory nuclear factors.

- DNA replication and transcriptional regulation are interrelated and coordinated processes.

- Approximately 5% of the mammalian genome is evolutionarily highly conserved, and experimental assays demonstrate functionality for 60% of these sequences. However, many important functional sequences are not conserved and may be relatively highly variable even among human populations.

Functional Annotation and Clusters of Gene Families

One of the first goals of any genome sequencing project is to broadly classify as many genes as possible into putative functional families. Answers to the questions "Are there any genes that are conspicuous by their absence?" and "Which genes are overrepresented relative to other genomes?" fuel hypotheses about gene function that may lead biotechnologists to probe new ways of attacking pathogens or harnessing microbes for useful purposes.

More generally, functional annotation is an essential step toward understanding how genes and gene products interact, using gene expression and proteomic information as discussed in Chapters 4 and 5. As with the structural annotation of genes, however, it is essential to realize that any functional annotation that depends solely on alignments and comparison is incomplete and prone to error. Classical genetic, biochemical, and cell biological methods must all be used in order to dissect the true function of genes.

First-pass classification is achieved by sequence-similarity searches, using software such as BLAST. Often protein similarities can be identified by conceptual translation of similar nucleotide sequences detected with BLASTn, but due to the redundancy of the genetic code, amino acid sequences tend to be more highly conserved than nucleotide sequences. BLASTp can be used to detect amino acid similarities, or BLASTx, which translates a nucleotide sequence into all possible reading frames and then scans these against a protein database. The theoretical basis for such searches was described in Box 2.2, including a description of the E-value (Expectation) score that is used to evaluate the significance of a match. A common result of BLAST procedures is that between one-third and one-half of all of the predicted proteins do not match a protein for which any functional data is available; hence these genes are classified as "unknown function" or "orphans." This is as true of multicellular eukaryotic genomes (human, fly, worm, weed) as it is of yeast and prokaryotes. Protein structure determination and other proteomic methods (discussed in Chapter 5) will gradually bring this number down.

It is believed there is a finite number of structural protein domains in the combined proteome of all organisms, and that once these domains have been identified, it ought in principle to be possible to at least cluster all proteins. Many genes evolve at a sufficiently fast rate that alignment based on sequence similarity alone is unlikely to be successful, and in these cases clustering will depend on structural predictions.

EXERCISE 2.3 *Perform a BLAST search*

Using the gene that you identified in Exercise 1.1 perform a BLASTn search for homologs in other species. What conclusions can you reach regarding whether this gene is part of a gene family, or regarding its phylogenetic distribution?

ANSWER: *Continuing with our example of the* IL13 *gene from Chapter 1, copy the transcript sequence of this gene either from the Ensembl or GenBank page. (For Ensembl, click on the "Transcript info" link and the sequence appears at the bottom of the page. For GenBank, first select "Nucleotide"; then type "IL13" into the search box and follow the link to the mRNA annotation for NM 002188.) Return to the NCBI home page, click on "BLAST," bring up the "Nucleotide-nucleotide (blastn)" page, and paste your sequence into the search box. Choose some search options (for example, "all RefSeq mRNA sequences") and submit the request. Depending on the number of queries in the queue, this could take a minute or two. In the case of* IL13, *alignments are shown for human, chimp, macaque, canine, and equine matches to the 1280 nt search. If you select optimization for somewhat similar sequences rather than highly similar ones, very high match scores are also seen for several other mammalian IL13 loci. The closest match for the Drosophila genome, however, has an E-value of just 0.17, and even this score is due to a single perfect match—a 23-bp element— that is probably not indicative of a homologous gene. Thus* IL13 *appears to be unique to vertebrates (and possibly even to mammals, since no fish or bird sequences have significant E-values), and to be present in a single copy per genome.*

Clustering of Genes by Sequence Similarity

Another common result obtained with protein BLAST searches is that each query sequence matches multiple proteins from one or more species. This can happen either because one domain in the query is present in a family of proteins, or because multiple domains match different proteins. These possibilities are readily distinguished because the alignment software reports a series of keywords associated with each positive hit and indicates the location of the stretch of similarity, as seen in Figure 2.19. When comparing closely related species, a query sequence containing multiple domains will typically identify a protein or proteins with the same domain structure.

However, over a period of divergence of tens of millions of years, domains shuffle, and over hundreds of millions of years different domains are found clustered in different combinations. There is some biological logic to domain clustering in that multiple domain matches tend to classify the gene product in the same broad category, such as transcription factor or receptor. That is, two different DNA-binding domains may be combined on the same protein, or a transmembrane region may be linked to a range of different intracellular and extracellular domains.

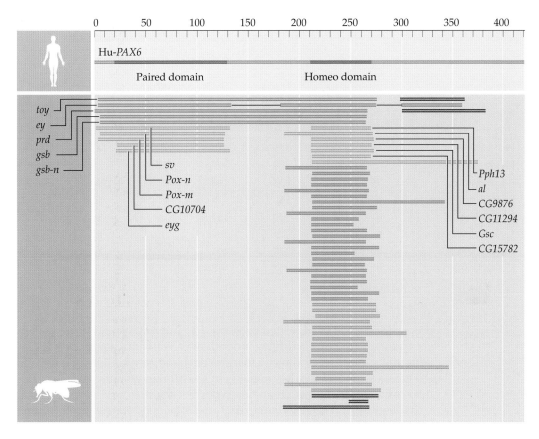

Figure 2.19 Alignment of human *PAX6* against the *Drosophila melanogaster* genome.   The basic protein BLAST search identifies 103 proteins in the fly proteome, with varying degrees of similarity to two portions—the Paired domain and the Homeo domain—of human *PAX6*. Sixty of the matches are shown graphically; mouse-over provides the identity of each match as shown for the 16 closest matches. Note that *toy, ey, prd, gsb,* and *gsb-n* match to both domains.

The output of a protein BLAST search is thus not so much a match to another gene as a match to one or more protein domains. Numerous databases have been established that classify protein domains according to criteria agreed upon by groups of experts. One of the first to emerge was the Enzyme Commission (EC) hierarchical classification of enzymes, in which each enzyme is assigned a number that reflects the subclassification of function. For example, EC1.1.1.1 is alcohol dehydrogenase. Similar classification schemes for nonenzymatic proteins are not so obvious, since most of these proteins perform functions whose biochemical or biological consequences are dependent on the context in which the protein is used. Nevertheless, a general scheme devised for the classification of bacterial proteins has attained wide acceptance.

Structural biologists have assembled protein databases such as PFAM that allow researchers to immediately access data concerning biochemical

TABLE 2.3 *Number of InterPro Protein Domains in Some Eukaryotic Genomes*

Domain type	InterPro ID	Number of domains				
		Human	Fly	Worm	Yeast	Weed
Immunoglobulin	IPR003006	765	140	64	0	0
C$_2$H$_2$ zinc finger	IPR000822	706	357	151	48	115
Protein kinase	IPR000719	575	319	437	121	1,049
Rhodopsin-like GPCR	IPR000276	569	97	358	0	16
P-loop motif	IPR001687	433	198	183	97	331
Reverse transcriptase	IPR000477	350	10	50	6	80
Rrm domain	IPR000504	300	157	96	54	255
G-protein WD-40 repeat	IPR001680	277	162	102	91	210
Ankyrin repeat	IPR002110	276	105	107	19	120
Homeodomain	IPR001356	267	148	109	9	118

Source: From IHGSC 2001, Table 25.

properties of a set of predicted proteins. The InterPro classification system goes a step further in classifying individual protein domains. Gene annotations now typically link directly to InterPro classifications. Table 2.3 lists the ten most common InterPro domains in the human genome and their frequencies in other multicellular eukaryotic genomes.

It follows that it is important to draw a distinction between protein function prediction and classification of genes as members of the same family. The latter involves distinguishing between paralogs and orthologs, and is addressed in the next section. The former allows us to make some very general conclusions about the content of genomes based on the association between protein domains and function.

The major classes of protein molecular function are enzyme, signal transduction (including receptors and kinases), nucleic acid binding (including transcription factors and nucleic acid enzymes), structural (including cytoskeletal, extracellular matrix, and motor proteins), and channel (voltage and chemically gated). Several other important categories include immunoglobulins, calcium-binding proteins, and transporters. Subclasses within each group vary widely in frequency among genomes, as do the absolute numbers and relative proportions of each major category. For example, the human genome shows a marked increase by comparison with the fly and worm in proteins involved in neural development, signaling during development, hemostasis, and apoptosis.

Clusters of Orthologous Genes

Sequence alignment serves the useful purpose of identifying which genes to include in a phylogenetic analysis. Prior to genome sequencing, many genes were cloned by hybridization to cloned genes from another species, or by degenerate PCR. These methods were not guaranteed to identify all the potential relatives of a particular gene, and in fact a typical procedure was to focus on the one or two clones that either gave the strongest hybridization signal or happened to be identified first.

The postgenomic equivalent of this bias is simply to assume that your gene is the same as the one that shows the closest sequence match in another genome. If similar genes from a number of different species all have the closest match, all the better. The problem arises when the query gene matches multiple members of a family of genes that have discrete functions. In this case, the closest match may well be very misleading in terms of function.

Consequently, classification of molecular function starts with the identification of as large a set of possible family members as possible. The PSI-BLAST (Altschul et al. 1997) and PHI-BLAST (Zhang et al. 1998) algorithms and related procedures have been developed for this purpose. The basic idea of PSI-BLAST is to align the sequences obtained in an initial protein database search and use this to construct a profile (see Box 5.1), which is then used to initiate a fresh search. The process is iterated until no further matches are identified. A true family of genes ought to be bounded by a significance cut-off, so that there is a limit as to which proteins will be included in a family, as long as realistic parameters are chosen. The iterative procedure has the effect of reducing the degree of sequence similarity required for a significant match.

A slightly different approach to protein classification has been to assemble **clusters of orthologous genes**, or **COGs** (Tatusov et al. 2001). COGs are created by identifying the best hit for each gene in complete pairwise comparisons of a set of genomes. Comparison of the 185,500 proteins encoded by the genomes of 66 microbes led to the identification of 4873 COGs as of September 2003. These represent about 75% of all predicted microbial proteins and can be accessed at http://www.ncbi.nlm.nih.gov/COG. Similarly, just over 50% of 110,000 eukaryotic proteins from fly, nematode, human, *Arabidopsis*, two yeasts, and a microsporidian assemble into 4,852 eukaryotic COGs (Tatusov et al. 2003), 20% of which form a conserved core of proteins.

As shown in Figure 2.20, triangles are drawn linking genes that have the same best hit in at least one direction (for example, the genes that are the best hit in both species B and C for a gene in species A are also the best hit for at least one of the comparisons of B against C). These triangles are then merged together to assemble a cluster of putative orthologs.

COGs will typically consist of both orthologs and paralogs. A **paralog** is a duplicate copy of a gene that arose subsequent to the split between the two lineages that are being compared. An **ortholog** is a gene in another lineage that is derived from the same ancestral gene that was present prior to the lineage split (Mindell and Meyer 2001). Thus, using Figure 2.21A as a guide, say there are two copies of a human gene *HuA*, the two copies (*HuA* and *HuA'*) are paralogs, and both are orthologs of a mouse gene, *MmA*. Distinguishing which of these has retained the ancestral function is not necessarily possible simply by comparing the levels of sequence similarity. To see why, suppose (trivially) that *HuA* and *MmA* differ at five amino acids, none of which affect

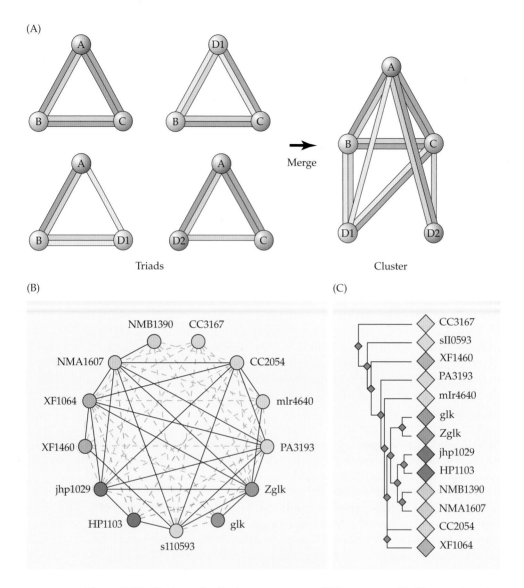

Figure 2.20 Clusters of orthologous genes. COGs are assembled by merging triads of genes from different species that are the best match to one another. (A) Each triad shows one gene from each of four species (genes A, B, C, and D) linked to the gene that is the best match in the other species. Thus genes A, B, and C are the best matches to one another in the respective species' genomes, and although A is the best match for both D1 and D2, D1 is not the best match for A, so these genes are joined by a single line. (B) COG0837, identified as glucokinase, as viewed at the NCBI Web site. In the merger of the triads on the left, dashed lines indicate examples where the best match occurs only in one direction; solid lines imply that each gene is the best match in the genome of the other. The tree at the right (C) shows a more standard representation of the relationship among the genes.

the function of the protein, while *HuA′* and *MmA* differ at just four amino acids, one of which changes a critical residue involved in substrate binding. In this case, clustering by similarity would erroneously identify *HuA′* and

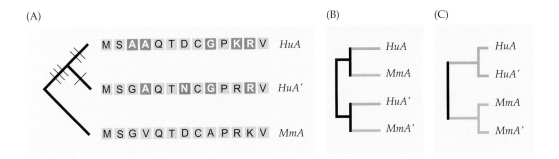

Figure 2.21 Orthologs and paralogs. When any three sequences are contrasted, usually two will cluster together. (A) In this example, the two human sequences differ from the mouse sequence at three common sites, and *HuA* has two further changes compared to a single unique change in *HuA'*. However, the difference in charge introduced by changing an aspartate [D] to an asparagine [N] may confer a different function to the *HuA'* gene, highlighting the fact that degree of sequence similarity does not always indicate which proteins are most similar functionally. (B) Duplicate genes may be identified as pairs of orthologs if the sequences between species (*HuA* and *MmA*; *HuA'* and *MmA'*) are more related than within species. (C) If *HuA* and *HuA'* are more similar to one another than to *MmA* and *MmA'*, they are likely to be recently duplicated paralogs.

MmA as most likely to be functionally equivalent. The situation is additionally complicated by the fact that such a high proportion of genes have duplicates that arose prior to the divergence of the two species being compared.

It cannot be emphasized strongly enough that COGs provide at best only a hint as to gene function. Errors are increasingly likely to arise at more and more refined levels of a clustering hierarchy. This is because, particularly in animals and plants, genes tend to duplicate and diverge in function in lineage-specific manners, but clustering by overall sequence divergence does not reflect key events involving single amino acid changes that shift the functional specificity of enzymes, kinases, and transcription factors. Phylogenetic methods are considered to provide a much more reliable indicator of functional subclassification. Nevertheless, the **InParanoid** Web site of the Stockholm Bioinformatics Center (http://inparanoid.sbc.su.se/cgi-bin/index.cgi; Remm et al. 2001) provides a very useful tool for looking up putative orthologs between species pairs, using a database of more than 600,000 predicted protein sequences from 35 organisms.

Phylogenetic Classification of Genes

Molecular evolutionists have developed sophisticated tools for the assembly of gene phylogenies, as described in Box 2.4. The methods are the subject of numerous complete textbooks (e.g., Graur and Li 2000; Hall 2007). Different procedures for phylogenetic analysis can yield very different answers, but procedures exist for attempting to ascertain the likelihood that a partic-

BOX 2.4 Phylogenetics

A concept that has arisen several times in this discussion is that of homology, or *shared descent from a common ancestral sequence.* There are many reasons why we may want to use an aligned set of DNA or amino acid sequences to propose an **evolutionary tree** or **phylogeny** that displays the hierarchical ancestral relationships among the sequences. We may simply wish to understand the pattern of relatedness of a group of species. There are also less obvious objectives. Many proteins are members of multigene families, and we may want to classify a newly discovered gene into the appropriate subfamily. We may want to test predictions about the phylogenetic placement of one or more sequences, perhaps inferring the source of a viral infection. In any case, there is a large and growing literature on the inference of phylogenies using sequence data. We will provide only a brief overview of some of the more common methods. For further details, see the excellent presentation by Swofford et al. (1996) or the handbook by Hall (2007).

General Principles

Before describing the leading phylogenetic reconstruction methods, a few technical issues must be introduced. First is the distinction between rooted and unrooted tree topologies (Figure A), which is important because most phylogenetic methods can only produce unrooted trees. Additional information regarding evolutionary rates or the most ancient relationships is needed to root the inferred trees.

A second important concept is that of branch length, which in molecular terms is defined as the average number of nucleo-tide substitutions per site. For example, along a branch of length 0.2, the average nucleotide site has undergone 0.2 changes. Clearly, that number is not achievable in nature; it represents the average value over sites, most of which will have changed either 0 or 1 times.

The primary methods of phylogeny reconstruction are **parsimony, distance,** and **likelihood.** There are many variants within each of the three broad classifications. The shared thread among all of the methods is an attempt to *identify the topology that is most congruent with the observed data.*

The methods differ in their mechanisms for measuring this congruence. Some methods (parsimony, likelihood, and some distance methods) define a metric between topology and data and require (in principle) an exhaustive search through all possible tree topologies. Unfortunately, the number of topologies increases explosively. The number of unrooted tree topologies for n sequences is $3 \times 5 \times 7 \times \ldots \times (2n - 5)$—a value that exceeds 2 million when n reaches 10. In practice, heuristic search methods are used to search through a likely subset of the vast collection of possible topologies. The commercially available software PAUP* (Swofford 2002; http://paup.csit.fsu.edu) incorporates many of the algorithms and is widely used to construct phylogenies.

Parsimony Methods

Parsimony methods are probably the most widely used and most intuitive methods of phylogenetic inference. The underlying idea is a simple one: The best topology is the one requiring the smallest number of

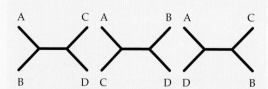
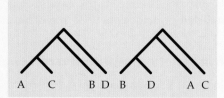

FIGURE A Rooted and unrooted trees. For any four taxa (the tips of the branches) there are three distinct unrooted trees. Each unrooted tree can be rooted on any of its five branches; two of the five possible rooted trees for the center unrooted tree are shown.

changes to explain the observed data. In Figure B we see a short alignment of four DNA sequences, along with the three possible unrooted topologies for the four sequences. Consider position 1 in the alignment. If the true topology were topology 3, a bit of thought reveals that at least two nucleotide substitutions (for example, one on the branch leading to sequence A, and a second on the branch leading to sequence B) would be required. At position 2, the data can be explained with a single C→T (or T→C) change on the interior branch of topology 3.

Likelihood Methods

In its mechanics, likelihood reconstruction of phylogeny is similar to parsimony. A search through tree topologies is necessary, with the congruency measure being the maximum probability of the data given the topology. The **maximum likelihood estimate** of the true tree is the topology providing the highest such probability. In order to compute the probability, one must first select a probabilistic model describing the change of sequences over time. Many factors might be included in such a model, including base frequency biases and unequal probabilities for different types of nu-

cleotide substitutions. This feature should be contrasted with most parsimony methods, where all changes are treated equivalently, regardless of the type of substitution or the amount of time on a given branch. Likelihood methods tend to be very powerful tools for phylogeny reconstruction, but are typically limited to relatively small data sets (say, fewer than 50 sequences) because of their computational expense.

Distance Methods

Distance methods are the oldest family of phylogenetic reconstruction methods. As do likelihood methods, they rely on a probabilistic model of sequence evolution. However, distance methods use the model only to calculate pairwise *evolutionary distances* for all pairs of sequences—a task that is much faster than the likelihood calculations. Given a matrix of pairwise evolutionary distances, there are many distance methods to choose from. While many require topology searching, some of the more popular ones do not, relying instead on a predetermined algorithm for obtaining the estimated phylogeny.

One of the most effective distance methods is the **neighbor joining (NJ)** method, which is very fast and tends to have a high probability of reconstructing the correct tree (when used with a proper choice of pairwise distance).

Evaluating Reconstructed Trees

As with any statistical inference, an estimate without some notion of its reliability is essentially useless. For the phylogeny problem, we would like a measure of data support for particular groupings, or **clades,** in an estimated tree. The most common approach for measuring support is through the use of **bootstrapping,** as introduced by Felsenstein (1985). Numerical resampling techniques are used to compute bootstrap support levels for every node in the tree topology. Bootstrap values near 100% indicate clades that are strongly supported by the data, while lower levels indicate reduced support. What the precise bootstrap level to indicate a statistically significant clade should be is the source of tremendous controversy, but values greater than 70–80% are often taken to indicate fairly strong support for the clade.

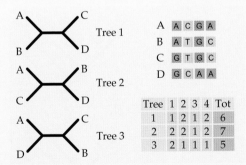

A	A	C	G	A	
B	A	T	G	C	
C	G	T	G	C	
D	G	C	A	A	

Tree	1	2	3	4	Tot
1	1	2	1	2	6
2	2	2	1	2	7
3	2	1	1	1	5

FIGURE B Maximum parsimony. The table within the figure shows the minimum number of changes at each site required by each topology, along with the total number of changes obtained by summing across sites. Applying the maximum parsimony principle to these data, tree 3 would be our best estimate of the "true" phylogeny because it requires the smallest number of nucleotide changes (five) to explain the aligned data.

ular phylogenetic hypothesis is correct. Whichever method is used, one of the goals of phylogenetic analysis is to ascertain how groups of similar genes isolated from a set of species are related by descent. In the preceding example, assume now that the mouse gene is also duplicated. Then if *HuA* and *MmA* are found to cluster in a phylogenetic analysis, as do *HuA'* and *MmA'*, then it is reasonable to infer that genes *A* and *A'* duplicated and diverged in function prior to the divergence of the mouse and human lineages (Figure 2.21B). However, if *HuA* and *HuA'* cluster more closely together, as do *MmA* and *MmA'* (Figure 2.21C), then it is likely that two independent duplications have occurred, and hence neither pair of genes are orthologs and functional assignments should be treated with more caution.

The above caveats notwithstanding, the most likely function of unknown genes can be inferred by phylogenetic analysis so long as the functions of a subset of similar genes or proteins are known, as shown in Figure 2.22. Application of these methods to large families of genes in multiple taxa is sometimes referred to as phylogenomics (Eisen 1998).

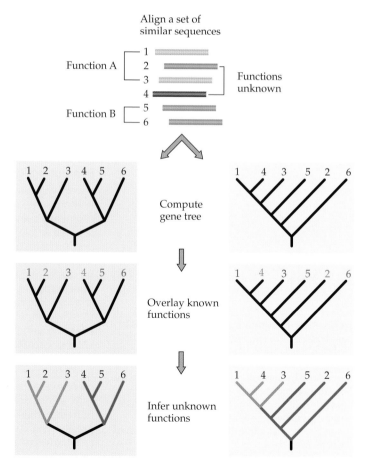

Figure 2.22 Inferring gene function from phylogenetic analysis.   In the example on the left, an ancient duplication event is inferred to have resulted in the evolution of two types of function, and the unknown function of protein 2 is most likely the same as that of proteins 1 and 3. By contrast, in the example on the right, no deep branch separating two clades is observed, but the function of the unknown protein 2 is most likely the same as that of the ancestral function (protein 6), since the more recently derived protein 5 retains the same function. (After Eisen 1998.)

EXERCISE 2.4 *A simple phylogenetic analysis*

A few years ago, a pressing question in the area of human evolution was the exact phylogenetic relationship between humans, chimpanzees, and gorillas. Fossil, morphological, and early molecular data provided conflicting and inconclusive results as to which pair of organisms were most closely related. Suppose the following four short DNA sequences are available, including a sequence from the outgroup organism, orangutan:

Human	CGAAATGCAT
Chimpanzee	CGGACTGTAT
Gorilla	CAAGCTGTAC
Orangutan	TAAGACGTAG

Carry out a phylogenetic analysis to infer the rooted evolutionary tree of human, chimpanzee, and gorilla.

ANSWER: *There is only one possible unrooted tree for three species, so we cannot answer the question using only sequences from human, chimpanzee, and gorilla. By adding the outgroup sequence, we can infer the unrooted tree for the four sequences and then root the tree on the branch leading to orangutan. The result of this rooting will also provide a root for the human-chimpanzee-gorilla subtree.*

Although distance or likelihood methods would also be acceptable, we will use the parsimony method as discussed in Box 2.4. The three possible unrooted trees for the four species are:

Following the method in Box 2.4, the required number of changes for each tree are:

	1234567890	Total
Tree A	1111210102	10
Tree B	1212210102	12
Tree C	1212110102	11

Thus, Tree A is chosen as the best estimate. Since we know that the orangutan is the outgroup, we can place the root of the tree on the branch leading to the orangutan, as follows:

These data suggest that human and chimp are closest relatives.

Gene Ontology

It has become clear that annotation on the basis of molecular function alone is insufficient to describe or predict biological function. Three examples illustrate the general point. Perhaps the best known example of the evolution of a novel function is the reuse of enzymes such as lactate dehydrogenase as lens crystallins, which has occurred multiple times in vertebrate lineages. In these cases, an enzyme is converted into a structural protein. Just as dramatically, the *Drosophila* ortholog of the mammalian dioxin receptor is encoded by the *spineless-aristapedia* gene, which is one of the key regulatory genes that control antenna differentiation. The only clear functional similarity here is the involvement of the protein product in the olfactory system. In both of these cases, knowledge from one species is not directly transferable to another.

Less dramatic but no less important are examples where the *molecular* function has remained the same but the *physiological* function has evolved. In some cases, apparently divergent functions (such as those of the Hox genes in patterning both the cranial nerves of vertebrates and the appendages of invertebrates) can be traced to a shared common function in patterning along the body axis. In others, one or more biological functions are clearly derivative, such as the use of the *Toll-dorsal* pathway in the embryonic patterning of flies as opposed to innate immunity in both vertebrates and invertebrates. In still other cases, functions performed by a single gene in one taxon are split between two genes in another.

Despite these reservations, there is considerable conservation of gene function across the Metazoa, much of which also extends to the fungi and plants. Just over 1,500 genes have unambiguous orthologs among the fly, worm, and human genomes—meaning that there is a single closest match in each genome. Although plants have novel families of transcription factors and seem to use different mechanisms for intercellular communication, most of their core physiological functions are the same as those of animals. Consequently, while developmental roles may diverge, it is well accepted that knowing the cell biological and molecular functions and/or the subcellular localization of each protein is a useful aspect of gene annotation.

To this end, the Gene Ontology (GO) consortium (Ashburner et al. 2000) has established a project to annotate all these features for the complete set of genes identified by genome projects. Several thousand genes from each of the major model organisms are already annotated in GO, which is accessible on the Web at http://www.geneontology.org or through links on the main genome project pages.

The GO site classifies protein function into one of three major categories, as shown in Figure 2.23. *Cell biological function* refers to the nature of the process that is regulated or affected, such as cell growth and division, respiration, signal transduction, and cAMP biosynthesis. *Molecular function* corresponds to the biochemical activities discussed in a previous section, such as enzyme, nucleic acid binding, DNA helicase, or tyrosine kinase. *Cellular component* refers to the place in a cell where the gene product is active, such as the cell surface, Golgi apparatus, or spliceosome. Each of these terms is

(A)

Biological process	Molecular function	Cellular component
▶ Behavior	▶ Antitoxin	▶ Cell fraction
▶ Cell communication	▶ Anticoagulant	▶ Cell wall
▶ Cell growth and maintenance	▶ Antioxidant	▶ Extracellular
▶ Death	▶ Apoptosis regulator	▶ Intracellular
▶ Developmental processes	▶ Cell cycle regulator	▶ Membrane
▶ Perception of external stimulus	▶ Cytoskeletal regulator	▶ Unlocalized
▶ Physiological processes	▶ Defense/immunity protein	
▶ Viral life cycle	▶ Vitamin transporter	
	▶ and many more	

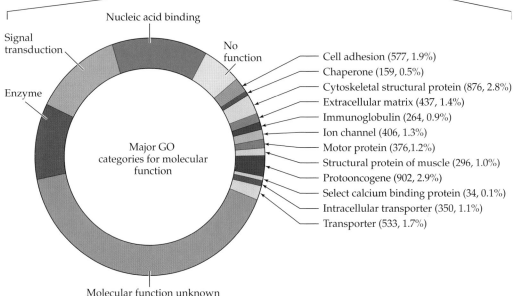

(B)

Evidence codes

▶ IMP, inferred from mutant phenotype
▶ IGI, inferred from genetic interaction
▶ IPI, inferred from physical interaction
▶ ISS, inferred from sequence/structural similarity
▶ IDA, inferred from direct assay
▶ IEP, inferred from expression pattern
▶ IEA, inferred from electronic annotation
▶ TAS, traceable author statement
▶ NAS, non-traceable author statement
▶ NR, not recorded

Figure 2.23 Gene Ontology categories. The Gene Ontology (GO) Web site categorizes genes by their probable function. (A) The three main functional rubrics are shown, with a partial list of the specific functions within each category. The graph shows the number and proportion of genes in each GO molecular function category in the human genome. (Graph after Venter et al. 2001, Figure 15.) (B) The evidence codes of the GO site indicate how functional information was obtained.

BOX 2.5 Gene Ontologies

In the not-so-distant past, a *Drosophila* geneticist might have queried a sequence database using the partial sequence of a gene named *mip*, and found very strong matches to the genes GAP, gap1, rI533, sextra, CG6721, and Ras-GAP. Upon brief inspection of the gene sequences, she would have reached the puzzling conclusion that all of these genes were identical in sequence. Might this have been an interesting example of sequence conservation? No, it was merely an example of the same gene being named multiple times in different locations in the literature. To make matters more confusing, once she solved the nomenclature issue she would most likely have found a bewildering array of functions assigned to this gene by various researchers. This state of affairs made it all but impossible to carry out automated annotation of genes, and it was a major obstacle to creating unified databases that span multiple organisms and use varying vocabularies.

The Gene Ontology(GO) consortium (www.geneontology.org) is the most well-known attempt to provide a standard vocabulary for describing the functions of genes. GO assigns a single name to each gene product, keeps track of known synonyms to that name, and provides a hierarchical annotation of the gene's functions. In the GO system, a gene product is considered to have one or more molecular functions, be involved in at least one biological process, and be associated with any number of cellular components. For each of these three major aspects, the GO team is developing a controlled vocabulary for use in gene annotation. For instance, the *sextra* gene is found to be a synonym of the Gap1 gene product, which is involved in the biological processes "mitosis" and "sister chromatid cohesion," among others. Gap1/sextra is associated with the cellular component "plasma membrane," and has the molecular function "Ras GTPase activator activity." Each of these terms is itself a member of a hierarchical system. "Plasma membrane" is found to be in the class "membrane," which in turn is a member of the class "cell." If the cellular component were localized even further or limited to structures of specific organisms, finer categories would have been provided (e.g., "apical plasma membrane" or "caveolar membrane").

The figure illustrates the structure of the GO classification scheme for an *Arabidopsis* gene named *INNER NO OUTER*. Notice that the central hub corresponds to a single gene product. Splitting from this hub are three separate branches, which correspond to the three GO ontologies: biological process, cellular component, and molecular function.

The use of GO is rapidly becoming a universal standard for gene annotation. It eliminates confusion arising from nonstandard nomenclature, and enables databases to interact more effectively. Consequently, for example, if a student discovers that a previously uncharacterized gene in her organism is homologous to an annotated gene in another species, the GO categories allow her to formulate hypotheses about what the function of the gene may be. Her collaborator might then initiate a bioinformatic search for potential interacting partners, starting by scanning through the lists of genes in other organisms with similar gene ontologies. Both the hierachical structure of GO and the uniformity of the vocabulary facilitate rapid hypothesis formulation.

The GO project arose in 1998 as a solution to barriers that inhibited communication between FlyBase (*Drosophila*), the Mouse Genome Database, and the *Saccharomyces* database (Ashburner et al. 2000). Since that time, many other groups have joined the consortium, including the databases at TIGR, Gramene (Monocots), WormBase (*C. elegans*), and TAIR (*Arabidopsis*). NCBI's Entrez Gene tool allows users to query for GO terms. GO consortium members and GO users have developed tools to encourage the widespread use of GO. The web-based AmiGO browser

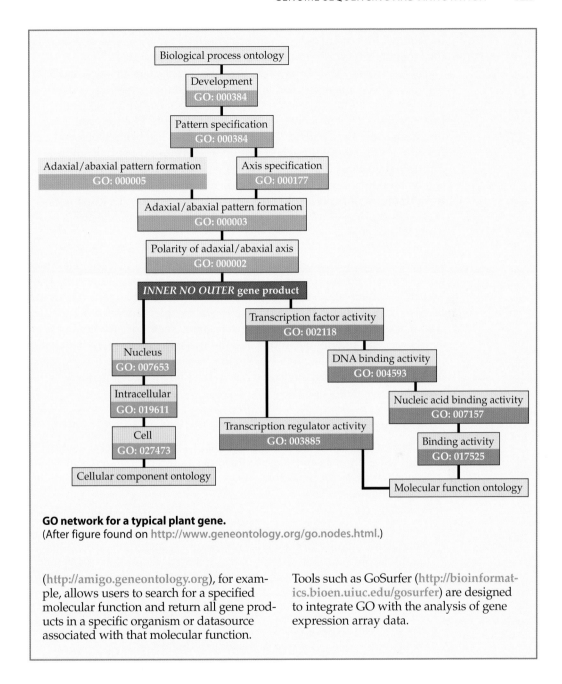

GO network for a typical plant gene.
(After figure found on http://www.geneontology.org/go.nodes.html.)

(http://amigo.geneontology.org), for example, allows users to search for a specified molecular function and return all gene products in a specific organism or datasource associated with that molecular function.

Tools such as GoSurfer (http://bioinformatics.bioen.uiuc.edu/gosurfer) are designed to integrate GO with the analysis of gene expression array data.

arranged hierarchically, from general to specific function, and attributes are continually updated as new information becomes available. The current organization of GO assumes function in an idealized general eukaryotic cell type, so does not attempt to convey information as to which tissue or cell type is affected, nor does it capture the pleiotropy of gene function (namely the ability of the same gene to perform multiple functions). However, such information is typically linked to online literature resources.

There are over 7,000 terms in GO describing molecular function and 5,000 describing biological process, with some annotations having as many as twelve levels within a hierarchy of terms. This is too deep for efficient computational searches and development of statistical models of protein families. Consequently, an allied PANTHER database has been developed that has a simplified ontology schema built on HMMs and manual curation of multiple sequence alignments (http://www.pantherdb.org). This database can be used to analyze the conservation of amino acid sequence segments in subfamilies of proteins, to rank the likelihood that individual replacements affect protein function, and to annotate predicted proteins from novel genome sequences.

Other modes of functional classification are conceivable. In reality, the Web resources for each organism portray different attributes in their annotation. Wormbase, for example, allows a nematologist to pull up images of which particular cells express a gene product, while FlyBase offers comprehensive links to allelic information and genetic interactions, among other data. The essential focus of emerging annotation methods is in linking gene expression and regulatory information to the static properties that can be inferred from a gene sequence. Dynamic modeling of gene function will depend on successfully integrating as much knowledge of genes as databases can reasonably and sensibly portray.

Summary

1. Almost all genome sequencing to date has been performed using automated fluorescence-based dideoxy chain-termination (Sanger) sequencing on machines that can generate in excess of a megabase of raw sequence per day.

2. Sequence traces are converted to raw sequences using algorithms that assign confidence scores to each base according to the spacing, height, and quality of the peaks representing each nucleotide. Phred quality scores are then used in the automated generation of sequence assemblies, or contigs, based on alignment of overlapping sequence reads.

3. The next generation of sequencing methods utilize novel technologies and will make it possible to sequence an entire human genome in a couple of days for just thousands of dollars.

4. Two main strategies are used to assemble whole genome sequences: hierarchical sequencing and shotgun sequencing. The former relies on prior ordering of clones to build the genome sequence in megabase-length pieces, while the latter uses computational power to assemble the entire sequence simultaneously.

5. Repetitive sequences are a major obstacle to sequence assembly, but can be overcome by masking the repeats from initial alignment and

using mate pairs to provide information on spacing and orientation of sequences that are interrupted by repeats.

6. First-draft genome sequences assembled from between five- and ten-fold redundant sequences have an error rate greater than 1 in 10,000 bases and short sequence gaps every few hundred kilobases. "Finishing" refers to the process of filling the gaps, improving the accuracy, and resolving ambiguities. Finishing is labor-intensive and can be more time-consuming than preparation of the draft.

7. Sequence alignment uses an algorithm that imposes penalties for mismatches and indels and allows the most likely alignment to be inferred from the scores of each possible alignment. This is a computationaly intensive process, so tools such as BLAST and FASTA use a heuristic method to focus searches only on the subset of plausible matches.

8. Annotation of genes using software that searches for open reading frames, splice sites and promoters using hidden Markov models (HMMs) is efficient overall, but can miss genes and often fails to identify intron-exon boundaries with accuracy.

9. Estimates of the number of genes in a genome based on the number of distinct ESTs in a set of libraries tend to greatly overstate the true number of genes, due to alternative splicing of complex transcripts and to errors that occur during cDNA cloning. Alignment of EST with genomic sequence can collapse families of ESTs into single genes.

10. Conservative annotation of gene locations utilizes two or more pieces of information, generally including de novo prediction, cDNA evidence, and similarity to sequences in other genomes. These approaches are likely to underestimate the true gene content, since many transcripts are tissue-specific or novel.

11. Comparative analysis of genome sequences is expected to assist in the identification of conserved regulatory sequences required for control of transcription and possibly of functions of the chromatin.

12. Numerous structural features of genomes such as the distribution of repetitive DNA, sequence content of centromeres and telomeres, and variation in GC content, are of biological interest in addition to the identification of genes. The ENCODE Project aims to define the function of all DNA sequences.

13. COGs are clusters of orthologous genes identified by connecting best-hit matches of each gene in the complete genomes of other organisms. They provide a draft picture of the distribution of gene families in each completed genome.

14. Phylogenetic methods should be used to investigate the likely functional relationships among a group of aligned gene sequences. Orthologs are sequences in two lineages that are derived from a common ancestor, while paralogs are duplicate sequences in the same lineage.

15. The Gene Ontology and Panther projects seek to categorize genes not just according to predicted function, but also location of the gene product in a cell, and the biological process or processes that it regulates.

16. There are numerous ways of grouping and categorizing proteins, so it is generally advisable to contrast the predictions made by several methods and to integrate predictions made directly from sequence comparisons with whatever molecular biological information is available in the species of interest. The internet facilitates online comparisons across diverse species.

Discussion Questions

1. By what criteria can we judge the accuracy of a draft (or complete) genome sequence? If 50% of the base pairs are apart by more than 100 kb in two different drafts, what areas of genome research are most likely to be adversely affected?

2. Should the fact that both of the draft human genome sequences were derived predominantly from individual males be cause for concern?

3. Do you think that the gene counts derived from first pass annotation of complete genome sequences are more likely to be underestimates or overestimates of the true number of genes? What are the criteria used for gene discovery?

4. Different tools for predicting gene function often give very different predictions. How useful are bioinformatic methods for functional annotation, and what can be done to improve their accuracy?

5. What are some of the applications for emerging DNA sequencing methods such as single molecule sequencing and sequencing by hybridization?

Web Site Exercises

The Web site linked to this book at http://www.sinauer.com/genomics provides exercises in various techniques described in this chapter.

1. Edit two sets of overlapping sequence trace files with phred, then merge them into a contig using phrap and consed.

2. Perform a BLAST search with the consensus sequence derived from Exercise 2.1.

3. Use Clustal to align the sequences obtained in Exercise 2.2.

4. Run GenScan on a scaffold sequence containing the full-length gene.

5. Perform a phylogenetic analysis of the aligned sequences in Exercise 2.3.

Literature Cited

Altschul, S., W. Gish, W. Miller, E. Myers and D. J. Lipman. 1990. Basic local alignment search tool. *J. Mol. Biol.* 215: 403–410.

Altschul, S. et al. 1997. Gapped BLAST and Psi-BLAST: A new generation of protein database search programs. *Nucl. Acids Res.* 25: 3389–3402.

Ashburner, M. et al. 2000. Gene ontology: Tool for the unification of biology. The Gene Ontology Consortium. *Nat. Genet.* 25: 25–29.

Bentley, D. S. 2006. Whole-genome re-sequencing. *Curr. Opin. Genet. Dev.* 16: 545–552.

Bernardi, G. 2000. Isochores and the evolutionary genomics of vertebrates. *Gene* 241: 3–17.

Blanchette, M. and M. Tompa. 2002. Discovery of regulatory elements by a computational method for phylogenetic footprinting. *Genome Res.* 12: 739–748.

Boffelli, D., J. McAuliffe, D. Ovcharenko, K. Lewis, I. Ovcharenko, L. Pachter and E. M. Rubin. 2003. Phylogenetic shadowing of primate sequences to find functional regions of the human genome. *Science* 299: 1391–1394.

Boguski, M., T. Lowe, and C. Tolstoshev 1993. dbEST: Database for "expressed sequence tags." *Nat. Genet.* 4: 332–333.

Burge, C. and S. Karlin. 1997. Prediction of complete gene structures in human genomic DNA. *J. Mol. Biol.* 268: 78–94.

Calabrese, P. P., R. T. Durrett and C. F. Aquadro. 2001. Dynamics of microsatellite divergence under stepwise mutation and proportional slippage/point mutation models. *Genetics* 159: 839–852.

Carninci, P. et al. 2000 Normalization and subtraction of cap-trapper-selected cDNAs to prepare full-length cDNA libraries for rapid discovery of new genes. *Genome Res.* 10: 1617–1630.

Cummings C. and H. Zoghbi 2000. Trinucleotide repeats: Mechanisms and pathophysiology. *Annu. Rev. Genomics. Hum. Genet.* 1: 281–328.

Dayhoff, M. O. et al. 1972. *Atlas of Protein Sequence and Structure*, Vol. 5. National Biomedical Research Foundation, Silver Springs, MD.

Deamer, D. and M. Akeson 2000. Nanopores and nucleic acids: Prospects for ultrarapid sequencing. *Trends Biotech.* 18: 147–151.

Dermitzakis, E. T. and A. G. Clark. 2003. Evolution of transcription factor binding sites in mammalian gene regulatory regions: Conservation and turnover. *Mol. Biol. Evol.* 19: 1114–1121.

Dermitzakis, E. T. et al. 2003. Evolutionary discrimination of mammalian conserved non-genic sequences (CNGs). *Science* 302: 1033–1035.

DeSalle, R., J. Gatesy, W. Wheeler and D. Grimaldi. 1992. DNA sequences from a fossil termite in Oligo-Miocene amber and their phylogenetic implications. *Science* 257: 1933–1936.

Eisen, J. A. 1998. Phylogenomics: Improving functional predictions for uncharacterized genes by evolutionary analysis. *Genome Res.* 8: 163–167.

ENCODE Project Consortium. 2007. Identification and analysis of functional elements in 1 percent of the human genome by the ENCODE pilot project. *Nature* 447: 799–816.

Ewing, B. and P. Green 1998. Base-calling of automated sequencer traces using phred. II. Error probabilities. *Genome Res.* 8: 186–194.

Ewing, B., L. Hillier, M. Wendl and P. Green 1998. Base-calling of automated sequencer traces using phred. I. Accuracy assessment. *Genome Res.* 8: 175–185.

Felsenstein, J. 1985. Confidence limits on phylogenies: An approach using the bootstrap. *Evolution* 39: 783–791.

Gopal, S. et al. 2001. Homology-based annotation yields 1,042 new candidate genes in the *Drosophila melanogaster* genome. *Nat. Genetics* 27: 337–340.

Gordon, D., C. Abajian and P. Green. 1998. Consed: A graphical tool for sequence finishing. *Genome Res.* 8: 195–202.

Gordon, D., C. Desmarais, and P. Green. 2001. Automated finishing with autofinish. *Genome Res.* 11: 614–625.

Graur, D. and W.-H. Li. 2000. *Fundamentals of Molecular Evolution*. Sinauer Associates, Sunderland, MA.

Green, P. 1997. Against a whole-genome shotgun. *Genome Res.* 7: 410–417.

Griffiths-Jones S. 2004. The microRNA Registry. *Nucl. Acids Res.* 32 (DB issue): D109–D111.

Hall, A. E., K. C.Keith, S. E. Hall, G. P. Copenhaver and D. Preuss. 2004. The rapidly evolving field of plant centromeres. *Curr. Opin. Plant Biol.* 7: 108–114.

Hall, B. G. 2007. *Phylogenetic Trees Made Easy: A How-To Manual for Molecular Biologists*, 3rd Ed. Sinauer Associates, Sunderland, MA.

Henikoff, S. and J. Henikoff. 1992. Amino acid substitution matrices from protein blocks. *Proc. Natl. Acad. Sci. (USA)* 89: 10915–10919.

IHGSC (International Human Genome Sequencing Consortium). 2001. Initial sequencing and analysis of the human genome. *Nature* 409: 860–921.

Imanishi, T. et al. 2004. Integrative annotation of 21,037 human genes validated by full-length cDNA clones. *PLoS Biology* 2: 856–875.

Larsen, F., G. Gundersen, R. Lopez and H. Prydz. 1992. CpG islands as gene markers in the human genome. *Genomics* 13: 1095–1107.

Lowe, T. and S. Eddy. 1997. tRNAscan-SE: A program for improved detection of transfer RNA genes in genomic sequence. *Nucl. Acids Res.* 25: 955–964.

Margulies, M. et al. 2005. Genome sequencing in microfabricated high-density picolitre reactors. *Nature* 437: 326–327.

Mindell, D. and A. Meyer. 2001. Homology evolving. *Trends Ecol. Evol.* 16: 434–440.

Moore, D. S. and G. P. McCabe. 1998. *Introduction to the Practice of Statistics*, 3rd Ed. W.H. Freeman, New York.

Needleman, S. and C. Wunsch. 1970. A general method applicable to the search for similarities in the amino acid sequence of two proteins. *J. Mol. Biol.* 48: 443–453.

Ovcharenko, I., D. Boffelli and G. G. Loots. 2004. eShadow: A tool for comparing closely related sequences. *Genome Res.* 14: 1191–1198.

Penn, S., D. Rank, D. Hanzel and D. Barker. 2000. Mining the human genome using microarrays of open reading frames. *Naure Genet.* 26: 315–318.

Petrov, D., T. Sangster, J. Johnston, D. Hartl and K. Shaw. 2000. Evidence for DNA loss as a determinant of genome size. *Science* 287: 1060–1062.

Rajewsky, N., M. Vergassola, U. Gaul and E.D. Siggia. 2002. Computational detection of genomic *cis*-regulatory modules applied to body patterning in the early *Drosophila* embryo. *BMC Bioinformatics* 3: 30.

Reboul, J. et al. 2001. Open-reading-frame sequence tags (OSTs) support the existence of at least 17,300 genes in *C. elegans. Nat. Genet.* 27: 332–336.

Remm, M, C. Storm and E. Sonnhammer. 2001. Automatic clustering of orthologs and in-paralogs from pairwise species comparisons. *J. Mol. Biol.* 314: 1041–1052.

Rogic, S., A. Mackworth and F. Ouellette. 2001. Evaluation of gene finding programs on mammalian sequences. *Genome Res* 11: 817–832.

Sanger, F., J. Donelson, A. Coulson, H. Kossel and D. Fischer. 1974. Determination of a nucleotide sequence in bacteriophage ?1 DNA by primed synthesis with DNA polymerase. *J. Mol. Biol.* 90: 315–333.

Smith, T. and M. Waterman. 1981. Identification of common molecular subsequences. *J. Mol. Biol.* 147: 195–197.

Shoemaker, D. D. et al. 2001. Experimental annotation of the human genome using microarray technology.

Nature 409: 922–927.

Sonnhammer, E., S. Eddy, E. Birney, A. Bateman and R. Durbin. 1998. Pfam: Multiple sequence alignments and HMM profiles of protein domains. *Nucl. Acids Res.* 26: 320–322.

Swofford, D. L. 2000. *PAUP*: Phylogenetic Analysis Using Parsimony and Other Methods* (software). Sinauer Associates, Sunderland, MA.

Swofford, D. L., G. J. Olson, P. J. Waddell and D. M. Hillis. 1996. Phylogenetic inference. *In* D. M. Hillis, C. Moritz and B. K. Mable (eds.), *Molecular Systematics*, 2nd Ed., pp. 407–514. Sinauer Associates, Sunderland, MA.

Tatusov, R. et al. 2001. The COG database: new developments in phylogenetic classification of proteins from complete genomes. *Nucl. Acids Res.* 29: 22–28.

Tatusov, R. et al. 2003. The COG database: an updated version includes eukaryotes. *BMC Bioinformatics.* 4: 41.

Venter, J. C., M. Adams, G. Sutton, A. Kerlavage, H. Smith and M. Hunkapiller 1998. Shotgun sequencing of the human genome. *Science* 280: 1540–1542.

Venter, J. C. et al. 2001. The sequence of the human genome. *Science* 291: 1304–1351.

Waterman, M. S. 1995. *Introduction to Computational Biology*. Chapman and Hall, London.

Weber, J. and H. Myers. 1997. Human whole-genome shotgun sequencing. *Genome Res.* 7: 401–409.

Zhang, Z., et al. 1998. Protein sequence similarity searches using patterns as seeds. *Nucl. Acids Res.* 26: 3986–3990.

3 Genomic Variation

The central place that the study of genomic variation occupies in modern biology was highlighted with its recognition by the journal *Science* as the "Breakthrough of the Year" for 2007. The elucidation of the nature and distribution of genetic variation within and among populations has applications across a wide range of biological investigation, from biomedical research to ecology and evolutionary biology. Most variation at the DNA level comes in the form of small insertions and deletions (**indels**), copy number variation (**CNV**), and **single nucleotide polymorphisms** (**SNPs**, pronounced "snips"). The majority of indels are just one or a few bases in length, while CNVs are typically several kilobases long and encompass the loss or duplication of whole genes or stretches of genes. A SNP is any site at which two or more different nucleotides are segregating within a population.

In this chapter, we start by considering the nature and distribution of SNPs and the theory of SNP applications, then move on to a description of current methods for SNP discovery and, finally, SNP genotyping. Thorough treatment of basic theoretical population and quantitative genetics can be found in textbooks by Falconer and Mackay (1996), Lynch and Walsh (1998), and Hartl and Clark (2007), while the primary literature is the best reference for emerging methods of SNP detection.

The Nature of Single Nucleotide Polymorphisms

Classification of SNPs

Single nucleotide polymorphisms are naturally occurring variants that affect a single nucleotide. They are most commonly changes from one base to another—transitions and transversions—but single-base indels are also SNPs. Some authors also regard two-nucleotide changes and small indels (up to a few nucleotides) as SNPs, in which case the term **simple nucleotide polymorphism** may be preferred. Microsatellites, longer simple sequence repeats, and all other classes of molecular polymorphism (including transposable element insertions, deletions from several bases to megabases in length, chromosome inversions and translocations, and aneuploidy) are not SNPs. In the euchromatic portion of most genomes, SNPs outnumber large deletions and insertions (CNVs), but CNVs typically involve a much greater number of nucleotides. For example, in the diploid genome of the first fully sequenced human, J. Craig Venter, there are 3.2 million SNPs and 0.9 million CNV and repeat polymorphisms, but the SNPs constitute only 26% of the 12.3 Mb of polymorphic DNA (Levy et al. 2007). The chromosomal locations of some of the SNPs that may affect various attributes of this individual are shown in Figure 3.1.

SNPs are classified according to the nature of the nucleotide that is affected. **Noncoding SNPs** may be located in a 5′ or 3′ nontranscribed region (NTR), in a 5′ or 3′ untranslated region (UTR), in an intron, or they may be intergenic. **Coding SNPs** may be **replacement polymorphisms** (i.e., they change the amino acid that is encoded) or **synonymous polymorphisms** (they change the codon but not the amino acid). **Regulatory polymorphisms** include synonymous and noncoding polymorphisms that may affect gene function through effects on transcriptional or translational regulation, splicing, or RNA stability.

Alternatively, SNPs can be classified as transitions or transversions. **Transitions** change a purine to a purine or a pyrimidine to a pyrimidine—A to G or C to T, and vice versa. **Transversions** change a purine to a pyrimidine and vice versa—A or G to C or T, and C or T to A or G. Even though there are twice as many possible transversions as transitions, transitions tend to be at least equal to transversions in frequency, and often are more prevalent. This phenomenon is broadly true of both coding and noncoding SNPs, and in part is the result of differences in the mechanisms by which particular types of mutations arise and are repaired. In addition, due to the nature of the genetic code, transitions are less likely to affect amino acids than are transversions and thus are thought to have a higher probability of retention in coding regions. As a result, the transition/transversion ratio is affected by constraints on sequence evolution and must be accounted for in the computation of genetic distances based on the amount of sequence divergence (Graur and Li 2000).

Confusion exists over the distinction between polymorphisms and mutations, largely due to the dual usage of the term *mutation*. All SNPs arise as mutations, in the sense that the conversion of one nucleotide into another is a mutational event. But by the time a sequence variant is observed in a population, the event that created it is usually long past, so the observed SNP is no longer a mutation—it is just a rare sequence variant or a polymorphism. However, *mutation* is also used to describe any allele that deviates from the majority type ("wild type"), particularly where the aberrant allele is thought to affect some phenotype or disease status. Since this distinction applies to only a small fraction of all SNPs, the term *polymorphism* is more general.

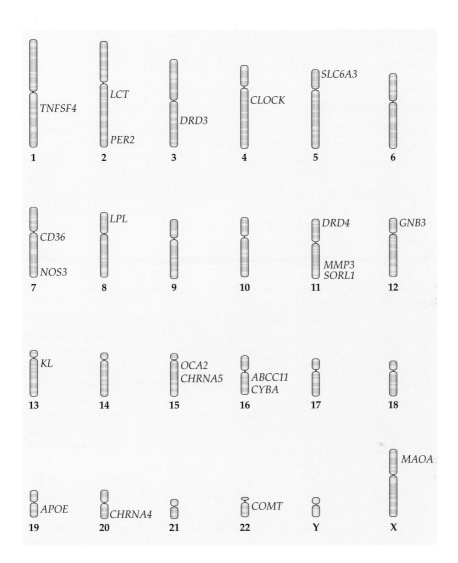

Figure 3.1 Location of polymorphisms that have been associated with various traits in one human. The complete sequence of the first diploid human genome (that of J. Craig Venter) uncovered more than 4 million SNPs, several of which have been implicated in various traits. See Levy et al. (2007) for a brief description of the phenotypes and disease susceptibility associated with each gene. Note that merely having variation at the loci shown in the figure does not mean that this polymorphism causes the individual's particular phenotype.

Further, most loci in natural populations harbor multiple haplotypes (a **haplotype** is a distinct combination of single nucleotide types on a single chromosome at a locus; in highly polymorphic species, there are generally many different haplotypes per locus), so the wild type is actually a set of different haplotypes. Even where one or more haplotypes are associated with a quantitative phenotype, it is not trivial to make the distinction between wild type and mutant—and the more rare form may in fact be ancestral—so the term polymorphism is again preferred.

Distribution of SNPs

The study of the distribution of genetic variants, including SNPs, lies within the domain of population genetics, and the study of the relationship between SNPs and phenotypic variation lies in the domain of quantitative genetics. Genomic methods have led to a renaissance of interest in both fields of enquiry. Whereas the earliest applications of genomics were in the mapping of Mendelian loci, the emerging importance of SNPs lies in the mapping and identification of **quantitative trait loci** (**QTL**), which are loci that contribute to polygenic phenotypic variation. On the one hand, SNPs provide the wherewithal to scan genomes for linkage to QTL; on the other, the vast majority of QTL effects are almost certainly due to as yet unidentified SNPs. For both reasons, characterization of the distribution of SNP variation is a major goal of most genome projects.

According to the **neutral theory** of molecular evolution, most SNPs are maintained in natural populations as a result of a balance between mutation and genetic drift. That is to say, the rate at which mutations are introduced into a population is in equilibrium with the rate at which polymorphisms are lost as a result of random sampling effects. Most mutations, whether deleterious, advantageous, or neutral in effect, are lost within a few generations, but by chance some attain appreciable frequencies and can even move toward fixation. To this theoretical formulation must be added the effect of selection, which is believed to be most often slightly deleterious when present, but on occasion tends to favor a new allele (**positive selection**) or to maintain two or more polymorphisms (**balancing selection**) at some loci.

Population structure is also a crucial determinant of allele frequencies within and among populations, as migration is a potent source of diversity, isolation affects the rate at which variation is lost due to drift, and rapid population growth alters the spectrum of haplotype frequencies. While the dynamics of genotypic variation are described by the balance of mutation, drift, migration, and selection, neutral theory supplies the null hypothesis by which observed patterns of molecular variation are assessed, and for the most part is regarded as a successful explanation for the distribution of SNP variation in genomes. However, the concepts of population genetics are often counterintuitive and seemingly self-contradictory, and textbooks such as Hartl and Clark (2007) should be consulted for a full treatment.

Three key concepts are important in characterizing SNP variation:

• Allele frequency distribution

• Linkage disequilibrium

• Population stratification

While a common body of theory describes each of these attributes, the proximate reasons for differences in each quantity among species and within genomes are not yet well characterized. Among the prominent causes of variation in SNP diversity between species are population size (including bottlenecks during speciation), mating system, population structure, and migration, including admixture. Within a genome, SNP diversity is also heavily impacted by genetic factors impinging on the neutral mutation rate, such as variation in recombination rate along and between chromosomes, and structural and functional constraints on genes.

As a result of the balance between mutation and drift, the vast majority of SNPs are rare, and in fact the frequency distribution of SNPs is expected to be L-shaped (Figure 3.2A). In a sample of several hundred alleles, the most common class of SNPs are typically singletons (those that appear only once in the sample), followed by doubletons, tripletons, and so on up to two equally common variants. Only between one-third and one-half of all SNPs are "common," meaning that the rarer allele is present in more than 5% of the individuals. This in turn means that a SNP that is useful for mapping in one pedigree or cross is not necessarily going to be useful in another set of comparisons. Similarly, population structure affects the relative frequencies of alleles among populations and, although "private alleles" unique to a single population are

(A)

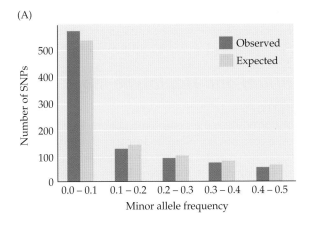

Figure 3.2 Nucleotide diversity in natural populations. (A) Observed and expected distributions of SNP frequencies for 874 SNPs from 75 candidate human hypertension loci. Rare alleles are the most frequent, and the number of SNPs in each frequency class declines as the more rare allele becomes more common. (B) Linkage disequilibrium (D') decays with time (number of generations) in proportion to the recombination rate r. (C) As a result, the level of nucleotide diversity is a function of recombination rate, and hence chromosomal position, as in this example for *Drosophila melanogaster*. (A after Halushka et al. 1999; B after Hartl and Clark 2007; C after Begun and Aquadro 1992.)

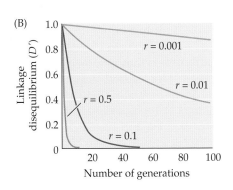

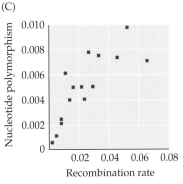

useful for many applications, differences in allele frequency can be prevalent and greatly complicate quantitative genetic inference.

The second relevant aspect of the frequency distribution is **nucleotide diversity**, which can be defined as the average fraction of nucleotides that differ between a pair of alleles chosen at random from a population. Nucleotide diversity is a function of the number of segregating sites and the frequency of each allele at these sites. Humans have relatively low nucleotide diversity, with an average difference of one SNP every kilobase between the two chromosomes of any individual. Both *Drosophila* and maize have more than an order of magnitude greater polymorphism, with an average of one SNP every 50–100 bases. These numbers vary over a wide range within genomes; nucleotide diversity exceeds 10% in some parts of the human major histocompatibility complex, but several kilobases can be monomorphic in other parts of the genome.

Wide variation in nucleotide diversity is also seen within loci, often between exons and introns. Some regulatory sequences in introns and 5′ NTRs are among the least polymorphic of all sequences, despite the general trend for coding sequences to be more highly constrained. Nucleotide diversity is strongly affected by recombination rate, as a consequence of its effect on the duration over which positive or negative selection affects linked sites: the faster recombination separates two SNPs, the less effect selection on one will have on the frequency of the other. As a result, diversity tends to be lower around centromeres and at the telomeres, and more generally to vary according to chromosomal location, as shown in Figure 3.2C (Begun and Aquadro 1992).

Linkage Disequilibrium and Haplotype Maps

Linkage disequilibrium (**LD**) refers to the nonrandom association of alleles. If two SNPs assort at random, then the expected fraction of each of the four possible two-locus genotypes can be calculated simply by multiplying the respective SNP allele frequencies. If the proportion of, for example, double homozygotes is greater than predicted, then there is linkage disequilibrium between the SNPs. As described in Box 3.1, statistical tests have been devised to detect LD. As will be described later in this chapter, linkage disequilibrium is important in that it allows mapping of disease loci in large populations; however, it may also obscure correct inference of the meaning of association between a SNP genotype and a phenotype. Consequently, characterization of the extent of LD is crucial for most quantitative applications of genomics.

The primary cause of linkage disequilibrium is thought to be historical contingency. When a mutation arises, the SNP it creates is in complete LD with every other site on the chromosome. As the new SNP rises in frequency in a population, recombination breaks up the LD in proportion to the genetic and hence the physical distance between sites. For this reason, LD is generally expected to decay monotonically on either side of each SNP (Figure 3.2B). However, this expectation is only observed approximately. It is not uncommon for two sites several kilobases apart to be in statistically signif-

BOX 3.1 Disequilibrium between Alleles at Two Loci

For a variety of reasons, genome scientists are interested in studying the independence (or *lack* of independence) of allelic combinations. A deviation from independence among alleles at a single locus is called **Hardy-Weinberg disequilibrium.** When alleles at two or more loci do not segregate independently, the phenomenon is called **gametic phase disequilibrium** or, more commonly, **linkage disequilibrium.**

Linkage Disequilibrium

As described in the text, linkage disequilibrium (LD) may arise from a variety of sources including nonrandom mating, population admixture, and epistatic selection. The last is of particular relevance for genome scientists, since linkage disequilibrium may indicate a functional interaction between loci associated with a phenotype of interest. For this reason, we now discuss LD and statistical tests for its presence.

Consider two loci, A and B, and for simplicity suppose they each segregate two alleles, A_1 and A_2 at the A locus, B_1 and B_2 at the B locus. Let the allele frequencies at A be p_1 and p_2, while q_1 and q_2 denote the frequencies at locus B (implying $p_1 + p_2 = 1$, $q_1 + q_2 = 1$). If the alleles at the two loci segregate randomly to form gametes, then gametes of the form A_iB_j occur with frequency p_iq_j. If this is the case, loci A and B are said to be in linkage equilibrium. If, on the other hand, the loci are in linkage disequilibrium, the frequency of A_iB_j gametes can be described with the introduction of a linkage disequilibrium coefficient, D (Lewontin and Kojima 1960), as shown in Table A. Those with a mathematical inclination should verify that the marginal totals are correct.

Table A

	B_1	B_2	Total
A_1	$p_{11} = p_1q_1 + D$	$p_{12} = p_1q_2 - D$	p_1
A_2	$p_{21} = p_2q_1 - D$	$p_{22} = p_2q_2 + D$	p_2
Total	q_1	q_2	1

Other measures of linkage disequilibrium have been suggested (see Hedrick 1985 for a review). Recognizing that the range of D depends on the allele frequencies at A and B, Lewontin (1964) proposed the measure:

$$D' = \frac{D}{D_{max}}$$

where D_{max} is the maximum possible value of D given the allele frequencies. To illustrate this concept, consider the following 3 sets of 10 haplotypes in which $p_1 = 0.6$, $p_2 = 0.4$, $q_1 = 0.6$, and $q_1 = 0.4$:

Equilibrium	Coupling	Repulsion
A_1B_1 A_1B_2	A_1B_1 A_1B_1	A_1B_2 A_1B_2
A_1B_1 A_1B_2	A_1B_1 A_1B_1	A_1B_2 A_1B_2
A_1B_1 A_1B_1	A_1B_1 A_1B_1	A_1B_1 A_1B_1
A_2B_1 A_2B_1	A_2B_2 A_2B_2	A_2B_1 A_2B_1
A_2B_2 A_2B_2	A_2B_2 A_2B_2	A_2B_1 A_2B_1

The left-hand group has what looks to be a random assortment of alleles such as would be expected under Hardy-Weinberg equilibrium; the middle group has *all* A_1B_1 and A_2B_2 haplotypes; and the right-hand group has *no* A_2B_2. You should be able to calculate that D for these three groups is 0.04, 0.24, and –0.16, respectively, and to see that in fact there is no way that the absolute value of D could be any greater for the latter two populations. We say that the middle group, where the two common alleles are always on the same chromosome, exhibits complete linkage disequilibrium in coupling phase, whereas the right-hand group, in which the rare allele of one locus is always with the common allele of the other, exhibits complete LD in repulsion phase. The quantity D' takes on values of 0.17, 1.0, and –1.0 for these three groups, which is useful because, ranging from –1 to 1, it allows us to compare the relative amounts of LD in either phase. Furthermore, unlike D, D' is independent of allele frequencies, so it allows us to compare LD at different combinations of

(Continued on next page)

BOX 3.1 (*continued*)

loci. For a similar reason, Hill and Robertson (1968) suggested the use of the squared correlation coefficient,

$$r^2 = \frac{D^2}{p_1 p_2 q_1 q_2}$$

For discussions of the relative merits of these and other measures, see Clegg et al. (1976) or Devlin and Risch (1995). Weir (1996) provides a theoretical treatment of statistical estimation of linkage disequilibrium coefficients.

Statistical Tests for Linkage Disequilibrium

We now focus on statistical tests for detecting the presence of linkage disequilibrium. The basic objective is a two-way test of independence, leading naturally to the use of traditional contingency table procedures. Consider a random sample consisting of $2n$ gametes. Let n_{ij} denote the number of $A_i B_j$ gametes in the sample, with $n_i.$ and $n_j.$ being the observed number of A_i and B_j alleles, respectively:

	B_1	B_2	Total
A_1	n_{11}	n_{12}	$n_1.$
A_2	n_{21}	n_{22}	$n_2.$
Total	$n._1$	$n._2$	$2n$

If alleles at the two loci segregated independently, the expected frequency of $A_i B_j$ gametes is $p_i q_j$, and the expected number of $A_i B_j$ gametes in a sample of size $2n$ is $e_{ij} = 2n p_i q_j$. If we estimate the unknown allele frequencies with the observed frequencies (e.g., $p_1 = n_1./2n$), we can then construct a chi-squared test in the typical fashion:

$$\chi^2 = \sum_{i,j} \frac{\left(n_{ij} - e_{ij}\right)^2}{e_{ij}}$$

The χ^2 statistic (the summation of the squared differences between observed and expected divided by the expected values) is compared to a chi-squared distribution with 1 degree of freedom to determine significance.

Suppose we have an observed data set with the following counts:

	B_1	B_2	Total
A_1	9	1	10
A_2	2	4	6
Total	11	5	16

The estimated value for p_1 is

$$\hat{p}_1 = {}^{n_1.}\!/_{2n} = {}^{10}\!/_{16} = 0.625$$

Similarly, the estimate for q_1 is

$$\hat{q}_1 = {}^{n._1}\!/_{2n} = {}^{11}\!/_{16} = 0.6875$$

The disequilibrium coefficient is $D = 0.5625 - 0.625 \times 0.6875 = 0.1328$. To carry out the chi-squared test we must first compute the expected counts. As an example, $e_{11} = 16 \times 0.625 \times 0.6875 = 6.875$, and the corresponding term in the χ^2 formula is $(9 - 6.875)^2 / 6.875 = 0.657$. The complete value is $\chi^2 = 0.657 + 1.445 + 1.095 + 2.408 = 5.605$, which is greater than the 3.84 critical value. Thus, we conclude that loci A and B are in linkage disequilibrium. The data contain an excess of $A_1 B_1$ and $A_2 B_2$ gametes (i.e., a paucity of heterozygotes).

An alternate procedure, particularly useful for small samples, is to perform a Fisher exact test. The idea of the Fisher exact test is to enumerate all possible samples of size $2n$ having the same allele counts as the observed sample. For each of these possible samples, the value of D is computed along with the probability of observing that sample if alleles segregated independently. Finally, a *p*-value is calculated by ranking the possible samples according to the value of D and, then tabulating the cumulative probability of samples with D greater than the value actually observed in the data. Weir (1996) shows that the conditional probability of a sample with counts n_{11}, n_{12}, n_{21}, and n_{22} given marginal alleles totals $n_1.$ and $n._1$ is

$$\frac{n_1.!n_2.!n._1!n._2!}{n_{11}!n_{12}!n_{21}!n_{22}!(2n)!}$$

Using that formula, the Fisher exact test for linkage disequilibrium can be computed as shown in Table B. The observed data (shaded) produce an exact test with a *p*-value of 0.035, suggesting that the two loci are in linkage disequilibrium.

The discussion above treats the case of two loci with two alleles, and assumes that data on gametes are available. Treatments accounting for multiple alleles and for unknown linkage phase of double heterozygotes, along with a comprehensive discussion of tests for linkage disequilibrium, may be found in Weir (1996). Lewontin (1995) discusses a different approach to detection of linkage disequilibrium across a large sample of rare alleles such as are typically encountered in the pooled genome sequences of many individuals.

Table B

n_{11}	n_{12}	n_{21}	n_{22}	D	Probability	Cumulative probability
10	0	1	5	0.195	0.001	0.001
9	1	2	4	0.133	0.034	0.035
8	2	3	3	0.070	0.206	0.241
7	3	4	2	0.008	0.412	0.653
6	4	5	1	−0.055	0.288	0.941
5	5	6	0	−0.117	0.058	1.000

icant LD, while numerous SNPs between them are in linkage equilibrium with one another and with the sites that are in LD (Stephens et al. 2001). Thorough sampling across the full extent of a locus is required to characterize the extent of association among polymorphisms; an example from the human *LPL* gene is shown in Figure 3.3.

The extent of LD varies within and among species. In mammals, LD is commonly observed for several tens, and in many cases hundreds, of kilobases on either side of a SNP. Figure 3.4 shows that, on average, detectable LD extends out to 160 kb between randomly sampled sites in the human genome, but that there is wide variation in how quickly it decays. In flies, by contrast, LD is rarely observed other than sporadically more than one or two kilobases away from a site and often decays within a few hundred bases. Furthermore, the haplotypes that exist in the human genome as a result of LD have a somewhat block-like, or clustered, distribution. To a broad approximation, our genome consists of tens of thousands of blocks, each up to a hundred kilobases long, containing three to five common haplotypes formed by tens or even hundreds of SNPs in LD.

The extent of linkage disequilibrium is also a function of local recombination rates. LD is greater in non-recombining regions of the genome, including the Y chromosome and parts of the X chromosome, as well as centromere-proximal regions of autosomes. Even over just a few centimorgans, recombination rates are known to vary widely, and hotspots just kilobases in length along the chromosome are thought to influence the boundaries of haplotype blocks (Coop et al. 2008).

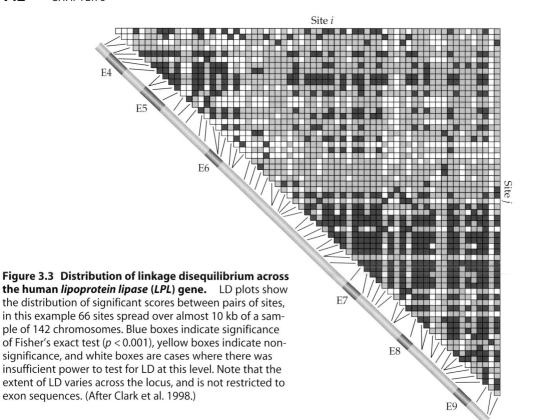

Figure 3.3 Distribution of linkage disequilibrium across the human *lipoprotein lipase* (*LPL*) gene. LD plots show the distribution of significant scores between pairs of sites, in this example 66 sites spread over almost 10 kb of a sample of 142 chromosomes. Blue boxes indicate significance of Fisher's exact test ($p < 0.001$), yellow boxes indicate nonsignificance, and white boxes are cases where there was insufficient power to test for LD at this level. Note that the extent of LD varies across the locus, and is not restricted to exon sequences. (After Clark et al. 1998.)

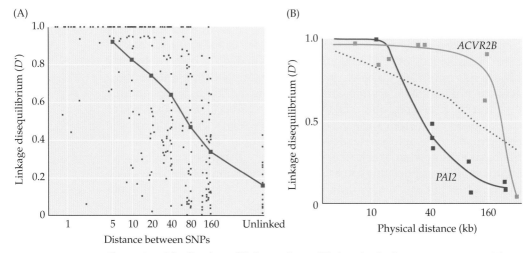

Figure 3.4 Distribution of linkage disequilibrium in the human genome. (A) Plot of average level of LD against distance in kilobases for 19 randomly sampled regions of the human genome. Linkage disequilibrium typically decays to background levels after more than 100 kb, although individual loci can show very different patterns (B), as exemplified by the loci ACVR2B and PAI2. The dotted line in (B) is the same overall plot shown in green in (A). (After Reich et al. 2001.)

EXERCISE 3.1 *Quantifying heterozygosity and LD*

Using the following ten sequences:

```
 1 gctgcatcag aagaggccat caagcgcatc actgtacttc tgccatggcc
 2 gctgtatcag aacaggccat caagcgcatc actgtacttc tgccatggcc
 3 gctgtatcag aacaggccat caagcacctc actgtacttc tgccatggac
 4 gctgcatcag aagaggccat caagcacatc actgtccttc tgccatggcc
 5 gctgcatcag aagaggccat caagcacatc actgtccttc tgccatggcc
 6 gctgtatcag aacaggccat caagcgcatc actgtccttc tgccatggcc
 7 gcggcatcag aagaggcgat caagcacatc actgtccttc tgccatggac
 8 gctgcatcag aagaggccat caagcacatc actctacttc tgccatggcc
 9 gctgtatcag aacaggccat caagcgcatc actgtccttc tgccatggcc
10 gctgcatcag aagaggccat caagcgcatc actctccttc tgccatggcc
```

(a) Count the number of segregating sites.

(b) Calculate the expected average heterozygosity per nucleotide.

(c) Determine the level of linkage disequilibrium between the common polymorphisms.

ANSWER: *(a) There are 9 segregating sites, at positions 3, 5, 13, 18, 26, 28, 34, 36, and 49.*

(b) The minor allele frequencies at these 9 sites are, respectively, 0.1, 0.4, 0.4, 0.1, 0.5, 0.1, 0.2, 0.4, and 0.2. Since the expected heterozygosity of each nucleotide is given by 2pq where p is one allele frequency and q (= 1 – p) is the other allele frequency, the expected average heterozgosity H (including the 41 nonpolymorphic sites) is

$$H = [(41 \times 0) + (3 \times 0.18) + (3 \times 0.48) + (2 \times 0.32) + 0.50]/50 = 0.0624$$

This means that, on average, just over three site differences are expected between any pair of randomly chosen alleles.

(c) There are four common polymorphisms, at positions 5, 13, 26, and 36.

```
cgga
tcga
tcaa
cgac
cgac
tcgc
cgaa
cgaa
tcgc
cggc
```

If we extract these sites, it is easier to see how they are related. Following the procedure in Box 3.1, draw a table of haplotype frequencies for each pairwise combination. For example, for sites 5 and 13:

Site 5	Site 13	
	c $(q_1 = 0.4)$	**g** $(q_2 = 0.6)$
t $(p_1 = 0.4)$	$p_{11} = 0.4$	$p_{12} = 0.0$
c $(p_2 = 0.6)$	$p_{21} = 0.0$	$p_{22} = 0.6$

Since D = $p_{11} - p_1 q_1$, for this pair $D_{5,13} = 0.4 - (0.4 \times 0.4) = 0.24$.

(Continued on next page)

EXERCISE 3.1 (continued)

This is also the maximal value, D_{max}, since all of the less common alleles at both sites always segregate together. Consequently, D' is equal to 1. You should be able to calculate the following table for the other linkage disequilibrium estimates. D_{max} is just the maximum value p_{11} could take, minus p_1q_1.

Allele pair	p_1	q_1	p_{11}	D	D_{max}	D'
5t, 13c	0.4	0.4	0.4	0.24	0.24	1.00
5t, 26g	0.4	0.5	0.3	0.10	0.20	0.50
5t, 36a	0.4	0.4	0.2	0.04	0.24	0.17
13c, 26g	0.4	0.5	0.3	0.10	0.20	0.50
13c, 36a	0.4	0.4	0.2	0.04	0.24	0.17
26c, 36a	0.5	0.4	0.2	0.00	0.20	0.00

This means that there is complete LD between the first two sites, but linkage equilibrium between the last two sites, with partial LD for all other comparisons.

Figure 3.5 illustrates the concept of **tagging SNPs**, which are SNPs that capture most of the nucleotide variation included in haplotypes. Ninety-nine percent of all common human SNPs are tagged with an LD correlation statistic $r^2 > 0.8$ by 1 million SNPs, implying that genotyping of this number of markers will generally be sufficient for the purposes of disease or phenotype association mapping as described in more detail below. The International HapMap Project is an effort to map all of the common haplotypes in the human genome for the purposes of facilitating discovery of disease-promoting alleles. Phase II of this project, reported in January 2007, documented the allele frequencies of 3.2 million SNPs in 270 individuals from four populations (Europeans, Han Chinese, Japanese, and Yoruban Africans). The data is available to the public at http://www.hapmap.org, including a browser that allows users to visualize the density of SNPs, the LD structure, and the haplotype structure in a particular region of the genome, as well as select tagging SNPs, call up the sequence surrounding particular SNPs, and link to literature that refers to phenotype associations.

An important parameter affecting the utility of haplotype data is the degree of **population stratification**, which is the partitioning of genetic variation among populations within a species. Statistics for the identification and quantification of population structure have been in use by population geneticists for well over a half a century. Outbreeding species show wide variation in the degree of population structure, reflecting constraints on migration due to geographic, ecological, and behavioral factors, as well as historical contingency and local adaptation. Self-fertilizing and asexual

Sequences

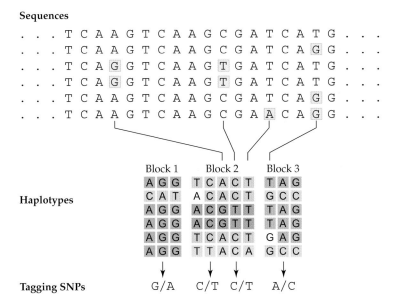

Figure 3.5 Tagging SNPs and haplotype blocks. Extraction of the polymorphisms from a set of sequences typically reveals a blocklike pattern of haplotypes. In this hypothetical example, Block 1 has two classes of haplotypes, one rare and one common; Block 2 has three classes of haplotypes; and Block 3 has two classes of haplotypes. Note that the boundaries between blocks are relatively sharp. The tagging SNPs can be used to define most of the variation in the sample.

species are likely to exhibit very high levels of population structure, which can have a marked effect on plant and parasite diversity in particular. Other species (including humans) have more limited geographic variation.

The majority of all human SNPs are shared among the major racial groups. The few detailed studies of human loci, however, are starting to reveal significant levels of population structure at the level of haplotype frequencies, as shown for the human *LPL* locus in Figure 3.6. A survey of sequence diversity in at least 2 kb of 313 loci in 82 unrelated humans found that only 8% of all haplotypes were unique to one of the four ethnic groups surveyed, and that 71% were common to all groups, even in this moderate sample (Stephens et al. 2001). A more extensive survey of 377 microsatellites in 1,056 individuals showed that differences in allele frequency among six major population groups account for at most 5% of human variation, yet with this number of markers it is possible to assign most humans to a genetic cluster that broadly reflects geographic distribution (Figure 3.7; Rosenberg et al. 2002). Similarly, typing of hundreds of microsatellites allowed individual dogs to be assigned with high probability to the appropriate breed, also resolving the genetic relatedness among breeds (Parker et al. 2004). To the extent that SNPs with quantitative effects on traits, particularly disease sus-

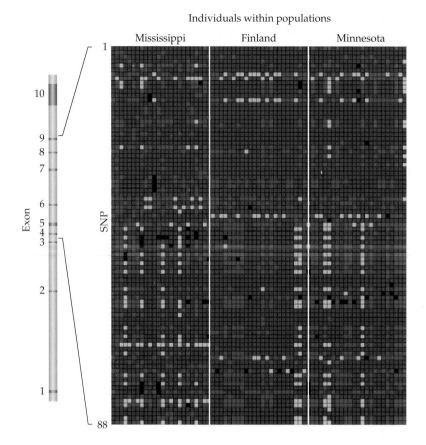

Individuals within populations

Figure 3.6 Haplotype structure in the human lipoprotein lipase (LPL) gene.
Plot of 9.7 kb from each of 71 individuals from three populations (Jackson, MS; North Karelia, Finland; Rochester, MN) were sequenced. The figure shows the distribution of genotypes at each of 88 SNPs, with homozygotes for the more common allele in blue, homozygotes for the more rare allele in yellow, heterozygotes red, and missing data black. Two major classes of haplotype are visible in the data; these are found in each of the populations, possibly at different frequencies and against a background of diversity. (After Nickerson et al. 1998.)

ceptibility, are embedded in haplotype structure, variation in haplotype frequencies among subpopulations might also have a considerable effect on inference in regard to genotype-phenotype associations.

Applications of SNP Technology

Population Genetics

High-throughput DNA sequencing has numerous applications in classical population genetic research. First among these is the study of population

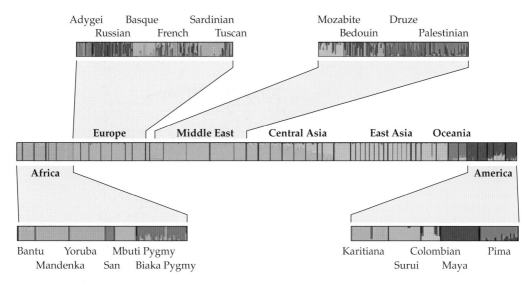

Figure 3.7 Human diversity and population structure. Assuming six different human population groups, the Structure algorithm distinguishes Africans, Europeans, Middle Easterners, Central and East Asians, Oceanics, and Americans. In the central bar, each individual within a population is represented by a thin vertical line, and colors identify what proportion of the microsatellite alleles is attributable to each population group. More detailed analysis of each population clearly differentiates populations within groups in Africa and America, but there appears to be more mixture in Europeans and Middle Easterners, which adds ambiguity to those classifications. (After Rosenberg et al. 2002.)

structure at greater resolution than can be achieved with microsatellites. As DNA sequencing throughput increases, sequence comparison using markers dispersed throughout a genome becomes a practical option for characterizing population genetic variation for a wide range of species. The only limitation lies in the identification of polymorphic markers, but the next-generation automated sequencing platforms (see Figure 2.5) enable rapid and cost-effective SNP discovery. With SNPs in hand, it is also possible to generate unambiguous genotype data for hundreds of organisms quickly and affordably. Box 3.2 introduces the "coalescent" framework underlying much of the modern population genetic theory that allows inference from polymorphic sequence data.

The utility of DNA sequence comparisons for inferring demographic history is well illustrated by certain findings of molecular anthropology. Studies of mitochondrial DNA have established that human genetic diversity is greatest in Africa, suggesting that *Homo sapiens* emerged in Africa between 150,000 and 200,000 years ago (Cann et al. 1987), in contrast with the alternate view that humans evolved multiregionally across the globe from a species that emerged closer to a million years ago. Studies of ancient fossil specimens, as well as Y-chromosomal markers and numerous autosomal

BOX 3.2 The Coalescent

The ability to collect the DNA sequence of a specific gene or genomic region from multiple individuals in a species cheaply and quickly has led to a rebirth of the field of population genetics. The data offer the potential to study evolution at the finest level of detail, and the promise of identifying individual nucleotide mutations that influence variation at the phenotypic level. In order to make use of DNA sequence surveys, it was necessary to develop a new body of population genetic theory and a collection of data analysis tools based on that theory. The driving force behind this effort has been the emergence of *coalescent theory*.

The most fundamental notion behind the coalescent is the idea that if one takes a random sample of DNA sequences from a population, then the number of generations one must trace back to find the **most recent common ancestor (MRCA)** of all the sampled sequences is a random variable. Figure A illustrates the case of two samples of $n = 3$ taken from a very small population of size $N = 6$ diploid individuals, which have a total of $2N = 12$ alleles. For a simple random mating population with constant population size, it is of no concern which individual carries an allele, and thus, the study of a diploid population of N individuals is the same as that of a haploid population with $2N$ individuals. Tracing the history of the three sampled blue sequences, we see that two of them shared a common ancestor two generations ago and that the MRCA of the sample of all three was six generations in the past. By contrast, when a second sample of size three is taken from the same population, such as the red sample, the time back to the MRCA is likely to be different than six (in this case, the three different red individuals coalesce to a common ancestor four generations in the past, and any other sample will also have a slightly different history). In other words, the number of generations back to the MRCA of the n sampled sequences, t_{MRCA}, is a random variable with a probability distribution.

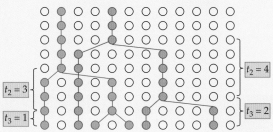

Figure A

A large body of literature has been devoted to understanding the probabilistic behavior of t_{MRCA}. Upon reflection, one sees that t_{MRCA} has a significant impact on the amount of variability observed in the sample data: When t_{MRCA} is large, there are more opportunities for mutations, which in turn result in more sample variation. It is this link between variation and t_{MRCA} that make understanding its properties so important. When dealing with simple random mating populations it is quite simple to show that t_{MRCA} is described well by properties of exponential distributions.

Returning to the example from the figure, consider the probability that none of the sampled sequences share a common ancestor in the previous generation. This event occurs only if all of the $n = 3$ sampled alleles have unique ancestors in the previous generation, and has probability

$$\text{Pr (3 unique ancestors in previous generation)} = 1 \times \frac{2N-1}{2N} \times \frac{2N-2}{2N}$$

The form of the expression reflects the fact that the first individual can trace back to any of the $2N$ sequences in the previous generation. In order to have unique ancestors, the second sampled sequence must be a descendant of one of the $2N - 1$ remaining sequences. Likewise, the third sampled sequence must trace to one of the remaining $2N - 2$. In general, if n sequences are sampled from a population of $2N$, the probability that none of the sequences share a common ancestor in the previous generation is:

$$P_n = \Pr (n \text{ unique ancestors} \\ \text{in previous generation})$$

$$= \prod_{i=1}^{n} \frac{2N - (i - 1)}{2N}$$

This expression can be used to find the probability that the first ancestor of any pair of individuals in the sample occurred t generations ago:

$$\Pr (\text{first ancestor of any pair was} \\ t \text{ generations ago}) = p_n^{t-1} (1 - p_n)$$

This formula arises from the requirement of $t - 1$ consecutive generations with no common ancestors, followed by a single generation with two sequences tracing to a common ancestor, and it can be used to show that the times between coalescent events follow exponential distributions.

Once the first coalescent event has occurred it is as if the entire process restarts with a sample size of $n - 1$, because two original lineages have been combined into one. Indeed, this recursive nature continues after each coalescent event until all lineages converge into a single one at the MRCA. Figure A shows the lengths of the periods of time between each event for the blue sample, $t_n, t_{n-1}, \ldots, t_2$, with time t_k denoting the length of time during which the sample is represented by k lineages. Using properties of the exponential distribution it is simple to verify that the expected value of t_k is:

$$E(t_k) = \frac{4N}{k(k - 1)}$$

and that the expected value of t_{MRCA} is:

$$E(t_{MRCA}) = E(t_n) + E(t_{n-1}) + \ldots + E(t_2)$$

$$= 4N\left(1 - \frac{1}{n}\right)$$

Consider next the total length of the genealogy. Noting that during the kth interval there are k lineages of length t_k, the total length is:

$$T = \sum_{k=2}^{n} k t_k$$

The value of T is intimately related to the amount of variability in a sample. Because it measures the total amount of time "captured" in a sample of sequences, the genealogy length provides the number of opportunities for mutations to occur. Note, however, that T is also a random quantity, its distribution being described by sums of exponential random variables as described above. If mutations arise at constant rate μ, then the expected number of mutations on a genealogy of length T is simply μT. If one assumes the infinite sites population genetic model, then careful thought reveals that the number of mutations on the genealogy is exactly the same as the number of variable, or segregating sites in the sample of DNA sequences, S. Some algebra shows that the expected number of segregating sites in a sample of size n taken from a population of N diploid individuals with per-site mutation rate μ is:

$$E(S) = \mu E(T) = \theta \sum_{k=1}^{n-1} \frac{1}{k}$$

where $\theta = 4N\mu$.

This formula is a cornerstone of coalescent theory, because it provides a direct relationship between a quantity observable in sampled data (S) and a population parameter that governs the accumulation of variation (θ). The relationship also offers a statistical estimator for θ,

$$\hat{\theta} = \frac{S}{\displaystyle\sum_{k=1}^{n-1} \frac{1}{k}}$$

Only the product of mutation rate and population size is estimable. However, if additional information is available, the two components can be separated. If, for example, an empirical estimate of the mutation rate is available, the estimation formula can also be used to estimate the population size.

Consider the following three sequences that differ at five segregating sites:

CGCTACAAAGTCATACCTGGTCAGATAGA

CGTTACAAAGTCACACCTGGTCTGATAAA

CGTTACAAGGTCACACCTGGTCAGATAAA
　　*　　　　*　　　*　　　　　*　*

(Continued on next page)

BOX 3.2   *(continued)*

The estimate of $\theta = 4N\mu$ is simply:

$$\hat{\theta} = \frac{S}{\sum\limits_{k=1}^{n-1}\frac{1}{k}} = \frac{5}{\frac{1}{1}+\frac{1}{2}} = \frac{5}{\frac{3}{2}} = 3.33$$

If available experimental data provided an estimate of the mutation rate to be 0.0005, then an estimate of the effective size of the population from which the sample is drawn is:

$$\hat{N} = \frac{\hat{\theta}}{4\hat{\mu}} = \frac{3.33}{(4 \times 0.0005)} = 1,666$$

The process and concepts described above lay the foundation for a large body of theory relating sequence variation to population and mutation processes. Coalescent theory in recent years has evolved to offer estimators of recombination rates, selection coefficients, migration rates, and rates of population growth. It has been the basis of methods for identifying genes undergoing adaptive evolution. In conjunction with rapid improvements in desktop computing power, the coalescent has become the most important theoretical tool in modern molecular population genetics.

loci and microsatellites, continue to lead to refinement of the "out-of-Africa" hypothesis, and to fill the gaps in our understanding of human migration patterns into Eurasia, the Americas, and Australasia over the past 50,000 years (Garrigan and Hammer 2006).

High-volume sequencing and genotyping has the potential to greatly expand the range and volume of traditional population genetic applications. One such area is the ascertainment of the breeding structure and degree of dispersal of soil microbes, pathogenic bacteria, parasitic nematodes, and numerous insect pests. Sequence comparisons can reveal the degree of haplotype divergence in microbial genomes, providing clues as to whether the species spreads clonally or sexually, by epidemics or pandemics. Whole-genome sequences provide hints as to how microbes have adapted to particular niches and may highlight genes that are likely to regulate parasitism, virulence, and toxicity. Comparing host and parasite phylogenies can indicate whether associations are ancient or recent, restricted or labile. Viral evolution can be studied at the whole-genome level, in the case of RNA viruses such as HIV even providing clues as to how the virus responds over a period of months to antiviral therapy. Numerous applications in conservation genetics will become increasingly feasible as costs decline and standard technologies emerge.

Another broad class of application of SNP diversity studies is the inference of the evolutionary factors experienced by individual loci. Numerous "tests of neutrality" have been developed that compare observed patterns of sequence variation within and among genes, with expectations derived from neutral theory (reviewed in Kreitman 2000). The simplest test, developed by Tajima, is based on the expectation that the number of segregating sites in a sample of alleles from a population should be proportional to the average pairwise distance between alleles. This test has low power and is often supplemented with more sophisticated analyses that are based on

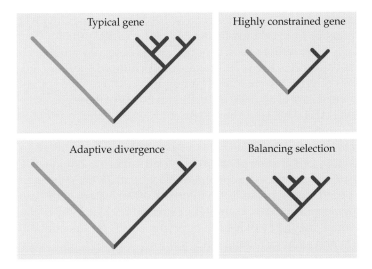

Figure 3.8 Basis for tests of neutrality. Four possible scenarios for comparing the amount of intraspecific polymorphism (blue; right-hand arm of each panel) with the extent of divergence to a sibling species (orange; left arm). Under neutrality, there is an expectation that highly constrained genes are less polymorphic and diverge less between species. After adaptive divergence, diversity is reduced within a species, whereas balancing selection is one possible explanation for the maintenance of polymorphism despite low divergence.

the notion diagrammed in Figure 3.8. The **HKA test**, for example, contrasts the ratio of polymorphic sites within a species to the number of fixed differences between species, between two or more loci, using an evolutionary model to predict the most likely relationship among these parameters. An excess of intraspecific polymorphism may indicate that balancing selection is acting on a site or sites and retaining linked neutral SNPs at the locus, whereas a deficit of intraspecific polymorphisms may under some circumstances indicate that a selective sweep recently removed the expected variation at the locus. Similarly, the **McDonald-Kreitman statistic** tests whether the ratio of polymorphism to divergence within a single locus is the same for synonymous and replacement sites, as it should be under neutrality.

Such tests are of theoretical importance in further delineation of the balance of mutation, drift, and selection across the genome. They may also have practical applications in the identification of target loci for advanced plant and animal breeding, as well as control of pathogens. Domestication of plants, notably maize (see Figure 1.18), has been shown to leave a "footprint" of reduced genetic diversity within a few kilobases of loci that experienced strong artificial selection (Doebley et al. 1997). Similarly, scans of sequence diversity in microbial parasite genomes contribute to the identification of strongly selected genes that are required for infectivity, immune system evasion, and pathogenicity.

Sequence analysis has allowed inference of selection on genes involved in human divergence from other primates. As many as 500 proteins have more amino acids substitutions than expected given observed numbers of synonymous substitutions between humans and chimps. These include many proteins involved in olfaction and hearing, amino acid catabolism, and early development (Clark et al. 2003), as well as an excess of those genes represented in the OMIM disease database. By contrast, the high nucleotide diversity of the human immune complex loci at the MHC and across the immunoglobulin superfamily is likely, at least in part, to reflect maintenance of variation due to a combination of heterozygote advantage and frequency-dependent selection during disease epidemics (Vogel et al. 1999). Yet another application of population genetics lies in inference of the history of disease-promoting polymorphisms. For example, a common regulatory SNP that affects expression of the IL4 cytokine has opposing effects on susceptibility to HIV infection and asthma, among other immune disorders, and occurs at such different frequencies in different populations that it seems as though different selective forces act on the gene in different parts of the globe (Rockman et al. 2003).

Recombination Mapping

The major application of SNPs in human genetics is in the mapping and identification of disease loci. **Recombination mapping**, also known as linkage mapping, has been used as the basis for positional cloning of Mendelian disease genes, namely single-gene disorders such as cystic fibrosis, Huntington's disease, and heritable long-QT syndrome. **Positional cloning** refers to the isolation of a gene that is responsible for a phenotype on the basis of its map position (as opposed to isolation via biochemical assay, for example); it relies on complete association between the mutation and the disease. Given a sufficiently high density of SNPs and a large enough set of pedigrees, identification of the disease-causing mutation is (in theory) simply a sampling and genotyping problem. For this reason, the completion of the human genome sequence has been hailed as the first step toward mapping every one of the thousands of known inborn errors of metabolism and other single-gene disorders, as it provides the framework for comprehensive SNP identification.

In practice, most disease genes are mapped in two steps: localization of the locus to several hundred kilobases, followed by sequencing of candidate genes within that interval to identify candidate disease-causing SNPs (such as premature stop codons). In humans, the mapping is performed in pedigrees (Figure 3.9). Each meiosis has the potential to result in recombination between the disease locus and any of the anonymous markers being used in the mapping. Failure of a marker to segregate with the disease in affected offspring indicates that a recombination event has occurred, and allows the investigator to deduce whether the disease locus lies proximal or distal to the marker on the chromosome. Similarly, segregation of the marker in a non-affected sibling provides mapping information. Clearly, the greater the number of affected

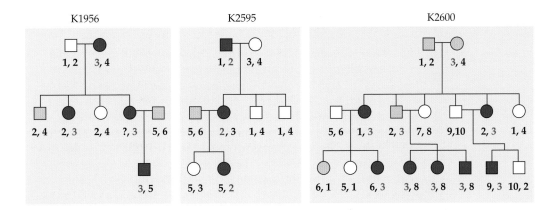

Figure 3.9 Pedigree mapping.   Portions of three of the nonrelated pedigrees that contributed to the mapping of the gene for a dominant form of long-QT syndrome are shown. In each case, one haplotype, numbered in red, co-segregates with the disease. (Note that the numbers refer to different haplotypes in each family; it is only identity by descent within each pedigree that is relevant.) Squares represent males, circles females. Dark blue indicates symptomatic individuals, light blue equivocal or unknown disease status. Each haplotype was inferred from five linked markers. (After Curran et al. 1995.)

individuals that can be screened, the more refined the mapping. Affected individuals can be identified in extended pedigrees covering multiple generations where DNA and medical records are available, or they can be contrasted across multiple distinct pedigrees. In the latter case, different markers might be used for each pedigree, but so long as a good genetic map is available, relative positional information can be extracted.

Numerous strategies now exist for jumping straight to candidate genes within mapped intervals. Several hundred kilobases may include tens of different genes. With the complete and annotated genome sequence, the sequences of predicted ORFs may provide an immediate clue. This proved to be the case with long-QT syndrome, a fatal heart arrhythmia condition for which sodium and potassium channels were identified as potential modulators of cardiac rhythmicity (Curran et al. 1995).

Synteny with other mammalian genomes may be used to identify known candidate genes with related disease phenotypes, as with a number of congenital diseases in dogs. Alternatively, each of the genes in the region of synteny of the mouse may be tested by Northern blots or in situ hybridization to assess whether the gene is expressed in the affected tissue, such as muscle, bone marrow, or liver. Even more divergent model systems such as zebrafish or invertebrates might also suggest functional associations that hint at a disease association, as described in Chapter 5. Or, microarray data may suggest that the one candidate gene is co-regulated with other genes that are implicated in the etiology of the disease. Failing such circumstantial evidence, it is now feasible simply to sequence each of the exons of the

EXERCISE 3.2 *Inferring haplotype structure*

*Suppose that a couple has three children. The genotypes at four sites in a
gene of interest for the parents and children are given below. Assuming
there was no recombination between the sites during transmission from
parent to child, write down the four possible haplotypes that each parent
could have, and then infer the haplotypes that the three children have,
using the parents' genotypes as a guide. Is there anything about this data
that might lead a forensic scientist to suspect that the father is not the
biological father of one of the children? If the mother's genotype had
been CT at the fourth site, would you be able to infer the haplotype
phases?*

Mother:	GG	AT	CA	TT
Father:	CC	AA	AC	CT
First child:	GC	AA	CC	CT
Second child:	GC	AT	AA	TT
Third child:	GC	AA	AC	CT

ANSWER: *The four possible haplotypes that each parent could have are
as follows:*

Mother:	G	A	C	T	*(M1)*
	G	T	A	T	*(M2)*
	G	T	C	T	*(M3)*
	G	A	A	T	*(M4)*
Father:	C	A	A	C	*(F1)*
	C	A	C	T	*(F2)*
	C	A	A	T	*(F3)*
	C	A	C	C	*(F4)*

*The first child must have received haplotypes M1 and F4, the second child
M2 and F3, and the third child either M1 and F1 or M4 and F4. Since each
parent can only transmit two haplotypes, but three haplotypes from each
parent are required to explain the three children's genotypes, something is
amiss. There is a further constraint that the mother must be M1 and M2 or
M3 and M4, and similarly the father F1 and F2 or F3 and F4. From this we
can infer that the third child must have received either M4 from another
mother, or F1 from another father (or was adopted).*

*If the mother were heterozygous at the fourth site, she would have eight
possible haplotypes, and inferring the linkage phases in the parents would
be ambiguous.*

genes in the candidate region to search for candidate causative SNPs, on the assumption that Mendelian mutations are likely to completely knock out or otherwise dramatically alter the protein product.

Recombination mapping has also been used to map major susceptibility factors for a small number of more complex disorders. Examples include the *LDL* receptor gene for heart disease, *APC* for colon cancer, and the *BRCA1* and *BRCA2* genes for heritable forms of breast cancer. The first two cases were aided by clear candidate roles of the mutated candidate genes—in one case a receptor for the low-density lipoproteins that are a major risk factor for heart disease, and in the other a DNA repair enzyme. Neither of the *BRCA* genes, by contrast, were obvious candidates, and in fact evidence that the identified genes are the true causative loci was debated until multiple independent mutations were associated with breast cancer in different pedigrees. Formal genetic proof that a mutation causes a particular phenotype consists either of reversion of the mutation, or rescue of the defect with a wild-type transgene, neither of which are practicable in humans.

Several countries have taken steps to coordinate public health databases with collection of blood samples from tens of thousands of individuals with the aim of accelerating disease gene discovery. The first example of this was an agreement between the Government of Iceland and the deCODE company, which received access to generations of health records in return for their investment in high volume genotyping of anonymyzed samples. Successes in relation to psoriasis and schizophrenia have led to similar commercial/public collaborative projects in populations with particularly interesting genetic or epidemiological attributes.

QTL Mapping

Techniques for recombination mapping in plants and nonhuman animals are more flexible and in many cases offer increased power to map quantitative trait loci (QTL)—polygenes affecting complex traits. The key features of successful **QTL mapping** designs are the ability to control the starting genetic variance; to reduce the environmental variance (in order to *increase* the proportion of overall phenotypic variance that is due to each QTL); and to increase the number of meioses as desired. Most designs start with two inbred parents that differ with respect to the trait of interest, either by chance or as a result of divergent artificial selection. Recombination is then allowed to break up linkage disequilibrium between markers over succeeding generations. In F_2 or backcross designs, two F_1 offspring of a cross between the parents are themselves crossed (or one or more F_1 individuals are back-crossed to a parent), and several hundred F_2 grandchildren are scored for both the phenotype and genotypes at SNPs distributed throughout the genome (Figure 3.10).

Alternatively, F_2 offspring can be sib-mated (that is, crosses set up between siblings) for 10 or more generations to create a set of recombinant inbred lines (RIL), each of which contains a random but nearly homozygous set of chromosome segments derived from each parent. The advantages of RIL are that

(A) F₂ design   (B) Backcross design   (C) RIL

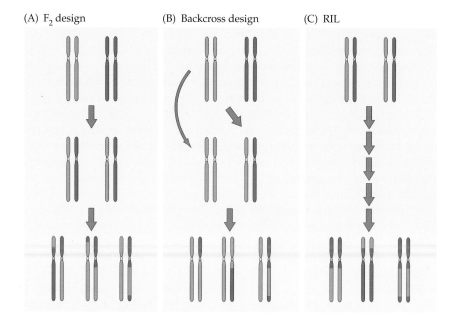

Figure 3.10 F₂, BC, and RIL experimental designs. Controlled mapping crosses are designed to segregate variation that is fixed in two divergent parental lines. (A) In an F₂ experiment, combinations of all three possible single-locus genotypes are generated by sib-mating of F₁ progeny. (B) In a backcross, F₁ individuals are crossed back to one parent, generating heterozygotes and homozygotes of a single class. (C) Recombinant inbred lines are produced by repeated sib-mating for at least 15 generations, which results in homozygous lines that allow testing of multiple genetically similar individuals.

multiple genetically identical individuals can be scored for each line, decreasing the environmental contribution to the trait measure; and that the same lines can be used by different investigators to characterize numerous traits.

Breakpoints between chromosomal segments can be mapped using SNPs or other molecular markers. A further variant on this strategy is to introgress one parental haplotype into the genetic background of the other parent, either by **marker-assisted selection** in each generation or by artificial selection on the trait of interest. In the case of marker-assisted selection, the investigator substitutes the chromosomal region of one strain into the genetic background of another and then asks whether the introgressed portion of the genome affects the trait. In artificial selection, by contrast, selection on the phenotype fixes chromosomal regions that influence the trait, and then marker genotyping can be used to delimit the portion of the chromosome that is responsible for the quantitative effect. Subsequently, transgenesis can be used to confirm that a particular gene is responsible for the effect, though such a study is not trivial for quantitative traits. A well known example of such mapping is the cloning of a gene for small fruit size in tomato, which was performed by introgression mapping to several hundred kilobases followed by regeneration of the phenotype in transgenic plants with a cosmid that carried the QTL (Frary et al. 2000).

Relatively high-resolution QTL mapping was enabled in the 1990s by the development of new techniques for marker genotyping (including SNP detection) and by novel statistical methods. The core idea of QTL mapping is no different from that of Mendelian recombination mapping, except that multiple loci, none of which is solely responsible for the trait, are mapped simultaneously. Instead of looking for perfect association between a marker and the phenotype, the genome is scanned for statistically significant associations between markers and the phenotype. Initially, significance was judged by *t*-tests for a difference between the mean phenotypes of the two allele classes at each marker. This procedure underestimates the magnitude of QTL effects, is unable to separate closely linked QTL, and has low precision. The **interval mapping** procedure (Figure 3.11; Lander and Botstein 1989) improved resolution and power by estimating marker genotypes and association at each position in the interval between adjacent pairs of mark-

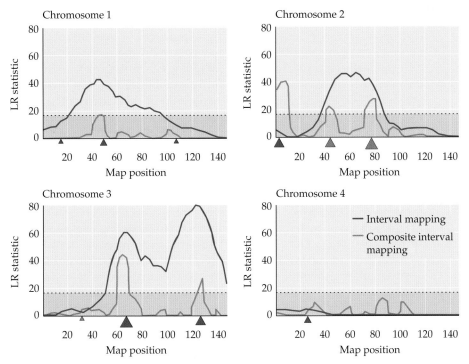

Figure 3.11 Interval mapping and CIM. The location, significance, and number of peaks detected in QTL analysis is a function of the analytical method used. For this simulation of 10 QTL of varying magnitude of the effect of the QTL (size of triangle) and sign (red triangles positive, green negative) distributed on four equal-length chromosomes with markers every 10 cM, composite interval mapping (CIM) resolved the six strongest QTL with more precise positions (red peaks that exceed the threshold dotted horizontal line for each chromosome) than did ordinary interval mapping (blue curves). Significance thresholds were determined by permutation testing. (After Zeng 1994.)

ers. Subsequent modifications include conditioning of marker effects on other significant markers in the genome (composite interval mapping; Zeng 1994) and simultaneous fitting of multiple QTL effects.

In these cases, statistical support is measured using a likelihood ratio, or by a logarithm of the odds (LOD) score, which is proportional to the logarithm of the likelihood ratio. For many purposes in human genetics, a LOD score of 4 (analogous to $p < 0.0001$) is taken as significant; however, given the number of tests performed, it is not always clear that this is appropriate, and permutation tests are now commonly used to define significance thresholds.

QTL are said to lie close to the significant peaks in a LOD distribution, but to have a confidence interval of ±2 LOD units—which even in a study involving more than 100 individuals can be anywhere from several to tens of centiMorgans. While useful for estimating the distribution and magnitude of allelic effects, such mapping is not usually sufficient to identify candidate genes, since the interval will span several megabases and hundreds of genes. For first-pass mapping, markers spaced between 10 and 20 cM are appropriate. Finer resolution can be obtained using a higher density of markers in the vicinity of each QTL and thousands of meioses. Even so, resolution of genes corresponding to QTL that individually account for less than 10% of the phenotypic variation has not yet been achieved in more than a handful of cases.

It should be emphasized that QTL mapping can be extremely useful even if it does not lead to cloning of the gene responsible for the effect (Lynch and Walsh 1998). The localization of a QTL to within 10 cM is sufficient to initiate marker-assisted breeding by introgression of the QTL interval flanked by markers into desired genetic backgrounds. For example, introducing tropical genetic material into a wide range of crop plants for the genetic improvement of traits such as yield, disease resistance, and fruit quality will depend on QTL mapping technology. With mapped QTL, it is possible to estimate the degree of dominance of segregating alleles. Mapped QTL also make it possible to measure **epistasis**, which is the effect of interactions between genotypes at two or more loci. Such data are fundamental to modeling evolution at the species level, and can also be utilized for agricultural breeding purposes.

Alternative methods for identification of QTL such as haplotype homozygosity mapping (Lander and Botstein 1987), transposable element insertional mutagenesis (Anholt et al. 1996), and mapping the haploinsufficient effects of small deletions (also known as **deficiency complementation mapping**; Long et al. 1996) are being employed alongside interval mapping methods and functional genomics, resolving the molecular basis for quantitative traits to levels that could barely be imagined in the pregenomic era.

Linkage Disequilibrium Mapping

Recent technological advances have made it feasible to detect common variants that contribute to specific diseases or phenotypes using a series of methods that are loosely categorized as **linkage disequilibrium mapping**, also known as **association mapping**. The idea behind each of these procedures is to ask whether a particular SNP is more commonly seen in affected, unrelated individuals than you would expect by chance. Significant results can arise either because the SNP contributes to the trait, or (more commonly) because the sampled SNP is in linkage disequilibrium with a site that contributes to the trait. In either case, the resolution of the approach is to the level of the single gene, and in some cases to the causal polymorphism. While this represents a significant advance on linkage mapping, the methods are expensive and require extensive SNP information, so to date are only available for a handful of species.

Linkage disequilibrium is the nonrandom association of alleles, as described in Box 3.1. If there were no linkage disequilibrium, it would be necessary to test every SNP in the genome for association. On the other hand, if every SNP were in LD with every other SNP, there would be no way to resolve the location of a significant effect, since all of the SNPs would show the association. If, however, there is local LD over tens of kilobases (as is often the case), in theory it is only necessary to sample one of a cluster of sites that tend to segregate together. Thus it should be sufficient to sample SNPs at a density similar to the level of local LD, or 50–100 kb in humans. Genome scans are predicated on this principle: that if there is LD, then it is only necessary to sample well-spaced tagging SNPs for association tests. As pointed out above, the caveat is that there is wide variation in levels of LD, so there is no guarantee that one SNP per 50 kb interval is sufficient to represent any particular interval.

Once a significant site is identified, more detailed sampling is employed to identify the gene and eventually QTN (quantitative trait nucleotides, the polymorphic nucleotides that are responsible for a QTL effect). Two general classes of procedure have been adopted: **case-control population sampling**, which is essentially looking for associations between SNPs and the disease in a large population; and pedigree-based **transmission disequilibrium testing**, which is essentially looking for unequal transmission of SNP alleles to affected and nonaffected siblings.

Theoretical basis of association studies.   At the heart of association mapping is the "CD-CV proposition": that common diseases are due to common variants. Whereas rare diseases are often due to rare Mendelian mutations, the idea is that complex diseases such as diabetes, asthma, stroke and depression might be due to polymorphisms in several genes that are likely be common, thus explaining the high heritability of these conditions. Each CD-CV polymorphism is expected to contribute a few percent of the total disease susceptibility. This proposition does not preclude at least two alternate models (that there are hundreds of common disease-promoting alleles, each of very small effect; or there are hundreds of very rare alleles of large effect, each giving rise to similar symptoms). Rare alleles or those of small effect will not be detected in genome association scans. If common variants of *moderate* effect exist, however, it should be possible to identify these common variants with sufficiently powerful study designs involving several thousand individuals.

Before giving a very general overview of these genome association scans, it is worth reviewing some of the hurdles that must be overcome. An important distinction must be made between linkage and association. *Linkage* means that the marker SNP is inferred to be within 50 cM of the disease allele by virtue of physical placement on the same chromosome, whereas *association* means that the SNP is significantly associated with the disease or trait. It is possible to have association without linkage—a chance false positive, for example, or less trivially as a result of population stratification or association between the SNP and an environment that contributes to the disease. It is also possible to have linkage without association, for example where there is insufficient evidence to detect the association between the disease allele and the disease (a false negative), or because the disease association is weak and limited recombination between the marker and the disease allele has been sufficient to break up the association. In general, significant results of association tests between genotype and phenotype will be due to both linkage and association, and disentangling these effects to demonstrate that a SNP is causal and not just linked to the causal site is nontrivial.

As reviewed in Zondervan and Cardon (2004), numerous factors will affect the success of a linkage disequilibrium mapping experiment. Among these factors are:

- *The heritability of the disease or trait.* If the genetic contribution to the disease is too small, no sample size or design will be sufficient to map the individual loci. A closely related concept is that of **relative risk**, which describes the increased likelihood that one group of individuals will get a disease relative to the risk factor for another group.

- *The number of genes affecting the disease or trait.* A ballpark estimate of the number of genes affecting quantitative traits can be obtained by comparing the variance among generations, but this is extremely difficult for threshold-dependent traits (probably including many diseases) and is hard to obtain from pedigree data. The larger the number of genes, the smaller will be their average contributions and the harder they will be to detect.

- *The penetrance and expressivity of the QTL effects.* **Penetrance** refers to the fraction of individuals with the QTL who show the trait, and **expressivity** to the severity of that locus' effects. The lower the penetrance or expressivity, the harder it is to map the gene. For human diseases, penetrance can be strongly age-dependent, leading to misascertainment; and expressivity can be modulated by the environment, which for psychological diseases in particular can lead to misdiagnosis or misclassification of severity.

- *The extent and uniformity of linkage disequilibrium between SNP markers, and between markers and the disease locus.* Variable LD can result in large sections of the genome remaining unscored. The major factors affecting LD are the age of the disease allele; the physical and hence genetic distance between the marker and disease allele; and the effective population size and stability of the population being typed.

- *Genetic heterogeneity.* Genetic heterogeneity describes the frequent observation that the same disease is affected by different loci in different pedigrees; for example, retinitis pigmentosum can be the result of mutation at more than a dozen different loci. A different type of heterogeneity that causes problems for association and linkage mapping is **allelic heterogeneity** within a locus (for example, nearly 100 different mutations in the *CFTR* gene can cause cystic fibrosis).

- **Population stratification.** Differences in allele and haplotype frequencies among populations can lead to false positive associations (Figure 3.12) or, in cases of cultural or environmental transmission, to false attribution of an association to probable causation. A hypothetical example of the latter would be the improper inference that a particular Y-chromosomal haplotype found uniquely in a particular Jewish lineage contributes to the holding of specific religious beliefs.

- **Admixture.** Recent contact between previously isolated populations can result in transient linkage disequilibrium and novel population structures that often go undetected. The effects of such admixture will change over successive generations as heterozygosity returns to normal and allele frequencies attain a new equilibrium.

- **Hidden environmental structure.** This is particularly relevant in relation to assessment of risk for physiological diseases such as obesity, diabetes, and heart disease. Undetected pathogens or incidence of exposure to pathogens and toxins may also have an enormous impact on human health and psychology.

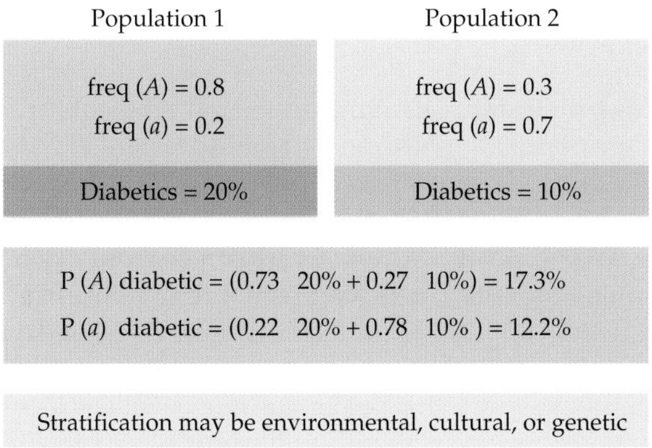

Population 1	Population 2
freq (A) = 0.8 freq (a) = 0.2	freq (A) = 0.3 freq (a) = 0.7
Diabetics = 20%	Diabetics = 10%

P (A) diabetic = (0.73 $\times$ 20% + 0.27 $\times$ 10%) = 17.3%

P (a) diabetic = (0.22 $\times$ 20% + 0.78 $\times$ 10%) = 12.2%

Stratification may be environmental, cultural, or genetic

Figure 3.12 Population stratification. Suppose two populations differ in both the frequency of a particular allele and the prevalence of a phenotype (such as diabetes). If the two populations are pooled (in equal ratios in this example), the measured association between the two genotypes and the disease prevalence may be significantly different, but not necessarily as a result of linkage to the marker. Since 73% [0.8/(0.8 + 0.3) = 0.73] of the *A* alleles are in population 1, which has a higher disease prevalence, the *A* allele would seem to be associated with disease, but it need not be linked to a true causative site. Environmental, cultural, or genetic factors may contribute to the apparent association, even when the population structure is not apparent to the observer.

- **Genotype by environment and genotype by genotype interactions.** Phenotypes are a function not just of genotype, but of the variable expression of genotypes in different environments (a plot of which is sometimes called a norm of reaction) and different genetic backgrounds (epistasis). These interactions are of essentially unknown magnitude, are difficult to detect statistically, and may affect the ability to detect main effects of susceptibility loci. Such effects may be particularly prevalent in relation to categorical and threshold-dependent traits, as well as in certain types of cancer.

- **Incomplete genotyping.** This category includes gaps in information due to missing and deceased individuals or to genotyping errors, as well as to factors such as undetected mispaternity and mismaternity (for example due to incomplete documentation of adoptions).

Despite all of these potential obstacles, association studies offer the best prospects for mapping of complex disease loci to the level of single genes (Hirschhorn and Daly 2005). Dozens of different tests have been proposed and applied with varying degrees of success. All of these generally require sample sizes in excess of 500 individuals to achieve over 80% power to detect a locus that accounts for at least 5% of the disease susceptibility (Long and Langley 1999). Studies of this magnitude are now routine, and in fact several journals now require independent replication on a second study population. Moreover, interesting associations are typically pursued by multiple investigators, and since there is a reasonable likelihood that any one study will fail to see an effect, proof that a locus affects disease susceptibility, along with estimation of the magnitude of the effect, generally follow multiple studies (Ioannidis et al. 2001). A useful online searchable database of human disease association studies can be accessed at http://geneticassociationdb.nih.gov.

Population-based case-control design. The most direct and efficient form of association mapping is the population-based case-control design. Marker frequencies are determined in a sample of affected individuals (the "cases") and compared with marker frequencies in an age, sex, and population-matched sample of unaffected "controls." The markers can be from a gene of interest (a candidate gene approach), a chromosomal region localized by linkage mapping (a sequential approach, outlined in Figure 3.13), or covering the whole genome (a genome scan).

Suppose a disease is thought to be affected by the status of a biallelic locus with alleles A and a. Assuming Hardy-Weinberg equilibrium, observed and expected frequencies can be contrasted most simply by a contingency chi-square analysis:

	Observed		Expected	
	Allele A	Allele a	Allele A	Allele a
Affected	24	278	49	253
Unaffected	86	296	61	321

$\chi^2 = 27.5$, $p < 0.001$

(A)

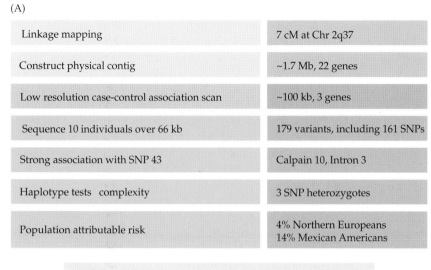

Linkage mapping	7 cM at Chr 2q37
Construct physical contig	~1.7 Mb, 22 genes
Low resolution case-control association scan	~100 kb, 3 genes
Sequence 10 individuals over 66 kb	179 variants, including 161 SNPs
Strong association with SNP 43	Calpain 10, Intron 3
Haplotype tests complexity	3 SNP heterozygotes
Population attributable risk	4% Northern Europeans 14% Mexican Americans

SNP affects binding to nuclear factor in human pancreatic extracts and transcriptional activation of a reporter gene in a cell line

Hypothesis: Calpain 10 protease involvement in NIDDM

(B)

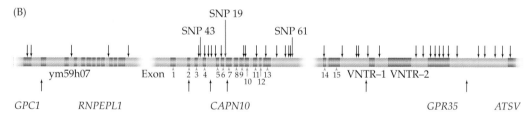

Figure 3.13 Positional cloning of a candidate complex disease gene. Though falling short of definitive proof, the study schematized here for the NIDDM1 type 2 diabetes susceptibility locus demonstrates one strategy for identifying candidate complex disease genes. After standard linkage analysis (A), low-resolution scanning of the 2-Mb region for SNPs focused the search on 100 kb (B) that included 3 genes and a handful of SNPs with replicated disease association across a number of case-control study populations. The strongest association was observed for heterozygotes for a particular haplotype involving three linked SNPs (red arrows). Follow-up biochemical studies are being used to test the hypothesis that variable transcription of the calpain10 protease contributes up to 15% of the population relative risk for the disease. (After Horikawa et al. 2000.)

In reality, continuity adjustments must be made for the fact that the alleles are not drawn from the same sample, so the distribution of expected alleles is not straightforward. Nevertheless, in this hypothetical example, the *A* allele appears to be significantly less frequent in affected individuals than you would expect given its frequency in the population, and one possible explanation is that this allele may be protective against the disease. Variants of this approach discussed in Box 3.3 include comparison of genotype (*AA*, *Aa*, *aa*) frequencies, or estimation of multimarker haplotypes and testing for differences in disease prevalence by analysis of variance.

More sophisticated case-control designs incorporate parametric model testing. Parameters describing the probable magnitude and dominance of the effect, as well as the recombination distance from the disease locus, can be estimated simultaneously using maximum likelihood methods. Prior knowledge of the inheritance patterns may also be incorporated, such as the degree of dominance of the trait. Significance is tested by assessing the likelihood ratio of the probability of obtaining the observed frequencies given the data and best-fit parameters, to the probability given a null hypothesis of no association.

Three major problems in establishing statistical significance are (1) adjusting for the number of statistical comparisons that are performed; (2) controlling for the non-independence of tests, and (3) dealing with the fact that power is a function of marker and disease allele frequencies. The traditional Bonferroni correction for multiple comparisons is to divide the nominal testwise significance threshold by the number of contrasts to obtain an experiment-wide acceptable false positive rate. This is a very stringent but essential threshold. It leaves many true associations undetected, as there is no way to distinguish them from the many tests that produce similar *p*-values

BOX 3.3 Case-Control Association Studies

In a sizeable population of individuals who tend not to mate with relatives, and who otherwise choose their partners at random, only very closely linked polymorphic sites tend to be in linkage disequilibrium. If we ask the question "Is a particular marker more or less frequent in individuals affected with a disease than in unaffected people?" the answer will almost always be independent of the answer obtained for another marker. That is to say, even if there is a significant difference for one marker, chances are that any marker more than about 100 kb away will not show a difference. Or, to put it the other way around, if you wish to demonstrate a statistically robust marker-disease association, you had better sample a site within 100 kb of any SNP that actually affects susceptibility to the disease.

For any disease, there are between a couple and about two dozen polymorphic loci that confer detectably elevated risk of disease occurrence. Detecting these sites among the millions of SNPs in the human genome is quite a challenge. Assuming that a very dense SNP map is available, then more than 25,000 tests must be performed to ensure that at least one marker in every 100-kb interval is sampled. This creates what is known as a multiple comparison problem, as follows.

We demand that there be less than a 1 in 20 chance that any one of the tests is positive before we accept the result as evidence for an association. But, since we would actually *expect* to see a significant result at the 0.05 level once in every 20 tests, we must further divide the testwise significance level of 0.05 by the number of tests performed (25,000), and hence seek a *p*-value of less than 1 in 500,000. The problem is that this is extremely stringent, so a risk factor would have to be extremely large to be detectable in a sample of even several hundred individuals. Alternatively, several thousand individuals must be sampled. For multifactorial diseases in which the environment interacts strongly with genotype, the task is even more daunting. The statistical issues associated with controlling for the number of tests performed are discussed in detail in Devlin and Roeder (1999).

So, while the technology for genome scans of thousands of individuals is on the horizon, it should be clear that statistical associations are just a first step toward identifying candidate genes. With multibillion-dollar markets as an incentive, the investment of hundreds of millions of dollars can be justified even though there is no guarantee of success, and even though some positive associations will turn out to be statistical artifacts. In practice, researchers may address the latter possibility by accepting an elevated false-positive rate in a first genome scan, in the hope that replication of the study focusing just on the suggestive regions will provide more robust evidence. The number of tests can also be reduced by initially focusing the study on a candidate gene that has been identified by linkage mapping in pedigrees or by biochemical or comparative genomic approaches. Positive results from such tests are likely to suggest molecular biological approaches to understanding why variation at the locus may affect the disease.

Association tests can be performed in several ways (Sasieni 1997). In all cases, the strategy is to collect genotype data from several hundred affected individuals and from several hundred age- and sex-matched unaffecteds from the same or a similar population. A table can then be drawn documenting the number of times in the case and control groups that each genotype class occurs (homozygotes with 0 or 2 copies of the allele of interest, or heterozygotes with 1):

Number of individuals in each genotype class

	0	1	2	Total
Case	r_0	r_1	r_2	R
Control	s_0	s_1	s_2	S
Total	n_0	n_1	n_2	N

Several χ^2-distributed test statistics with 1 degree of freedom have been developed. We describe one, known as a trend test, here. A trend test models the effect of an additive allele—that is, a similar increment in the likelihood of having the disease is observed when we compare heterozygotes to homozygotes *without* the allele as when we compare heterozygotes to homozygotes *with* the allele. The statistic is:

$$\chi^2 = \frac{N[N(r_1 + 2r_2) - R(n_1 + 2n_2)]^2}{RS[N(n_1 + 4n_2) - (n_1 + 2n_2)^2]}$$

For the example given in the text, if none of the 151 affected individuals were homozygous for the A allele ($r_0 = 0$, $r_1 = 24$, $r_2 = 127$), while 9 of the 191 unaffecteds were homozygous ($s_0 = 9$, $s_1 = 68$, $s_2 = 114$), then $\chi^2 = 26.4$. This is essentially the same result we obtain with allele counts alone when Hardy-Weinberg conditions apply.

If nonadditive effects are suspected, a genotype-case control statistic takes the form:

$$\chi^2 = \sum_i \left[\frac{(Nr_i - Rn_i)^2}{NRn_i} + \frac{(Ns_i - Sn_i)^2}{NSn_i} \right]$$

where the summation is over each of the three genotype classes ($i = 0$, 1, or 2). This example yields $\chi^2 = 27.8$ with 2 degrees of freedom.

Adjustments can also be made for multiallelic markers (for example, microsatellites as opposed to biallelic SNPs; Nielsen and Weir 1999) and for subpopulation structure if it is detected (Pritchard et al. 2000). An emerging approach is to perform haplotype association studies, in which multisite genotypes are reduced to haplotypes and treated as if they define a single multiallelic locus.

Much statistical theory is still being developed in relation to association studies, and human geneticists are just beginning to get a feel for the structure of large data samples. Sampling, variable patterns of linkage disequilibrium, population divergence, and cultural factors all complicate the analyses, but there is much excitement about the potential for dissecting the genetic components of complex diseases.

by chance. For example, if 100,000 tests are performed, then 10 are expected to have p-values less than 0.0001; even if 15 associations are observed at this level, it is impossible to say which are true positives and which are false. For a genome scan, where several hundred thousand markers are tested, each single marker p-value must typically be less than 0.00000005 ($5 ? 10^{-8}$) to be regarded as significant (Risch and Merikangas 1996).

The second and third problems are typically addressed by permutation testing. This is where the genotypes and phenotypes are randomized with respect to one another. The significance testing is performed on thousands of iterations, leading to an empirical estimate of the expected distribution of test statistics given the allele frequencies, phenotype distribution, and level of linkage disequilibrium in the dataset. The real data are then compared with the permutations.

Population-based association studies are susceptible to the effects of population stratification (see Figure 3.12). It is thus essential to assess whether there is evidence for mixing of two or more populations in the dataset. This can be done using software such as Structure (Pritchard et al. 2000a; see Figure 3.7) or Eigenstrat (Price et al. 2006) that look for signatures of population differentiation in hundreds of anonymous SNPs (Structure) or genome-wide (Eigenstrat). Even where just a single candidate gene is being evaluated, it is advisable to genotype a set of **ancestry informative markers** (**AIMs**) to ensure that hidden population structure is accounted for. Subsequently, the association tests can be conditioned on the estimates of genetic ancestry (Pritchard et al. 2000b), or genomic control methods can be used that incorporate an inflation factor into evaluation of the test statistic (Devlin and Roeder 1999). Ultimately, though, there is no substitute for independent replication—if possible, using a different population.

Pedigree-based analysis.   An alternative to population-based analysis is the use of family-based linkage disequilibrium methods such as **transmission disequilibrium tests** (**TDTs**), which have high resolution because they do not depend on meiosis in short pedigrees. The idea is to test whether heterozygous parents transmit either of their two alleles to a single affected child with equal probability, as shown for the following data from a study of the association between IDDM diabetes and a specific SNP in allele A for the gene that encodes insulin (Spielman et al. 1993).

	Transmitted allele	
	A	a
Observed	78	46
Expected	62	62

$\chi^2 = 8.2$, $p < 0.005$

Only one affected child is tested per family, and only heterozygous parents are informative. The reason for sampling only one affected child per family is to ensure that genetic correlations that occur due to linkage in pedigrees (Box 3.4) do not bias the analysis. And if a parent is homozygous AA,

BOX 3.4 Family-Based Association Tests

A difficult distinction to grasp in terms of disease mapping is that between linkage mapping and linkage disequilibrium (LD) mapping, particularly where family pedigrees are the source of data. Linkage mapping uses the small amount of recombination that occurs in each generation within a pedigree to locate small chromosomal regions containing hundreds of genes. It is like recombination mapping of quantitative trait loci in model organisms, except that many pedigrees are used instead of a single cross with hundreds of progeny. LD mapping, by contrast, uses the very large amount of recombination that has occurred during the history of a population to ensure that any associations that are detected only extend over one or a few loci. (Imagine recombination mapping where you were only typing the children a few thousand generations removed from when a cross was first established: recombination would have occurred between most pairs of markers, so the resolution would be very high, but you would have to sample at an extremely high density.) Case-control population-based LD mapping was explained in Box 3.2, and family-based association tests also utilize the low LD in populations.

Family-based linkage disequilibrium mapping of genetic diseases is based on unequal transmission of alleles from parents to a single affected child in each family. If only a single affected individual is typed in each family, and all of the families are unrelated genetically, then the only sites that can be associated statistically with a disease are those that contribute directly to the disease susceptibility, or markers in linkage disequilibrium with such sites in the general population. The families must be unrelated, as this ensures that the chromosomes are sampled without biases due to shared genetic factors. As we have seen, linkage disequilibrium among sites generally extends only 100 or so kilobases along a chromosome. Consequently, the resolution of LD mapping is at the level of individual genes, and in some cases it may be possible

to resolve which SNPs within a gene are likely to affect the disease susceptibility.

There are actually several forms of family-based linkage disequilibrium tests. The approach mentioned in the text is known as the **transmission disequilibrium test,** or **TDT.** Information is extracted from the ratio of transmission of the two alleles from heterozygous parents to the affected child. Genotypes are obtained for the "triad" of two parents and their child. If both parents are homozygous (whether or not for the same allele), offspring always receive the same genotype independent of disease status, so this situation is uninformative. If, by contrast, both parents are heterozygous, then both alleles in the affected child are informative. If only one parent is heterozygous, it is always possible to deduce which allele was transmitted, so long as the genotype of the second parent is known. Even if the second parent's genotype is unknown, if the child has a different heterozygous genotype for a multiallelic marker, it is still possible to infer which allele was transmitted.

With this information, a Z-test can be designed to test the association of each allele with the disease (Spielman et al. 1993). The Z-test statistic takes the form of the difference between the observed and expected number of transmitted alleles, divided by the variance of the observed number of transmitted alleles. Suppose that in a set of triads, the A allele is transmitted b times from heterozygous parents, and the other allele(s) c times. Then:

$$Z_{\text{TDT}} = \frac{b - c}{\sqrt{(b + c)}}$$

has a two-sided p-value that is chi-square distributed. The significance can be assessed by squaring the Z_{TDT} score and comparing this value with the chi-square table with 1 degree of freedom. Corrections for multiallelic markers, as well as for situations in which there is biased segregation of alleles in both affected and unaffected offspring (segregation distortion), have been employed, as have modifications to apply

(Continued on next page)

BOX 3.4 *(continued)*

the test to contrast extreme individuals for continuous (as opposed to discrete) traits (Allison 1997).

An alternative test, the **sibling-transmission disequilibrium test (S-TDT)** proposed by Spielman and Ewens (1998), uses information from one affected and one unaffected child with different genotypes in a set of unrelated families. In this case, the test statistic is

$$Z_{\text{S-TDT}} = \frac{Y - T}{\sqrt{V}}$$

where Y is the number of A alleles in the affected children in all of the families, T is half the total number of A alleles in all of the pairs of children (which is equivalent to the expected value of Y), and V is the variance of Y. Each of these quantities is easy to calculate given the observed number of homozygous and heterozygous affected and unaffected siblings. If there are r homozygotes and s heterozygotes for the allele, then $T = r + s/2$, and the variance V is the sum of the number of sibling pairs that are discordant homozygotes plus $1/4$ times the number of pairs in which only one of the siblings is heterozygous. An example of a situation in which the S-TDT may be preferable is for late-onset diseases, where parental genotypes are not always available.

To illustrate these tests, consider the data set in the table below, which shows the genotypes at a single biallelic site in two parents, an affected child, and an unaffected child for a hypothetical disease. Applying the TDT, 11 heterozygous parents in 8 of the triads transmit the A allele 9 times to affected children and the a allele twice. Note that two children provide no information because both parents were homozygous, while three children are counted twice each because both parents were heterozygous. Then:

$$Z_{\text{TDT}} = \frac{9 - 2}{\sqrt{(9 + 2)}} = \frac{7}{\sqrt{11}} = 2.11$$

Applying the S-TDT, the A allele is present 14 times in the affected siblings compared with an expected number of $(6 + 8/2) = 10$ times. The variance is $(1 \times 1 + 6 \times 0.25) = 2.5$, where the discordant homozygotes are found in the eighth family on the list. Thus,

$$Z_{\text{S-TDT}} = \frac{14 - 10}{\sqrt{2.5}} = 2.53$$

Neither of these measurements is significant with such a small sample size, but both suggest a trend toward biased segregation of the A allele. For biallelic markers, the same result is obtained for the alternate allele.

Parent 1	Parent 2	Affected	Unaffected	b	c	r	s	Y
AA	*Aa*	*AA*	*Aa*	1	0	1	1	2
Aa	*Aa*	*Aa*	*aa*	1	1	0	1	1
Aa	*aa*	*Aa*	*aa*	1	0	0	1	1
AA	*AA*	*AA*	*AA*	0	0	2	0	2
Aa	*Aa*	*AA*	*Aa*	2	0	1	1	2
Aa	*aa*	*Aa*	*aa*	1	0	0	1	1
aa	*Aa*	*aa*	*aa*	0	1	0	0	0
Aa	*Aa*	*AA*	*aa*	2	0	1	0	2
Aa	*AA*	*AA*	*Aa*	1	0	1	1	2
AA	*aa*	*Aa*	*Aa*	0	0	0	2	1
		Sum:		9	2	6	8	14

(A) Initial replication

Locus	T	U	Ratio	Significance
ABCC8	26	12	2.2	0.012
PPARγ	81	104	0.8	0.045
IRS1	30	26	1.2	0.30
ADRB2	96	124	0.8	NS
INS	104	115	0.9	NS
IRS1	14	21	0.7	NS
KCNJ11	138	154	0.9	NS
TNF	14	13	1.1	0.42

(B) Further replication

Study	N	Risk ratio	Significance
Sibships	1130	0.74	0.071
Case-control			
Scandinavia	481	0.88	0.10
Quebec	127	0.71	0.08
Total	2071	0.78	0.002

(C)

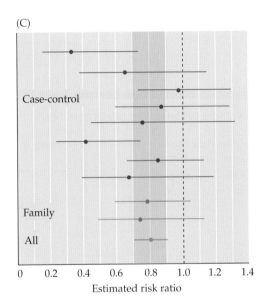

Figure 3.14 PPARγ alanine substitution and type 2 diabetes risk. Published reports had implicated common polymorphisms in eight different genes with type 2 diabetes susceptibility. (A) Initial replication studies on a set of 333 parent-offspring trios from Scandinavia confirmed just two of these associations (T = transmitted alleles, U = untransmitted; four nonsignificant results actually showed trends in the opposite direction to that initially reported). (B) Follow-up replication in three other cohorts confirmed only the association with the peroxisome proliferation-activated receptor-γ (PPARγ) Pro12Ala, at marginal levels. (C) Taken together, however, the consistent relative risk estimate and high frequency (~85%) of the susceptibility allele results in the SNP being associated with a population-attributable risk for the disease as high as 25%. (After Altschuler et al. 2000.)

all offspring receive the *A* allele and there is no basis for comparison; by contrast, if the parent is heterozygous, under normal circumstances both alleles should be transmitted with equal probability to affected and to unaffected children.

In some cases, there may be a bias for one allele to be transmitted more frequently than the other to all offspring, but adjustments can be made for such **segregation distortion**. A positive association between the frequency of one of the transmitted alleles and having the disease is expected to arise either because the SNP actually causes the disease, or because it is in linkage disequilibrium with it in the population from which the families were drawn. The only assumption is that the marker and disease loci are identical by descent in the population, as a result of which the method is sensitive to genetic heterogeneity, but is independent of population structure.

A battery of modified family-based tests is now available (Laird and Lange 2008) and are commonly used, particularly in replication studies. The combination of case-control and transmission disequilibrium tests may provide the best available evidence for association of a marker with a disease (Figure 3.14). Where an association is due to linkage disequilibrium (often reflecting proximity) rather than causation, it is likely that multiple SNPs in proximity to the locus will show association. Whether or not the strongest association is indicative of the closest linkage to the true causal SNP will depend on the precise LD structure at the locus. Since initial scans utilize SNPs at intervals of 10 kb or more, and SNPs typically occur every kilobase, a final survey of all of the SNPs in a locus is required to identify the most likely causal site.

Finally it must be noted that, given the high variance of background and environmental contributions to the effects of disease loci, there is in fact no guarantee that causal SNPs or clusters of SNPs can be identified. This fact is particularly sobering when it is realized that genes with antagonistic effects on a trait can lie adjacent to one another, and that regulatory and structural polymorphisms can have very different effects. Ultimately, suggestive statistical associations must be confirmed with biochemical, cell biological, and physiological methods.

EXERCISE 3.3 *Perform a case-control association test*

Two parallel association studies between a candidate gene and skin cancer are performed in New York City and Miami, Florida. The numbers of cases and controls who are homozygous for a 3-bp insertion or deletion polymorphism or who are heterozygous are as follows:

	New York				**Miami**		
	Insertion	Hetero-zygous	Deletion		Insertion	Hetero-zygous	Deletion
Case	221	198	55	Case	171	236	83
Control	279	212	35	Control	189	244	77

Calculate the probability of association between the deletion polymorphism and the disease for each population separately and for the combined population, using the formula:

$$\chi^2 = N \, [N(r_1 + 2r_2) - R(n_1 + 2n_2)]^2 / RS[N(n_1 + 4n_2) - (n_1 + 2n_2)^2]$$

Discuss the possible reasons for any discrepancy in the conclusions from the two studies.

ANSWER: *Following the method in Box 3.3, you should obtain the following values:*

Quantity	New York	Miami	Combined
r_1	198	236	434
r_2	55	83	138
n_1	410	480	890
n_2	90	160	250
N	1,000	1,000	2,000
R	474	490	964
S	526	510	1,036

Plugging these values into the formula results in χ^2 values of 7.6, 0.8, and 6.9 for New York, Miami, and the combined population, respectively. Prima facie this suggests little effect of the polymorphism in Miami, but suggests a significant effect at $p < 0.01$ in New York and across the two populations.

Overall, the New York frequencies closely match Hardy-Weinberg expectations assuming the insertion frequency is 0.3, whereas the Miami figures fit an insertion frequency of 0.4. Consequently, there is evidence for allele frequency differences between the two populations, and population stratification could affect the overall conclusion. Also, skin cancer is likely to be of higher prevalence in Miami, so genotype-by-environment interactions could occur. Since the insertion polymorphism is less frequent than expected in cases than controls in New York, it could be more protective in lower radiation.

Genome-wide association studies. With the development of genotyping panels that enable the interrogation of hundreds of thousands of SNPs and CNVs simultaneously at a cost of less than $1,000 per sample, it became feasible to scan entire genomes for association with diseases or phenotypes. The impact of the technology was felt in 2007, as dozens of **genome-wide association studies** (**GWAS**) reported novel associations with conditions as diverse as inflammatory bowel and restless leg syndromes, with breast and prostate cancer, as well as with normal variation for human body shape and pigmentation (McCarthy et al. 2008). The NIH maintains a catalog of published GWAS at http://www.genome.gov/26525384. As costs drop, the technology will likely be available for a variety of organisms of importance to the agricultural and evolutionary biology communities, and whole-genome sequencing options will also be considered alongside genotyping. The use of Q-Q plots to assess the distribution of genotype frequencies and associations is discussed in Box 3.5.

The British Wellcome Trust Case-Control Consortium's landmark seven-disease study (WTCCC 2007) summarized in Figure 3.15 illustrates the GWAS approach. For each of seven pathologies (bipolar disorder, coronary artery disease, Crohn's disease, hypertension, rheumatoid arthritis, and types 1 and 2 diabetes), DNA samples from 2,000 patients were compared with a common panel of DNA samples from 3,000 healthy controls. Genotyping involved over 500,000 common SNPs chosen to capture the majority of the variation in Caucasians. In Figure 3.15, the SNP associations for five of the diseases are plotted along the genome, from chromosome 1 through to the X, with more significant test statistics at the top. SNPs highlighted in green are significant at the genome-wide threshold of 10^{-7}: there are a dozen such loci for Crohn's disease, several for diabetes, and none that really stand out for two of the diseases (which thus do not appear in the figure). Replication studies by the WTCCC and other groups have confirmed

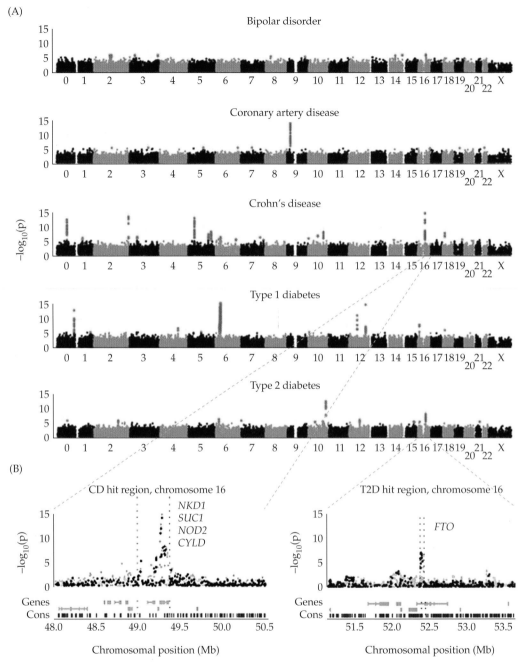

Figure 3.15 Genome-wide association mapping. These Manhattan plots show the magnitude of a test statistic for association between genotypes and each of five indicated diseases, walking along each chromosome in the human genome. The test statistic is the negative logarithm of the p-value, so small values appear as large numbers, for example 0.00001 (10^{-5}) is 5 on this scale. (A) Genome-wide significant SNPs are highlighted in green, while the vast majority of associations tests have $p > 0.1$ and do not appear individually on the plots. (B) Close-ups of two regions of association on chromosome 16. There are four strong candidate genes for Crohn's disease (left) and one strong candidate for type 2 diabetes (right). Yellow lines and boxes show known locations of exons and introns for genes in each region. Blue bars indicate the locations of highly conserved sequences in vertebrate genomes.

BOX 3.5 Genome-Wide Association Studies

Synergy between the fruits of the International HapMap Project and technological advances that now allow relatively inexpensive genotyping of over one million SNPs per individual have led to the emergence of genome-wide association studies (GWAS) as a major approach to discovery-based medical genetics. The objective of GWAS, simply stated, is to identify alleles that appear more frequently in individuals with a disease or condition than they appear in normal individuals (see Figure 3.15). For a variety of reasons (including mandates from funding agencies, requirements of scholarly journals, and the cost considerations of mistakenly reported disease-gene associations), standards for designing, executing, and analyzing GWAS are emerging at a much faster rate than has been seen for high-throughput methods in the past.

To a fairly good approximation, a GWAS consists of the following steps:

1. Experimental design
2. Data collection
3. Quality control and error checking of data
4. Identification of population stratification
5. Statistical analysis for disease-SNP association

In the following we will visit each of these steps except for data collection, which is treated in the text.

Experimental design

A typical GWAS consists of at least 1,000 "cases"—individuals who have the disease of interest—and at least 1,000 control individuals who are presumably disease-free. Samples of this size are necessary, first of all, because the analysis will eventually include a hypothesis test conducted at each SNP locus and large samples are needed to obtain the required statistical power. Very small p-values are necessary in order for a test to be deemed significant. A second factor is the assumption of the CD-CV model of disease underlying a GWAS: for multigene diseases, we do not expect to find dramatic differences in allele frequencies between cases and controls. Thus, large samples are required to provide precise estimation of those allele frequencies.

While the population case-control design is relatively simple to implement and analyze, an emerging design involves the use of controls at the individual level. In these experiments, a control individual might be carefully chosen to match characteristics of a specific case individual, including age, socioeconomic status, location, and/or known risk factors. Other designs, including trio and cohort studies, are also used in the appropriate settings.

The final, and perhaps most valuable, aspect of experimental design is replication. The identification of disease-associated SNPs in the same chromosomal region in samples from two or more populations is currently the gold standard for a GWAS. A replicated result adds considerable confidence that an identified association is the result of biological rather than stochastic factors. Short of functional verification, a replicated association is probably the closest we can expect to get to a **causal inference** using GWAS.

Quality control, error checking, and population stratification

Once data are collected from all individuals in the study, genotypes are called (usually by the proprietary algorithms of the array or chip manufacturer; see the text) at each SNP locus for each individual. The error rate of these calls appears to be impressively low. Individuals or loci might be excluded from analysis because of factors such as a high frequency of missing data, very low minor allele frequency, high heterozygosity (a potential indicator of sample contamination), departures from Hardy-Weinberg equilibrium, or evidence of population stratification.

The presence of nonrandom mating of individuals because of population structure can lead to artifactual associations. Thus a high priority is placed on screening for the presence of such stratification and subsequent corrections if evidence is found. The

(Continued on next page)

BOX 3.5 *(continued)*

Q-Q plot showing the effect of population stratification on gene associations for height. The red line graphs the association values expected if there were no effect of the surveyed genes on height. Points above the dashed line exceed the genome-wide threshold of significance. The arrowhead marks the point at which the number of observed associations exceeds the expected number. The blue curve lies below the black one as a result of adjustment for inflation of test statistics due to population stratification. (After Gudbjartsson et al. 2008.)

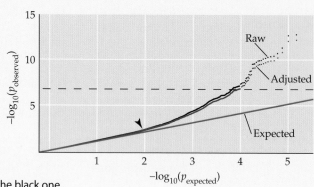

use of principal component analyses (PCA), such as those found in STRUCTURE or Eigenstrat, are currently preferred. If population stratification is found, the leading principal components of the PCA might be used as covariates in the subsequent analyses.

The figure above is a typical graphical demonstration of the effect of population stratification on the overall profile of associations. This example shows a so-called **Q-Q plot** of the observed against expected test statistics (as negative log *p*-values) for association of 304,226 SNPs with height in 32,223 adults adjusted for sex and age. The sample was mainly European with a small number of African Americans, which led to a slight inflation of *p*-values due to population structure. Comparison of the raw values with the adjusted curve suggests an approximate doubling of significance values across the range. There is such a large excess of associations starting at $p < 0.01$ ($-\log p > 2$; arrow) that this inflation is of little practical consequence, but for samples with just 1,000 cases the effect can be very important.

Statistical identification of associations

For basic population case-control studies, standard chi-squared tests for association

(see Box 3.3) are carried out at each SNP locus individually, with appropriate corrections made for the large number of multiple tests. However, these procedures do not allow for easy corrections to be made for factors such as population stratification or the presence of known clinical risk factors. It is becoming more popular to use the technique of **logistic regression** to identify associations.

Logistic regression is used widely to investigate the affect of variables (either quantitative or qualitative) on categorical response variables. In the GWAS setting, the logistic regression model relating the probability of disease, p_i, for individual i to the genotype (A or a) of that individual at SNP locus j takes the form

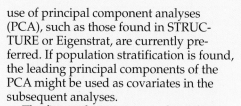

$$\ln\left(\frac{p_i}{1-p_i}\right) = \beta_0 + \beta_j x_{ij} + \underset{\sim}{\beta_s} s_i + \underset{\sim}{\beta_r} r_i + \varepsilon_i$$

where $\underset{\sim}{\beta_s}$ and $\underset{\sim}{\beta_r}$ are vectors of regression coefficients correcting for the presence of population stratification using PCA components and adjustments based on clinical or other potential risk factors, respectively. All regression parameters are estimated, and the null hypothesis $H_O{:}\beta_j = 0$ corresponds to the hypothesis that the allele at SNP locus j being tested has no effect on the probability of an individual having the disease. As with the case-control chi-squared tests, multiple testing procedures must be invoked to determine appropriate significance cutoffs. The free software PLINK is widely used to carry out these types of logistic regression analyses.

It is worth calling attention to the implicit assumption of an additive genetic model—that is, there are no epistatic interactions in the model. Current practice involves a two-step approach where the model above is used to identify associated SNP, followed by a second regression analysis that included potential interactions among the identified SNPs.

In the text it was noted that even after adjustments for multiple testing and identifying a set of, say, 15 "statistically significant" SNPs, there was no immediate way of determining which significant tests were most likely to be true positives. The use of **Bayes factors** is becoming popular for circumventing this problem, although not without some philosophical and theoretical controversy. A high Bayes factor for a locus provides increased evidence for that locus being associated with the disease, and the strength of evidence supporting individual SNPs can be ranked according to the magnitude of their Bayes factors. While proper interpretation of the evidence for a SNP conveyed by a p-value requires information about the power of the test, the Bayes factor for a SNP combines both of those aspects into a single metric.

most of these associations, and studies are under way to determine the identities of the causal SNPs and to decipher how they impact disease.

Two commercial platforms are available for whole-genome association studies. **Affymetrix GeneChips** rely on genotyping by hybridization (Wang et al. 1998). As shown in Figure 3.16, each SNP was originally represented by between 25 and 40 slightly different 25-nucleotide probes on a "variant detector array" (VDA) the size of a thumbnail, with millions of such probes printed on it. The latest versions instead have those probes that give the most consistent signal represented four times (McCarroll et al. 2008; see Chapter 4 for a description of how the chips are synthesized). Genomic DNA

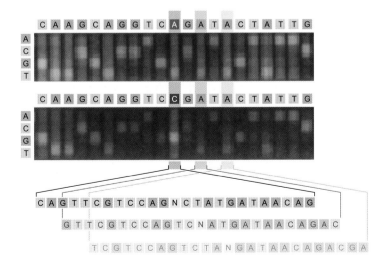

Figure 3.16 Sequencing by hybridization. Sequencing chips originally consisted of four rows of 25-mers in which the central position was varied. There were two sets of rows, one for each strand. Each successive column represents a shift of one nucleotide along the sequence. Hybridization of DNA from two different homozygotes for the central position results in differential hybridization. (After Wang et al. 1998.)

is fragmented, labeled, and hybridized to the array, and the signal intensity of probes that match the genomic DNA perfectly is compared with those that differ at a single base. The probes are identical except that they have A, C, G, or T in the central position. By comparing the signal from each array with the signals from hundreds of other individuals that are heterozygous or homozygous for the alternate alleles, a highly accurate readout of genotypes is obtained. The latest chips allow CNV to be genotyped at the same time.

The **Illumina Infinium** genotyping platforms have similar performance characteristics, providing whole-genome genotypes for from 100,000 to over 1 million SNP and CNV variants. This technology relies on extension of short oligonucleotides that hybridize immediately adjacent to each SNP on whole-genome amplified and fragmented genomic DNA (Figure 3.17). One version of the assay, Infinium I, uses allele-specific primers that will support polymerization only if the oligonucleotide matches the SNP perfectly at the terminal base of the primer (Gunderson et al. 2005); this assay requires different beads for each allele. By contrast, the Infinium II assay uses single-base extension of the polymorphic nucleotide adjacent to the terminal base of the primer, with two-color detection of biotin- or dinitrophenol-labeled nucleotides on a single bead for each SNP (Steemers et al. 2006). In either case, SNP identities are determined by decoding the location of the microbeads to which the primers are attached, by virtue of a genetic barcode associated with each bead. The assay is performed on an etched surface that harbors millions of beads while allowing fluorescence to be detected immediately beneath each bead.

Both companies report individual genotyping accuracies and success rates in the vicinity of 99% or better, which is remarkable and means that individual SNPs can be genotyped with confidence for about one-tenth of a cent each. However, systematic errors can occur that lead to mis-typing, and even if just a few percent of the individuals are affected (for example, if cases and controls are not randomly distributed with respect to assay performance) spurious results can arise. Stringent controls for Hardy-Weinberg equilibrium are thus used to filter potentially bad assays, and significant associations are typically double-checked with a second type of genotyping assay, at least in a subset of individuals.

Genome-wide scans can also be performed for association with continuous traits. Studies of pigmentation have uncovered half a dozen genes that contribute to hair, eye, and skin color in Europeans (Sulem et al. 2007); variation in one gene, *FTO*, accounts for around 3 kilograms of body weight comparing homozygote classes (Frayling et al. 2007); and at least 25 genes have been shown to influence height in a survey of more than 50,000 Caucasians, although taken collectively all 25 explain less than 5% of height variability (Weedon et al. 2008). Pooling of individuals with extreme trait values has been used to reduce costs and has led to the identification of a candidate gene for memory performance, based on estimates of allele frequencies directly from the genotype hybridization signal intensities (Papassotiropoulos et al. 2006).

(A) Infinium I

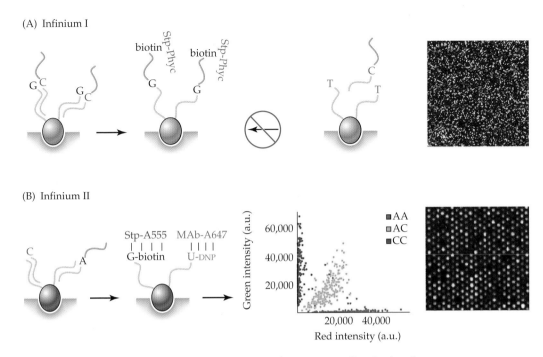

(B) Infinium II

Figure 3.17 The Illumina Infinium I and II genotyping assays. Illumina bead technology combines polymerase extension assays with hybridization of whole-genome amplified DNA to beads in etched wells to achieve high performance genotyping. (A) In the Infinium I assay, allele-specific primer extension (ASPE) occurs only where the terminal nucleotide of the primer is complementary to the SNP: in this example, the primer with a G on the left binds to the C polymorphism, but the one with a T on the right does not. Subsequently, biotinylated nucleotides are incorporated onto a growing chain and are detected by labeling with streptavidin-phycoerythrin, producing a grayscale image. A second bead with the alternate SNP allele on the primer must also be included on the chip, and the ratio of the signals provides the genotype. (B) In the Infinium II assay, a single primer is used on just one bead per genotype. The alleles are distinguished by the nature of nucleotide that is incorporated adjacent to the end of the primer: G or C are labeled with biotin, and A or T with dinitrophenol (DNP). These are detected, respectively, with streptavidin-Alexa555 or rabbit anti-DNP antibody conjugated to Alexa647, and the genotype is assessed from the ratio of red-to-green signal on the bead. G-C and A-T polymorphisms cannot be distinguished without additional beads. Both assays support parallel typing of up to a million SNPs in a single reaction without PCR amplification of the individual loci.

SNP Genotyping

SNP Discovery

SNPs are discovered by comparing sequences derived from different chromosomes. The probability of detecting a SNP for which the more rare of two alleles has a frequency of p in a study of n individuals is given by $\mathrm{Pr} = 1 - p^n - (1 - p)^n$ (Eberle and Kruglyak 2000). For an intermediate-frequency allele, this probability rises quickly from less than 50% for 2 chromosomes to more than 90% for a sample of 10 chromosomes (i.e., 5 individuals). Even for rare alleles with a frequency of 0.1, a sample of 10 chromosomes gives a reasonable probability of detection, but most such alleles will be missed when just 2 chromosomes are compared. Investigations that start with only a few individuals are thus heavily biased toward detection of common SNPs and underestimate the total number of segregating polymorphisms. For the purpose of finding SNPs for association studies or linkage mapping, however, this strategy is almost ideal, since alleles with frequencies in the range of 0.1 to 0.2 are optimal for linkage disequilibrium mapping purposes.

A number of different sources of chromosomes can be used to identify SNPs. The most obvious for many species is a comparison of the whole-genome sequence with existing sequences deposited in GenBank, which typically derive from different strains. For example, at least two-thirds of all discrepancies between the Celera whole-genome sequence of *Drosophila* and cDNA reports generated over the past two decades are true SNPs when verified in a dozen wild-type strains.

A more deliberate approach is to resequence clones from new individuals. Light shotgun reads of genomic DNA or of cDNA are an efficient way to identify sequence polymorphisms. Where there is a need to generate a high density of SNPs in a particular region of the genome, BAC clones can be isolated and resequenced, or the polymerase chain reaction (PCR) can be used to amplify the target DNA from a panel of individuals. These can be sequenced directly, or subcloned. For targeting SNPs to coding regions, or across the full length of a gene, it will often be preferable to sequence cDNA clones. However, the relatively high error rate of DNA polymerase and reverse transcriptase can cause the incorporation of mutations into cloned DNA fragments, so it is preferable to have each genotype represented by at least two independent reads.

SNP Genotyping

The ability to genotype large numbers of SNPs in large numbers of individuals rapidly and cost-effectively is essential for many applications, from linkage and association mapping to mutation detection and diagnosis. It is now feasible for single research groups, once they have invested the time and effort required to establish a method, to genotype thousands of SNPs per week, and for large consortia to genotype several hundred thousand SNPs per week. In this section, we first review the basic principles of some of the major methods in use based on traditional molecular biological tools and equipment (Mir and Southern 2000), and then consider three high-throughput commercial platforms that reduce costs to just cents per SNP.

Restriction fragment polymorphism. Most laboratory-based SNP genotyping methods depend on specific amplification of the DNA sequence surrounding the site to be genotyped, which is generally achieved using the PCR. By far the simplest way to detect a SNP within an amplified fragment is to monitor the cleavage of the fragment upon digestion with a 4-cutter or 6-cutter restriction endonuclease. These enzymes usually require a palindromic recognition site—that is, a site that reads the same sequence on the reverse strand, such as GATC or CCGG. Between one-fifth and one-half of all SNPs actually lie within just such a short palindromic sequence and, provided that a relevant restriction enzyme is commercially available, the SNP can be detected rapidly and for relatively little cost simply by running fragments out on agarose gels (Figure 3.18; Wicks et al. 2001). This **PCR-RFLP**, or **Snip-SNP**, method is a simplification of the restriction fragment length polymorphism method that was initially devised for detecting restriction site variants by whole-genome Southern blots.

It is also possible to turn most SNPs into **derived cleavable amplified polymorphic sequences (dCAPS)** by clever design of the PCR primers to introduce a restriction site associated with one allele, as shown in Figure 3.19 (Neff et al. 1998). The cleaved product typically is only 20 or so nucleotides shorter than the original amplified product, so small acrylamide

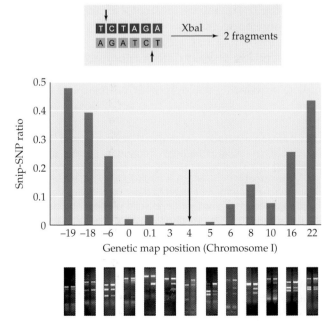

Figure 3.18 Bulked segregant mapping of mutations using Snip-SNPs. A polymorphic SNP restriction site in a PCR product, such as this XbaI site at position 20210 in clone T24B1 from *C. elegans* chromosome I, allows rapid genotyping of individuals with the two alleles. In bulked segregant analysis, pools of wild-type and mutant individuals from the F2 population of a cross between the two parental types are genotyped at multiple markers across a region. The relative intensity of the single higher molecular weight band, and the two bands produced by digestion of each marker, provides a ratio of wild-type to mutant alleles in the populations. In this example from Wicks et al. (2001), side-by-side comparison of wild-type (left) and mutant (right) pools led to rapid localization of a Mendelian mutation in the nematode to within 100 kb.

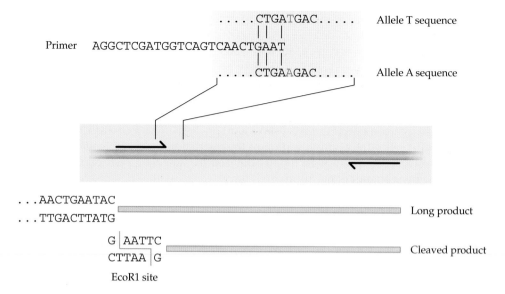

Figure 3.19 dCAPS.   A derived cleavable amplified polymorphic sequence (dCAPS) is generated by using a PCR primer that has a mismatch a few bases before the SNP. In this example, the second-to-last base is an A instead of a C, as a result of which the sequence of the amplified fragment includes a new restriction site for EcoR1 (GAATTC) for the A allele of the SNP (shown in blue at the top). By contrast, the T allele sequence (GAATAC) does not create a restriction site. Cleavage with EcoR1 results in a shorter fragment for the A allele.

gels are used to separate the fragments, which can be visualized with simple ethidium bromide staining. The method is ideal for screening a relatively small number of sites in a large population.

Minisequencing methods.   A philosophically different approach to genotyping is to actually *re-sequence* each allele in a sample of individuals. Instead of carrying out a complete sequencing reaction, which is relatively expen-

Figure 3.20 Single-base extension methods.   Multiplex PCR is performed to   ▶
amplify a number of different SNP-containing fragments from genomic DNA. Methods such as trapping a biotin-labeled strand with streptavidin-coated beads in microtiter-wells are then used to purify the target strand, which is hybridized to the allele-specific SBE probe. A variety of approaches can then be used to detect the incorporation of the appropriate fluorescently labeled single nucleotide. In solid-phase methods, the target is hybridized to the probe directly on a microarray, and the reaction is performed on the array. In liquid-phase methods, the SBE reaction is performed in solution. The products can then be separated by electrophoresis on the basis of size differences between the probes (possibly using new automated sequential injection techniques to stagger oligonucleotides, such as Molecular Dynamics' MegaBACE apparatus). Or they can be crosslinked to red/orange fluorescence-coded microbeads by way of a molecular "zip code" on the probe, then sorted by fluorescence-activated bead sorting. (After Chen et al. 2000.)

sive and takes hours rather than minutes, methods have been developed in which only one or a few bases are sequenced (Pastinen et al. 1997). Some of these are performed in microtiter plates using detectors that monitor fluorescence over a range of wavelengths and temperatures, such as the ABI Prism 7700 Sequence Detection System, and others are being adapted for microarray formatting, or even microcapillary electrophoresis. The cost effectiveness of this approach is a function of the density of SNPs under consideration: If each genotype costs a couple of dollars, and there are several SNPs within less than a kilobase, complete sequencing may be more efficient. However, if the SNPs are more widely dispersed, minisequencing will be cheaper once a particular method has been established in a particular laboratory.

The conceptually most direct minisequencing method is **single-base extension**, or **SBE** (Figure 3.20). In theory, DNA polymerase provides much higher specificity for detection of single-base polymorphisms than differential melting temperatures of hybridized sequences does (Pastinen et al.

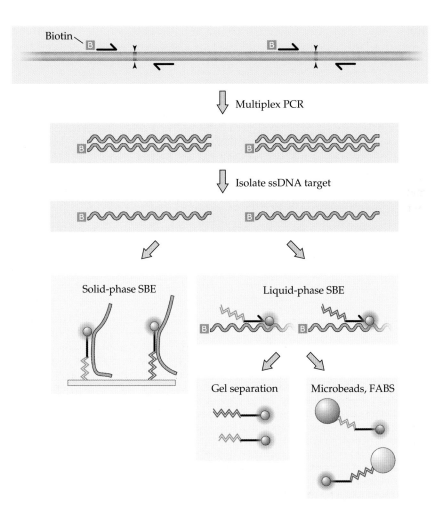

2000). The idea behind SBE is to incorporate a single dideoxy fluorescently labeled nucleotide at the SNP position immediately adjacent to the 3' end of the genotyping primer. Since only the base complementary to the SNP will be incorporated, SBE provides a direct readout of the genotype, including differentiation of homozygotes from heterozygotes, on the basis of the wavelength of the fluorescence. Both microarray and gel-based methods have been developed to facilitate multiplexing of the assay, so that dozens of genotypes can be determined in a single reaction.

Another highly automated minisequencing method is **pyrosequencing** (Figure 3.21; Ronaghi 2001). The idea is to perform 96 parallel minisequencing reactions in a microtiter plate, using fluorescence to detect incorporation of each nucleotide onto a primer in real time as nucleotides are cycled in and out of the microtiter wells. The fluorescence that is detected is actually derived from enzymatic cleavage of the pyrophosphate liberated from a dNTP when it polymerizes with a growing chain. The target polymorphism is amplified from genomic DNA and mixed with a genotyping primer that binds a few bases upstream of the SNP. The first couple of reaction cycles allow incorporation of the common bases between the 3' end of the primer and the SNP. In each cycle, the complementary nucleotide is added,

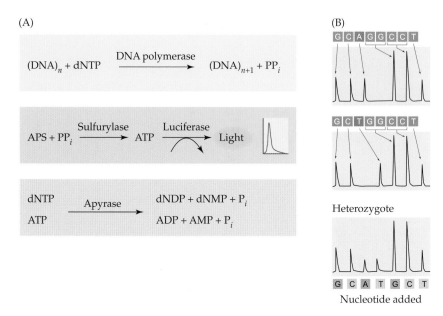

Figure 3.21 Pyrosequencing. (A) Pyrosequencing is based on cycling reactions, one of which generates ATP when a specific base is incorporated (top), one that converts ATP to light (center), and one that removes unincorporated nucleotides as well as excess ATP prior to the next cycle (bottom). (B) By controlling the order in which nucleotides are added, minisequencing traces known as pyrograms are produced over a matter of seconds. In this example, an A/T SNP is distinguished between homozygotes and heterozygotes. The height of each peak produced by the flash of light is proportional to the number of nucleotides incorporated.

DNA polymerase catalyzes incorporation and release of pyrophosphate, and the pyrophosphate is used by luciferase to convert luciferin to oxyluciferin, releasing a detectable flash of light. The enzyme apyrase continuously removes excess unincorporated nucleotides and ATP, allowing a fresh cycle to commence upon addition of the next nucleotide after a few seconds. At the SNP, both nucleotide types are incorporated in successive reactions. On the pyrogram (see Figure 3.21B), the first homozygote will produce a peak the same height as the preceding nucleotides, then a gap before the next common nucleotide is incorporated. The other homozygote will show a gap and then a peak, while heterozygotes will show two smaller peaks.

EXERCISE 3.4 *Designing a genotyping assay for a double polymorphism*

Design three different strategies for genotyping the two polymorphisms indicated in the following sequence:

GGTGCAGGCATGCAGAAGCGTG (G/A) CATG (T/C) GGAACAATGCTGAGTCCTAAT

ANSWER:

1. Use dCAPS Finder 2.0 (*http://helix.wustl.edu/dcaps/dcaps.html*) to generate primers and restriction sites that will recognize the two sites from either direction. For example, the primer GGTGCAGGCATGCA–GAAGGGTG terminates one base before the (G/A) polymorphism, and will produce a restriction site for HphI (GGTGA) with the A allele, but no restriction product with the G allele. The primer ATTAG–GACTCAGCATTGTTGC terminates one base before the (T/C) on the reverse strand, and will produce a restriction site for CviRI (TGCA) with the T allele, but no restriction product with the C allele.

2. Use PCR to amplify the whole sequence, and then a forward primer TGCAGGCATGCAGAAGCGTG to initiate a single base extension reaction for the G/A, and a reverse primer TTAGGACTCAGCATTGTTCC for the T/C polymorphism.

3. Use a single forward primer such as GGTGCAGGCATGCAGAAGCG to initiate a 7-nucleotide pyrosequencing reaction that will type both sites in the same reaction. You would add the nucleotides in the order T,G,A,C,A,T,G,T,C,G to generate a pyrogram.

High-throughput genotyping platforms

Short of whole-genome genotyping, a number of options are now available for rapid genotyping of up to several hundred SNPs in hundreds of individuals. Three of these are described briefly in this section and diagrammed in Figure 3.22, but this is by no means an exhaustive list of technologies. Analysis of genomic methylation profiles can also be performed by utilizing an

Figure 3.22 High-throughput genotyping platforms.   The Illumina and ABI geno- ▶
typing platforms are both based on allele-specific oligonucleotide ligation assays
(OLA). (A) In the GoldenGate assay, three unique primers are required for each SNP;
ligation of two of these by virtue of complementarity to the genomic DNA sequence
produces a template for universal PCR amplification of all of the genotyping reactions.
These primers hybridize to the bead array by way of a unique barcode address built
into P3, and the ratio of red to green fluorescence on either the U1 or U2 primer provides the genotype. (B) In the SNPlex assay, only two universal primers ("forward" and
"reverse") are used in the amplification step. The barcode is read by hybridization to a
unique ZipChute probe, which is linked to one of 96 molecular mobility modifiers and
a fluorophore. The ZipChutes are eluted and capillary electrophoresis allows readout
of the ratio of the two alleles for up to 48 SNPs simultaneously. (C) The Sequenom
iPLEX Gold assay utilizes single-base extension followed by mass spectrometry, based
on the different masses of each of the four nucleotides. Since the primers for each
SNP also have different masses, up to 40 SNPs can be genotyped.

extra experimental bisulfite treatment step that coverts non-methylated cytosine to uracil residues, which can then be detected on these platforms.

The **Illumina GoldenGate** genotyping assay (Shen et al. 2005) can be performed on 96,384 or 1,536 SNPs per sample on either a Sentrix Array Matrix
(SAM) or 16-sample BeadChips. It is a ligation-based assay that relies on the
amplification of short PCR fragments only when an allele-specific primer
matches the sequence of the genotype at the 3′ terminal nucleotide. When this
occurs, a brief extension reaction followed by ligation to another primer is supported, and this synthesized strand becomes a template for multiplex PCR.
The identity of the SNP is read by hybridization of the 5′ end of the allele-
specific primer to a molecular bar code on a bead. As with the Infinium assays,
the beads are deposited in etched wells and their identity is read by the manufacturer before the SAM or BeadChip is shipped to the user. The company
will custom design SNP genotyping chips for any species, and also offers a
variety of panels that can be used for linkage mapping, typing across the human major histocompatibility complex, and cancer research. Alternatively, Illumina's new VeraCode technology in conjunction with the BeadXpress reader,
allows greater flexibility in the number of SNPs and samples to be assayed.

Applied Biosystems' SNPlex genotyping system (Tobler et al. 2005) is
designed for 48 simultaneous assays and also uses an oligonucleotide ligation assay in conjunction with multiplex PCR amplification. However, the
SNPs are decoded by linking the allele-specific primers to one of 96 different dye-labeled "ZipChute" probes that have unique mobility as they migrate
through a capillary tube in an ABI Genetic or DNA Ananlyzer. Their SNaP-
shot Multiplex system provides for rapid genotyping of thousands of samples at just 10 SNPs, using a primer extension assay in conjunction with capillary electrophoresis. ABI also offers a completely different technology, the
TaqMan SNP Genotyping Assay (Holland et al 1991), for custom genotyping of any one of the millions of known human SNPs, or tens of thousands
of mouse or *Drosophila* SNPs. TaqMan is based on the release of a quenched
fluorescent tag during an allele-specific, real-time PCR and can also be
adapted for high-volume typing of most SNPs or indels in any organism.

(A) Ilumina GoldenGate

Allele-specific ligation   Universal amplification   Hybridization to bead array

(B) ABI SNPlex

Allele-specific ligation   Universal amplification   Electrophoresis of ZipChute

(C) Sequenom iPLEX

Hybridize primer   Single-base extension   Ratio of mass spectra

The **Sequenom MassARRAY iPLEX Gold** platform combines single base extension with high resolution mass spectrometry to support the simultaneous genotyping of up to 40 SNPs per reaction (Jurinke et al. 2001). Genotypes are distinguished on the basis of time-of-flight of oligonucleotides through a vacuum chamber, as described for protein fragment sequencing in Chapter 5. With parallelization, one machine can process 150,000 genotypes in a day. The system has seen broad application in association studies, assessment of copy number variation and loss of heterozygosity in cancer, as well as detection of viral and parasite loads.

Haplotype phasing methods

For many applications in quantitative genetics, it is important not just to have the genotypes of individual SNPs, but also to be able to assemble the haplotype of a series of linked SNPs. Suppose an individual is heterozygous at two adjacent SNPs, yielding the genotypes A/T and A/C. This information does not tell you which sites are in phase with one another; are the haplotypes AA and TC, or are they AC and TA? Establishment of the "linkage phase" over more than a couple of kilobases is a statistical problem discussed in Fallin and Schork (2000). Even within a short fragment, phase cannot be ascertained from the SNP genotypes alone (unless the PCR products are individually cloned prior to genotyping, in which case they necessarily derive from the same chromosome). For SNPs that are too far apart to be amplified in the same clone (aside from X or Y chromosomal haplotypes in males), there is no direct method for establishing linkage phase. The best solution is to monitor transmission of alleles from parents to offspring. Except in the case of double heterozygotes, co-transmission allows deduction of which SNP alleles must be derived from the same chromosome. For double heterozygotes, maximum likelihood estimates (Excoffier and Slatkin 1995) that fit the probability of gametic association dependent on individual allele frequencies and haplotype frequencies in the known sample can be used.

Summary

1. SNPs (single-nucleotide polymorphisms) are the most common form of segregating molecular variation in natural populations. Depending on the species and gene, they tend to occur at a density of between one in every 30 bases to one in every kilobase.

2. Population geneticists are interested in the distribution of SNPs, including their frequency in distinct populations, effect on encoded proteins, rate of divergence between species, and degree of linkage disequilibrium and haplotype structure.

3. A large proportion of SNPs are rare (occurring at a frequency of less than 5% of the alleles in a population), and the overall distribution of SNPs usually accords with the predictions of neutral molecular evolutionary theory.

4. Ultimately, most genetic variation is attributable to SNPs, small indel polymorphisms, and copy number variation (CNV), but there are complicated statistical difficulties in detecting the 10 or so variants that have the strongest genetic influence on any trait, among the millions of SNPs in a genome.

5. Genome-wide association studies (GWAs) allow complex traits and diseases to be dissected to the level of individual genes and, in some cases, causal polymorphisms.

6. Case-control disease mapping refers to procedures for detecting SNPs that are at a different frequency in unrelated affected compared with unaffected individuals. Detection of association is influenced by many factors including population stratification, admixture, variation in the level of environmental and cultural influence, disease penetrance and expressivity, the amount and distribution of linkage disequilibrium, and genetic heterogeneity.

7. Family-based association mapping procedures such as transmission-disequilibrium and sibling-disequilibrium methods are also available. Generally, more than 500 trios of parents and affected offspring must be studied to achieve statistical significance, but the advantage is that the tests are not biased by population stratification.

8. SNPs can be detected using next-generation, high-throughput sequencing approaches.

9. Sequencing by hybridization is a high-throughput method for detecting polymorphisms based on the specificity of hybridization to short oligonucleotides on a chip. Once a set of SNPs has been identified, they can be arrayed as oligonuleotides on a variant detector array. VDAs allow massively parallel genotyping of thousands of individuals at thousands of loci.

10. Low-technology methods for SNP genotyping include Snip-SNPs and dCAPS, which are SNPs that affect the ability of a restriction endonuclease to cleave a short PCR product.

11. A variety of new technologies for SNP screening are available, including minisequencing and single-base extension methods for microarrays, as well as several customizable commercial platforms.

Discussion Questions

1. Why is linkage disequilibrium so important for population and quantitative genetic analysis, and what level of sampling is required to quantify LD?

2. To what extent can the genes shown in Figure 3.1 be said to cause particular characteristics of a specific human individual? What is the relationship between genetic association and prediction?

3. Contrast the case-control and transmission-disequilibrium strategies for mapping disease loci.

4. Why is it essential that SNP association studies be replicated before it is concluded that a particular polymorphism is associated with a disease or clinical phenotype? Does failure to replicate a finding mean that the original study was flawed?

5. If you were asked to survey the population structure of an endangered species, which of the methods for SNP genotyping would you prefer to use?

Web Site Exercises

The Web site linked to this book at http://www.sinauer.com/genomics provides exercises in various techniques described in this chapter.

1. Use GDA to compute the pattern of linkage disequilibrium in a sequence dataset.

2. Carry out a hypothetical case-control association study.

3. Carry out a hypothetical transmission-disequilibrium association study.

Literature Cited

Altschuler, D. et al. 2000. The common PPARγ Pro12Ala polymorphism is associated with decreased risk of type 2 diabetes. *Nat. Genet.* 26: 76–80.

Anholt, R. R., H., R. Lyman and T. F. Mackay. 1996. Effects of single P-element insertions on olfactory behavior in *Drosophila melanogaster*. *Genetics* 143: 293–301.

Begun, D. and C. Aquadro. 1992. Levels of naturally occurring DNA polymorphism correlate with recombination rates in *D. melanogaster*. *Nature* 356: 519–520.

Cann, R., M. Stoneking and A. Wilson. 1987. Mitochondrial DNA and human evolution. *Nature* 325: 31–36.

Chen, J. et al. 2000. A microsphere-based assay for multiplexed single nucleotide polymorphism analysis using single base chain extension. *Genome Res.* 10: 549–557.

Clark, A. G. et al. 1998. Haplotype structure and population genetic inferences from nucleotide-sequence variation in human lipoprotein lipase. *Am. J. Hum. Genet.* 63: 595–612.

Clark, A. G. et al. 2003. Inferring nonneutral evolution from human-chimp-mouse orthologous gene trios. *Science* 302: 1960–1963.

Clegg, M. T. et al. 1976. Dynamics of correlated genetic systems. I. Selection in the region of the *glued* locus of *Drosophila melanogaster*. *Genetics* 83: 793–810.

Coop, G., X. Wen, C. Ober, J. K. Pritchard and M. Przeworski. 2008. High-resolution mapping of crossovers reveals extensive variation in fine-scale recombination patterns among humans. *Science* 319: 1395–1398.

Curran, M., I. Splawski, K. Timothy, G. Vincent, E. Green and M. T. Keating. 1995. A molecular basis for cardiac arrhythmia: HERG mutations cause long-QT syndrome. *Cell* 80: 795–803.

Devlin, B. and N. Risch. 1995. A comparison of linkage disequilibrium measures for fine-scale mapping. *Genomics* 29: 311–322.

Devlin, B. and K. Roeder. 1999. Genomic control for association studies. *Biometrics* 55: 997–1004.

Doebley, J., A. Stec and L. Hubbard. 1997. The evolution of apical dominance in maize. *Nature* 386: 485–488.

Eberle, M. A. and L. Kruglyak. 2000. An analysis for discovery of single-nucleotide polymorphisms. *Genet. Epidemiol.* 19 (Suppl 1): S29–S35.

Excoffier, L. and M. Slatkin. 1995. Maximum-likelihood estimation of molecular haplotype frequencies in a diploid population. *Mol. Biol. Evol.* 12: 921–927.

Falconer, D. S. and T. F. C. Mackay. 1996. *Introduction to Quantitative Genetics*, 4th Ed. Longman, Essex, England.

Fallin, D. and N. Schork. 2000. Accuracy of haplotype frequency estimation for biallelic loci, via the expectation-maximization algorithm for unphased diploid genotype data. *Am. J. Hum. Genet.* 67: 947–959.

Frary, A. et al. 2000. *fw2.2*: A QTL key to the evolution of tomato fruit size. *Science* 289: 85–88.

Frayling, T. M. et al. 2007. A common variant in the *FTO* gene is associated with body mass index and predisposes to childhood and adult obesity. *Science* 316: 889–894.

Garrigan, D. and M. F. Hammer. 2006. Reconstructing human origins in the genomic era. *Nat. Rev. Genet.* 7: 669–680.

Graur, D. and W.-H. Li. 2000. *Fundamentals of Molecular Evolution*. Sinauer Associates, Sunderland, MA.

Gudbjartsson, D. F. et al. 2008. Many sequence variants affecting diversity of adult human height. *Nat. Genet.* 40: 609–615.

Gunderson, K. L., F. Steemers, G. Lee, L. Mendoza and M. S. Chee. 2005. A genome-wide scalable SNP genotyping assay using microarray technology. *Nat. Genet.* 37: 549–554.

Hartl, D. and A. Clark. 2007. *Principles of Population Genetics*, 4th ed. Sinauer Associates, Sunderland, MA.

Hedrick, P. W. 1985. *Genetics of Populations*. Jones and Bartlett, Boston.

Hill, W. G. and A. Robertson. 1968. Linkage disequilibrium in finite populations. *Theoret. Appl. Genet.* 38: 226–231.

Hirschhorn J. N. and M. J. Daly. 2005. Genome-wide association studies for common diseases and complex traits. *Nat. Rev. Genet.* 6: 95–108.

Holland, P., R. et al. 1991. Detection of specific polymerase chain reaction product by utilizing the 5′–3′ exonuclease activity of *Thermus aquaticus* DNA polymerase. *Proc. Natl Acad. Sci. (USA)*. 88: 7276–7280.

Horikawa, Y. et al. 2000. Genetic variation in the gene encoding calpain-10 is associated with type 2 diabetes mellitus. *Nat. Genet.* 26: 163–175.

Ioannidis, J., E. Ntzani, T. Trikalinos and D. Contopoulos-Ioannidis. 2001. Replication validity of genetic association studies. *Nat. Genet.* 29: 306–309.

Jurinke, C. et al. 2001. Automated genotyping using the DNA MassArray technology. *Meth. Mol. Biol.* 170: 103–116.

Kreitman, M. 2000. Methods to detect selection in populations with applications to the human. *Annu. Rev. Genomics Hum. Genet.* 1: 539–559.

Laird, N. M. and C. Lange. 2008. Family-based methods for linkage and association analysis. *Adv. Genet.* 60: 219–252.

Lander, E. and D. Botstein. 1987. Homozygosity mapping: A way to map human recessive traits with the DNA of inbred children. *Science* 236: 1567–1570.

Lander, E. and D. Botstein. 1989. Mapping Mendelian factors underlying quantitative traits using RFLP linkage maps. *Genetics* 121: 185–199.

Levy, S. et al. 2007. The diploid genome sequence of an individual human. *PLoS Biol.* 5: e254.

Lewontin, R. C. 1995. The detection of linkage disequilibrium in molecular sequence data. *Genetics* 140: 377–388.

Lewontin, R. C. and K. Kojima. 1960. The evolutionary dynamics of complex polymorphisms. *Evolution* 14: 450–472.

Long, A. and C. Langley. 1999. The power of association studies to detect the contribution of candidate genetic loci to variation in complex traits. *Genome Res.* 9: 720–731.

Long, A. D., S. Mulleney, T. F. C. Mackay and C. H. Langley. 1996. Genetic interactions between naturally occurring alleles at quantitative trait loci and mutant alleles at candidate loci affecting bristle number in *Drosophila melanogaster. Genetics* 144: 1497–1510.

Lynch, M. and B. Walsh. 1998. *Genetics and Analysis of Quantitative Traits.* Sinauer Associates, Sunderland, MA

McCarroll, S. A. et al. 2008. Integrated detection and population-genetic analysis of SNPs and copy number variation. *Nat. Genet.* 40: 1166–1174.

McCarthy, M. I. et al. 2008. Genome-wide association studies for complex traits: consensus, uncertainty and challenges. *Nat. Rev. Genet.* 9: 356–369.

Mir, K. and E. Southern. 2000. Sequence variation in genes and genomic DNA: Methods for large-scale analysis. *Annu. Rev. Genomics Hum. Genet.* 1: 329–360.

Neff, M. M., J. D. Neff, J. Chory and A. E. Pepper. 1998. dCAPS, a simple technique for the genetic analysis of single nucleotide polymorphisms: experimental applications in *Arabidopsis thaliana* genetics. *Plant J.* 14: 387–392.

Nickerson, D. A. et al. 1998. DNA sequence diversity in a 9. 7-kb region of the human lipoprotein lipase gene. *Nat. Genet.* 19: 233–240.

Nielsen, D. and B. S. Weir. 1999. A classical setting for associations between markers and loci affecting quantitative traits. *Genetical Res.* 74: 271–277.

Papassotiropoulos, A. et al. 2006. Common *Kibra* alleles are associated with human memory performance. *Science* 314: 475–478.

Parker, H. G. et al. 2004. Genetic structure of the purebred domestic dog. *Science* 304: 1160–1164.

Pastinen, T., A. Kurg, A. Metspalu, L. Peltonen and A.-C. Syvänen. 1997. Minisequencing: A specific tool for DNA analysis and diagnostics on oligonucleotide arrays. *Genome Res.* 7: 606–614.

Pastinen, T., M. Raitio, K. Lindroos, P. Tainola, L. Peltonen and A.-C. Syvänen. 2000. A system for specific, high-throughput genotyping by allele-specific primer extension on microarrays. *Genome Res.* 10: 1031–1042.

Price A. L., N. J. Patterson, R. Plange, M. Weinblatt, N. Shadick and D. Reich. 2006. Principal components analysis corrects for stratification in genome-wide association studies. *Nat. Genet.* 38: 904–909.

Pritchard, J., M. Stephens, and P. Donnelly. 2000. Inference of population structure using multilocus genotype data. *Genetics* 155: 945–959.

Pritchard, J., M. Stephens, N. Rosenberg and P. Donnelly. 2000. Association mapping in structured populations. *Am. J. Hum. Genet.* 67: 170–181.

Reich, D. E. et al. 2001. Linkage disequilibrium in the human genome. *Nature* 411: 199–204.

Risch, N. and K. Merikangas. 1996. The future of genetic studies of complex human diseases. *Science* 273: 1516–1517.

Rockman, M. V., M. W. Hahn, N. Soranzo, D. B. Goldstein and G. A. Wray. 2003. Positive selection on a human-specific transcription factor binding site regulating *IL4* expression. *Curr. Biol.* 13: 2118–2123.

Ronaghi, M. 2001. Pyrosequencing sheds light on DNA sequencing. *Genome Res.* 11: 3–11.

Rosenberg, N., J. K. Pritchard, J. L. Weber, H. M. Cann, K. K. Kidd, L. A. Zhivotovsky and M. W.

Feldman. 2002. Genetic structure of human populations. *Science* 298: 2381–2385.

Sasieni, P. D. 1997. From genotypes to genes: Doubling the sample size. *Biometrics* 53: 1253–1261.

Shen, R. et al. 2005. High–throughput SNP genotyping on universal bead arrays. *Mutat. Res.* 573: 70–82.

Spielman, R. and W. J. Ewens. 1998. A sibship test for linkage in the presence of association: The sib transmission/disequilibrium test. *Am. J. Hum. Genet.* 62: 450–458.

Spielman, R., R. McGinnis and W. Ewens. 1993. Transmission test for linkage disequilibrium: The insulin gene region and insulin-dependent diabetes mellitus (IDDM). *Am. J. Hum. Genet.* 52: 506–516.

Steemers, F. J., W. Chang, G. Lee, D. Barker, R. Shen and K. L. Gunderson. 2006. Whole-genome genotyping with the single-base extension assay. *Nat. Methods* 3: 31–33.

Stephens, C. A. et al. 2001. Haplotype variation and linkage disequilibrium in 313 human genes. *Science* 293: 489–493.

Sulem, P. et al. 2007. Genetic determinants of hair, eye and skin pigmentation in Europeans. *Nat. Genet.* 39: 1443–1452.

Tobler, A. et al. 2005. The SNPlex genotyping system: A flexible and scalable platform for SNP genotyping. *J. Biomol. Tech.* 16: 398–406.

Vogel, T., D. Evans, J. Urvater, D. O'Connor, A. Hughes and D. Watkins. 1999. Major histocompatibility complex class I genes in primates: Co-evolution with pathogens. *Immunol. Rev.* 167: 327–337.

Wang, D. G. et al. 1998. Large-scale identification, mapping, and genotyping of single-nucleotide polymorphisms in the human genome. *Science* 280: 1077–1082.

Weedon, M. N. et al. 2008. Genome-wide association analysis identifies 20 loci that influence adult height. *Nat. Genet.* 40: 489–490.

Weir, B. S 1996 *Genetic Data Analysis II*. Sinauer Associates, Inc. Sunderland, MA.

Wicks, S., R. Yeh, W. Gish, R. Waterston and R. Plasterk. 2001. Rapid gene mapping in *Caenorhabditis elegans* using a high-density polymorphism map. *Nat. Genet.* 28: 160–164.

WTCCC (Wellcome Trust Case Control Consortium). 2007. Genome-wide association study of 14,000 cases of seven common diseases and 3,000 shared controls. *Nature* 447: 661–678.

Zeng, Z.-B. 1994. Precision mapping of quantitative trait loci. *Genetics* 136:1457–1468.

Zondervan, K. T. and L. R. Cardon. 2004. The complex interplay among factors that influence allelic association. *Nat. Rev. Genet.* 5: 89–100.

4 Gene Expression and the Transcriptome

After genome sequencing and annotation, the next major branch of genome science is analysis of the transcriptome, namely documenting gene expression on a genome-wide scale. The **transcriptome** is the complete set of transcripts and their relative levels of expression in a particular cell or tissue type under defined conditions. Several technologies have been developed for parallel analysis of the expression of thousands of genes, among which *cDNA microarrays* and *oligonucleotide arrays* are the best known. These methods are most suitable for contrasting expression levels across tissues and treatments of a chosen subset of the genome, but they do not provide data on the absolute levels of expression. A third method, called *serial analysis of gene expression* (SAGE), relies on counting of sequence tags to estimate absolute transcript levels, but is less suited to experimentation with large numbers of samples. Next-generation sequencing of cDNA preparations is also beginning to be used to profile absolute transcript abundance. In addition to describing these three methods, this chapter also discusses techniques for verifying differential gene expression on a gene-by-gene basis and describes some of the applications of comparative expression analysis. Since transcription is only one level of gene regulation, transcript levels do not necessarily translate into protein expression or activity. Methods for characterization of the proteome—the structure and expression of the proteins encoded in the genome—are described in Chapter 5.

Parallel Analysis of Gene Expression: Microarrays

The basic procedure of microarray analysis is to deposit a very small amount of DNA corresponding to each one of a collection of thousands of genes (the "probes") onto a 1- or 2-centimeter-square array surface, and then to interrogate these probes by hybridization to a "target" of mRNA that has been labeled with a fluorescent dye. The amount of target that sticks to each spot of probe is proportional to the abundance of the transcript in the sample, and is detected as the intensity of fluorescent signal. A change in abundance is measured as an increase or decrease in the signal, relative either to the signal from a control reference sample (a "ratio") or to the signals from the other probes on the array (a "relative intensity"). Estimates of the absolute abundance of each transcript in each sample are only approximate, so most inferences involve comparison of expression levels across treatments.

Applications of Microarray Technology

Gene expression profiling is much more than simply a method for rapidly finding genes that are differentially used under particular circumstances. Used properly, microarrays are a powerful tool for generating and testing hypotheses, classifying specimens, annotating genes, studying developmental and evolutionary processes, and performing clinical assessments. There are four basic steps in all microarray analyses: experimental design, technical performance, statistical analysis, and data mining (Figure 4.1). The first of these steps is arguably the most important. The starting point for any gene expression profiling experiment should always be the precise formulation of a biological question that will guide the implementation of an efficient and informative experimental design. Consultation with a statistician on issues such as the levels of replication that will be required to detect an effect or perform a contrast, and how to lay out the hybridizations, is always advisable, since the way an experiment is performed can have a large effect on statistical power and hence the conclusions that are reached.

There are at least half a dozen different common applications of microarray methodology, some of which are explored in the context of case studies at the end of this chapter. **Detection of candidate genes** remains a major objective of many studies and is based on the supposition that if a gene is transcribed under one set of conditions but not another, then its expression is likely to be required for, or at least to contribute to, any biological differences between the two conditions. Examples include implicating genes in disease states such as cancer or adverse immune responses; finding target genes for a regulatory factor; and describing the genetic response to drug exposure. An analogous usage is in **annotation of gene function**, generally following the principle of guilt by association. In any list of differentially expressed genes, there are likely to be several that encode proteins with unknown functions, and such genes can be tentatively ascribed a molecular or cellular function based on their pattern of shared regulation with genes whose function is known.

Microarrays are also used to dissect genetic mechanisms in a hypothesis-driven manner. An important application is **definition of genetic pathways** by a combination of precise temporal profiling and focused bioinformatic analysis. Time-series can be used to order the activation and repression of transcription, suggesting which genes regulate the expression of other genes. Inclusion of mutants and experimental manipulations in the analysis allows

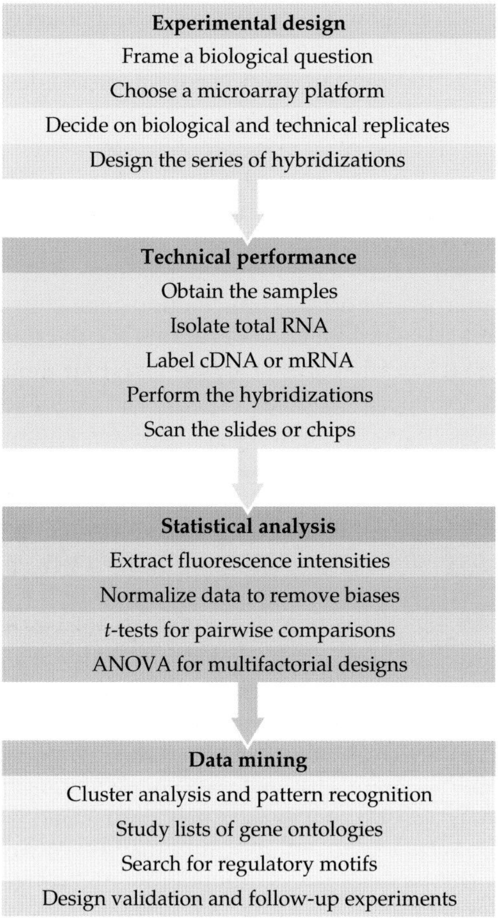

Experimental design

Frame a biological question

Choose a microarray platform

Decide on biological and technical replicates

Design the series of hybridizations

Technical performance

Obtain the samples

Isolate total RNA

Label cDNA or mRNA

Perform the hybridizations

Scan the slides or chips

Statistical analysis

Extract fluorescence intensities

Normalize data to remove biases

t-tests for pairwise comparisons

ANOVA for multifactorial designs

Data mining

Cluster analysis and pattern recognition

Study lists of gene ontologies

Search for regulatory motifs

Design validation and follow-up experiments

Figure 4.1 Flow diagram for gene expression profiling using microarrays.

causation to be distinguished from correlation. Investigation of the distribution of conserved sequence motifs in putative control regions of genes can lead to **dissection of regulatory mechanisms** that link pathways and networks of genes. In Chapter 6, we will discuss how the systems biology approach combines microarray analyses with proteomics, functional genomics, and metabolic profiling to arrive at a more complete description of complex cell biological processes.

Quantitative methodology has a central place in transcriptomics, in both basic research and translation into applications in agriculture and medicine. **Quantification of transcriptional variance** is an essential tool in evolutionary and quantitative genetics, allowing investigators to partition the differences among individuals, populations, and species under a variety of different environmental conditions. Rather than simply describing how a treatment affects a tissue, inclusion of individual variation in the experimental design enables assessment of how individuals respond differently to a treatment. Animal and plant breeders can attempt to select for specific phenotypes associated with this variation at the transcriptional level. Similarly, clinical applications of **molecular phenotyping** are being developed with regard for example to prognosis of cancer progression, classification of infection or toxin exposure, and prediction of pharmacological responses.

Experimental Design

The first experimental design decision is which microarray platform to use. As described below, there are currently three major types of platform: cDNA and long-oligonucleotide microarrays, short-oligonucleotide gene chips, and long oligonucleotide bead arrays. Major considerations involved in selecting a platform include cost (from $50 per homemade cDNA array to over $300 for some commercial whole-genome oligonucleotide arrays), coverage (from several hundred genes to multiple probes for over 20,000 annotated genes), availability (from having to make your own to having numerous options for model organisms), and quality (the relatively high cost of commercial arrays is generally offset by their very high repeatability and uniformity). There are numerous options for custom synthesis of microarrays for new organisms, or for representing a user-defined fraction of the transcriptome, all of which means that microarray technology is accessible to most biologists.

The second essential consideration is level of replication. It is generally necessary to choose a balance between available funding and experimental objectives. In the early days of microarray analysis, it was rare to see more than two replicates of any sample in an experiment, and most inferences were based on large (greater than twofold) changes in expression observed in multiple samples. It has since been recognized that subtle differences—down to 1.2-fold or even less—can be detected with moderate replication levels of 6–10 hybridizations per sample.

In addition, a fundamental distinction between technical and biological replicates has been drawn (Churchill 2002). Technical replicates are repeated samples of the same biological material, for example RNA preparations, labeling reactions, hybridizations, and duplicate spots of the same probe on each array. Biological replicates are independent samples of similar material, for example different individuals with the same genotype, leaves of a plant, trials of a treatment, or populations of a species. Because of cost and practicality, it is rarely possible to include all levels of replication in an experiment, so choices must be made based on which contrasts are of most interest. In some cases, pooling of samples allows the technical or biological source of variation to be included in the analysis without being measured explicitly. Examples include performance of two independent labeling reactions that are combined prior to hybridization in order to overcome stochastic noise in the labeling efficiency; and mixing a sample of individuals to gain a better representation of the response in a population.

The third major design issue is which samples to contrast on which arrays. For short-oligonucleotide Affymetrix gene chips and Illumina bead arrays, this is not an issue because only one sample is hybridized to each array. How-

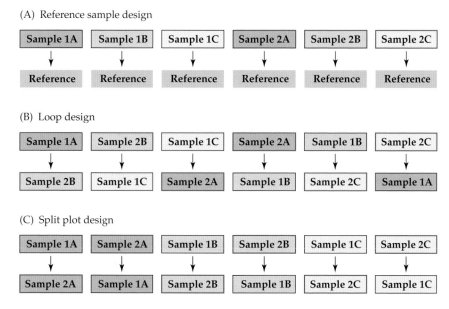

Figure 4.2 Common experimental designs for microarray analysis.

ever, most other microarrays involve competitive hybridizations of two samples labeled with two different colored dyes, and in these cases there are three basic types of experimental design to choose from, as shown in Figure 4.2:

- **Reference sample designs** contrast each experimental sample against a common reference sample, which is generally designed to include an average or intermediate level of transcript for every gene on the microarray. Though popular, reference sample designs waste expensive materials and generally are less statistically powerful than designs that employ more replicates of just the experimental samples.

- **Loop designs** are optimal where there are multiple biological samples of one or two treatment classes. A simple approach with two replicates of each sample is to contrast A → B, B → C, C → D, … G → A, where the tail of the arrow represents one dye and the head the other. In such a design, each sample is "balanced" with respect to the number of labeling reactions with each dye. Complete loop designs utilize multiple replicates so that each sample is contrasted at least once with each other sample.

- An alternative approach modeled after classical agricultural field trials is the **split-plot** design involving multiple factors (for example sex, drug, and population) in which the number of contrasts within and between factors is varied based on considerations of statistical power (Jin et al. 2001). The more direct contrasts there are, the more power there is for that contrast, but it is still possible to draw inferences involving contrasts that are not made directly.

EXERCISE 4.1 *Design a microarray experiment*

Suppose that you are setting up an experiment designed to see how gene expression in the liver of four strains of mice (A, B, C, and D) is affected by treatment with a particular drug, with the complication that you suspect that the two sexes respond differently. You have enough money to perform 16 two-color microarrays. Describe how you would set up the experiment in three ways: (a) using a reference sample; (b) maximizing your ability to find significant genes for each drug for a particular sex and strain combination; (c) maximizing your ability to detect interactions between the treatments by contrasting female against male, drug against control, and strain against strain as much as possible.

ANSWER: *There are 16 different samples to be contrasted: 4 strains × 2 treatments (drug vs. control) × 2 sexes (male vs. female). For design (a) it is simply a matter of setting up one array with each sample hybridized against the reference (which might, for example, be a mixture of each of the 16 experimental samples). Most often the reference is labeled consistently with just one dye. None of the samples are replicated.*

For design (b) you would be well advised to use a split-plot strategy, in which case you will be able to have four replicates of each sample, in eight mini-experiments as follows:

	Red dye	**Green dye**
Array 1:	*Female A, Drug*	*Female A, Control*
Array 2:	*Female A, Control*	*Female A, Drug*
Array 3:	*Female B, Drug*	*Female B, Control*
Array 4:	*Female B, Control*	*Female B, Drug*
Array 5:	*Male A, Drug*	*Male A, Control*
Array 6:	*Male A, Control*	*Male A, Drug*
Array 7:	*Male B, Drug*	*Male B, Control*
Array 8:	*Male B, Control*	*Male B, Drug*
Array 9:	*Female C, Drug*	*Female C, Control*
Array 10:	*Female C, Control*	*Female C, Drug*
Array 11:	*Female D, Drug*	*Female D, Control*
Array 12:	*Female D, Control*	*Female D, Drug*
Array 13:	*Male C, Drug*	*Male C, Control*
Array 14:	*Male C, Control*	*Male C, Drug*
Array 15:	*Male D, Drug*	*Male D, Control*
Array 16:	*Male D, Control*	*Male D, Drug*

Note that each pair of arrays involves the same set of treatments, in replicate, with the dyes flipped. The figure illustrates that every array is male against male (squares) or female against female (circles) for each strain, and every array contrasts drug (shaded) against control (unshaded).

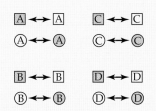

For design (c) you would choose a loop strategy in which each possible combination of strains is contrasted against each other, while all arrays contrast drug against control and male against female (see the figure at the end of this exercise). An example setup follows, but note that with three treatments it is not always possible to completely balance the design, so in this case arrays 8 and 16 are same-sex contrasts:

	Red dye	**Green dye**
Array 1:	*A, Drug, Female*	*A, Control, Male*
Array 2:	*A, Control, Male*	*B, Drug, Female*
Array 3:	*B, Drug, Female*	*D, Control, Male*
Array 4:	*D, Control, Male*	*C, Drug, Female*
Array 5:	*C, Drug, Female*	*C, Control, Male*
Array 6:	*C, Control, Male*	*D, Drug, Female*
Array 7:	*D, Drug, Female*	*B, Control, Male*
Array 8:	*B, Control, Male*	*A, Drug, Male*
Array 9:	*A, Drug, Male*	*D, Control, Female*
Array 10:	*D, Control, Female*	*D, Drug, Male*
Array 11:	*D, Drug, Male*	*A, Control, Female*
Array 12:	*A, Control, Female*	*C, Drug, Male*
Array 13:	*C, Drug, Male*	*B, Control, Female*
Array 14:	*B, Control, Female*	*B, Drug, Male*
Array 15:	*B, Drug, Male*	*C, Control, Female*
Array 16:	*C, Control, Female*	*A, Drug, Female*

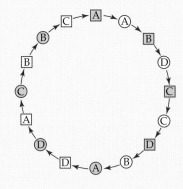

A fourth approach is not to adopt any specific design. This usually happens either because of lack of attention during planning or because an experiment grew or changed unexpectedly after it was commenced. It is generally possible to extract information from any design, but statistical power is strongly influenced by the way an experiment is set up.

Microarray Technologies

cDNA microarrays. The technology of cDNA microarrays provides a conceptually simple approach to monitoring the relative levels of expression of thousands of genes simultaneously (Schena et al. 1995). PCR-amplified cDNA fragments (ESTs) are spotted at high density (10–50 spots per mm^2) onto a microscope slide or filter paper, and probed against fluorescently or radioactively labeled target (Figure 4.3). The platform is flexible, since users print their own microarrays and are free to add or subtract clones as they refine their collection of ESTs. In recent years, similar genomic DNA microarrays have also been introduced, using PCR products generated with gene-specific primers that amplify predicted genes or gene segments from genomic DNA (Gilad et al. 2005).

The choice of which clones to array is influenced by the level of annotation of the cDNA library from which the clones were extracted and by the aims of the experiment. Ideally, each EST should be sequenced and should represent a unique gene or alternative splice variant, in which case the collection is called a **unigene set**. Because cDNA libraries are tissue-specific, multiple libraries must be sampled to produce an unbiased set of clones that is likely to be representative of the genes expressed under different conditions. It is also possible to generate a cDNA microarray simply from randomly chosen, unsequenced clones. This leads to overrepresentation of a small subset of highly expressed genes on the microarray; however, molecular subtraction techniques can be used to reduce redundancy. It is common practice to re-sequence the clones of interest at the end of an experiment to confirm their identity and ensure that no mistakes were made in the printing of the array.

For many organisms, the first generation of unigene sets was assembled by identifying unique clones in EST databases. Multigene families may be represented by several different clones, but in these cases there is some potential for cross-hybridization to occur, meaning that the same probe will be recognized by transcripts from different genes. This problem can be avoided to some degree by choosing probes that correspond to a nonconserved portion of the gene. Since cDNA microarrays typically have an upper limit of 15,000 elements (and often include fewer than 5,000 elements), they are unable to represent the complete set of genes present in higher eukaryotic genomes. Consequently, arrays have been developed that are specific for a certain developmental stage or tissue (for example gastrula or leaf), or class of gene (transcription factor or receptor).

The most expensive and time-consuming step in cDNA microarray analysis is the amplification of the EST set. While each spot on a microarray should

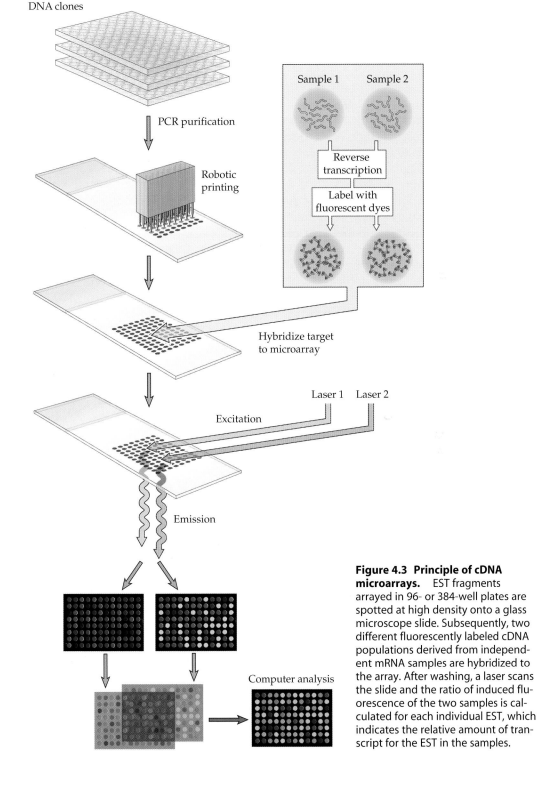

Figure 4.3 Principle of cDNA microarrays. EST fragments arrayed in 96- or 384-well plates are spotted at high density onto a glass microscope slide. Subsequently, two different fluorescently labeled cDNA populations derived from independent mRNA samples are hybridized to the array. After washing, a laser scans the slide and the ratio of induced fluorescence of the two samples is calculated for each individual EST, which indicates the relative amount of transcript for the EST in the samples.

contain just 10 ng of DNA, it is delivered as a few nanoliters of 5 mg/ml DNA solution. PCR-amplified fragments have been found to yield much stronger signals than concentrated plasmid DNA. PCR amplification is performed in 96- or 384-well plate format, typically in 100 μl volumes, and the products must be precipitated and re-suspended in a special spotting solution. Variation in clone length complicates amplification, and presumably affects signal strength, so clones in the 1–2 kb range are used preferentially.

Once amplified, the fragments may be spotted onto either a coated glass microscope slide or a nitrocellulose or nylon membrane. Membranes are most suited to applications where radioactivity is used to label the cDNA, while glass only supports fluorescence-based detection. Slides can be prepared in-house by coating them with a polylysine solution, but the uniformity of this surface is not guaranteed and a number of commercial suppliers have developed alternative coatings. Most solid support surfaces bond covalently with the sugar-phosphate backbone of spotted DNA after cross-linking with ultraviolet radiation at the end of the printing step.

Printing is performed with a robot that picks up samples of DNA from a 384-well microtiter plate and deposits aliquots sequentially onto a field of 100 or so slides. The spacing between spot centers is specified from 120–250 μm according to the density required (thus the robot must be accurate within < 20 μm with respect to where it deposits each spot). The entire microarray usually covers an area 2.5 × 2.5 cm, though longer grids can be printed when more clones are to be represented.

Commonly used print heads have from 4 to 32 individual printing pins spaced approximately 1 cm apart. The pins pick up DNA solution from nearby wells on the master plate. At a rate of a few deposits per second, it takes up to two days for the robots to prepare 100 microarrays containing at least 5,000 clones. Since only a few microliters of PCR product are used in each print run, each amplified batch of clones is sufficient for multiple print runs and hence at least 1,000 slides.

Several types of printing pin are in use. Most rely on capillary action to transfer solution from the pin to the slide. This method has the virtue that spot size can be controlled somewhat by adjusting the amount of time the pin touches the slide; with spot diameters close to 50 μm, extremely high-density microarrays can be produced. The pins are expensive and fragile, sometimes have a tendency to deliver doughnut-shaped spots (as the DNA spreads away from the tip), and can be prone to clogging. Ink-jet printers produce highly reproducible and efficient spot deposition, but are less popular in academic settings.

Long-oligonucleotide microarrays. Long oligonucleotides in the range of 50 to 70 bases have gained in popularity as probes since the late 1990s. The two principal differences relative to cDNA arrays are that the user is spared the step of having to use PCR to amplify thousands of clones (the oligonucleotides are ordered from a commercial firm); and a single strand of DNA of uniform length is deposited, affording greater control over hybridization specificity. Once the strand of interest has been identified in a cDNA or genomic sequence, freely available algorithms (for example, OligoArray; Rouillard et al. 2002) can be used to design long oligonucleotides with the desired base composition that promotes hybridization at the same temperature for all probes. Checks are also performed to eliminate the potential for cross-hybridization, and in general over 90% of genes yield suitable probes.

Spotting is performed with the same types of robotic arrayers described above. A concern is that if the probe DNA cross-links to the glass through its backbone, then only 20–30 bases of each individual molecule are likely to be exposed to the labeled target, which may be too short to ensure specific hybridization. Modified 5′ amino groups have been used to promote linkage of the oligonucleotides to an aldehyde on coated slides, so that the whole molecule is exposed.

Alternatively, the oligonucleotides can be synthesized *in situ* on the slide surface (Hughes et al. 2001). **Agilent Technologies** will synthesize arrays containing up to 244,000 long oligonucleotides (60-mers) using a combination of ink-jet printing to deliver the appropriate nucleotide in 60 consecutive steps and phosphoramidite chemistry to ensure the incorporation of a single base in each step. Agilent has commercial arrays for subsets of the human genome and for several model organisms, as well as unique products for interrogating microRNA expression and for various applications involving regulatory and intergenic DNA. Users may also provide a list of gene accession numbers that can be turned into a custom microarray, and options are available for generating "boutique" arrays with 11,000 probes, or whole genome arrays with 22,000, 44,000, or more probes. Hybridizations are performed using essentially the same methods as for cDNA microarrays.

Illumina produces high-density 50-mer oligonucleotide arrays in which the probes are linked to 3-μm diameter beads by way of a linker with a unique DNA-encoded address (Figure 4.4). The beads are distributed over

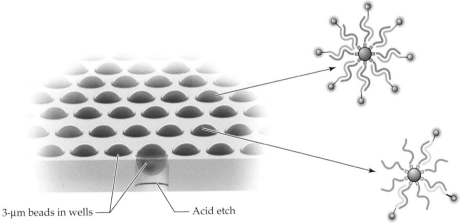

3-μm beads in wells —   — Acid etch

Figure 4.4 Illumina bead arrays. Tens of thousands of beads are distributed in etched wells of a silicon chip or optical fiber bundle, allowing the intensity of fluorescence due to hybridization of labeled cDNA to the long oligonucleotides on the bead to be read from each well. Each bead is pre-coded with the identity of the oligonucleotide according to the sequence of a linker (not shown).

an etched surface of either a silicon wafer or a bundle of optical fibers, where they are held in place by hydrostatic forces as solutions are washed over them during the hybridization steps. The addresses are read by the manufacturer before being shipped to customers, using a complicated series of hybridizations with a set of fluorescent dies (Gunderson et al. 2004). Each probe is represented by an average of 30 beads, providing technical replication and allowing precise estimation of transcript abundance. This is a one-color system, so there is no reference sample and the only major experimental design choices involve the number of samples and replicates to include in a study. The HumanRef8 and HT12 arrays provide coverage of 18,630 and 25,440 annotated genes respectively. The company also offers a slightly different "veracode" technology that allows for design of custom arrays with just hundreds of probes.

Short-oligonucleotide microarrays. The third general approach to parallel analysis of gene expression is the use of short-oligonucleotide microarrays, originally developed and marketed under the trademark Affymetrix GeneChip (Figure 4.5; Lockhart et al. 1996; Lipschutz et al. 1999). The unit of hybridization here is a series of 25-mer oligonucleotides designed by a computer algorithm to represent known or predicted open reading frames. Each gene is represented by a "probeset" of up to 22 different oligonucleotides to control for variation in hybridization efficiency due to factors such as GC content; this variation can be exacerbated due to the short length of the probes. The possibility of cross-hybridization with similar short sequences in transcripts other than the one being probed is controlled for by including a mismatch control adjacent to each oligonucleotide that has a single base change at the center of the oligonucleotide; this control should not hybridize under high-stringency conditions. However, in practice, perfect and mismatch intensities are often highly correlated, so many users ignore the mismatch data in their analyses.

The level of expression of each gene is calculated as a summary of the probe level data using automated procedures provided by Affymetrix with their Expression Console software. Data is generated from pixel intensities on the scanned chip as a .cel file, where each probe is associated with a measure of intensity. It is most simply output as a .chp file where each probeset is associated with a measure of gene expression. A third file type, the .cdf, is required to decode the position of each probe into its identity and probeset membership. The MAS5 and PLIER algorithms generate filtered "averages" of the difference between perfect match and mismatch probes, making adjustments for the fact that sometimes mismatch intensities are greater than perfect match ones, and dampening outlier effects. The Robust Multichip Analysis (RMA) algorithm (Irizarry et al. 2003) relies solely on the perfect-match probes and performs statistical normalization to control for background intensity. Several other statistical methods that take into account the variance among probes within a probeset are also in wide use (Gautier et al. 2004), and can be implemented using the R/Bioconductor open source software environment described later in this chapter.

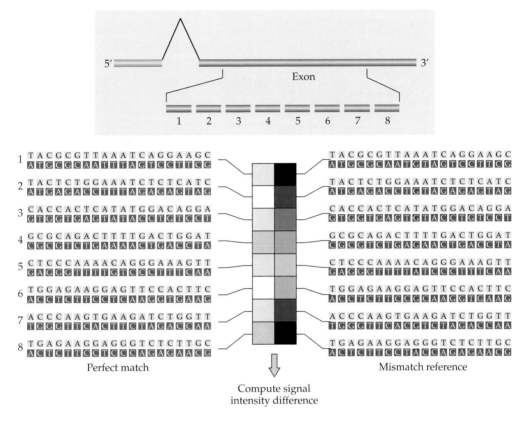

Figure 4.5 Principle of short oligonucleotide arrays. Up to twenty 20–25-mer oligonucleotide sequences are designed from an exon and printed on a chip adjacent to a mismatch oligonucleotide with a single base change (red) at the central position. When labeled RNA is hybridized to the chip, each oligonucleotide will produce a slightly different signal intensity. These are processed in conjunction with mismatch data to calculate a value for the expression level of the transcript, based approximately on the average difference between the perfect match and mismatch intensities.

High-density short-oligonucleotide arrays are constructed on a silicon chip by photolithography and combinatorial chemistry (Figure 4.6; McGall et al. 1996). For a 25-mer, 100 sequential nucleotide-addition reactions are performed across the surface of the chip in 25 cycles of A, T, G, and C. In each cycle, a localized flash of light "deprotects" the growing nucleotide chain just on that portion of the chip where the next nucleotide should be added. When a solution containing the nucleotide is added, a single nucleotide adds on to each deprotected chain, after which the remaining free nucleotides are washed off before the next cycle begins. The localization of the light is achieved by inserting a mask between the light and the chip, using technology developed by the microprocessor industry.

(A)

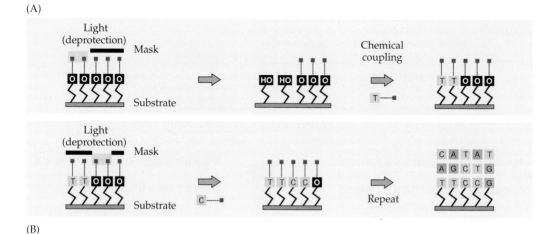

(B)

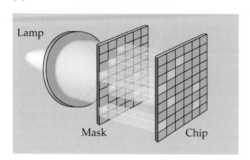

Figure 4.6 Construction of Affymetrix oligonucleotide arrays.   Oligonucleotides are synthesized in situ on the silicon chip, as described in the text. (A) In each step, a flash of light "deprotects" the oligonucleotides at the desired location on the chip; then "protected" nucleotides of one of the four types (A, C, G, or T) are added so that a single nucleotide can add to the desired chains. The light flash (B) is produced by photolithography, using a mask to allow light to strike only the required features on the surface of the chip.

Another company, **Nimblegen**, also produces short oligonucleotide arrays on glass slides but uses a maskless array synthesis method, employing miniature mirrors to focus a laser beam on each spot as the oligonucleotides are built. With either approach, several million oligonucleotides with their mismatch controls can be rapidly synthesized on thousands of identical chips, with extraordinarily high-quality reproducibility. The same technology is used to synthesize chips for sequencing by hybridization, as described in Chapter 2.

Both oligonucleotide and cDNA microarray technologies have their advantages and disadvantages. In brief, oligonucleotide arrays can accommodate higher densities of genes, including predicted genes not represented in cDNA libraries; have lower variability from chip to chip; incorporate mismatch controls; can be used by researchers without access to microarray construction facilities; and lend themselves to data comparison across research groups. Meanwhile, cDNA methods can be applied to any organism, irrespective of the status of genome sequencing efforts; and rely on hybridization over kilobases rather than tens of bases. This feature may reduce cross-hybridization artifacts and the effect of sequence polymorphism on hybridization.

Labeling and Hybridization of cDNAs

There are two common protocols for production of labeled target, both of which can be modified in several ways. These are direct labeling of cDNA prepared from mRNA by reverse transcription (Figure 4.7A), and linear amplification of labeled mRNA (Figure 4.7B). Total RNA—rather than the polyA fraction, which is generally less than 1% of total cellular RNA—is usually used as the template. If a radioactive label is used (^{33}P, ^{35}S or ^{3}H), it is incorporated directly on one of the nucleotides. More commonly, fluorescent dyes such as Cy3 and Cy5 conjugated to the base of a nucleotide are incorporated during the reverse transcription reaction. These dyes are bulky and do not incorporate efficiently, so an alternate approach is to use nucleotides with a small aminoallyl group that can be cross-linked to the

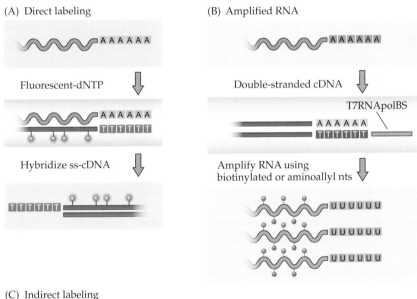

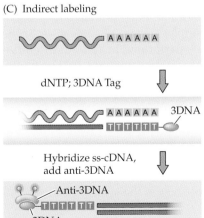

Figure 4.7 Three methods for labeling cDNA.
(A) Direct fluorescent dye incorporation during reverse transcription. (B) In aRNA synthesis, biotinylated, amplified RNA is produced by in vitro transcription from a T7 promoter at the 5′ end of the primer used in cDNA synthesis. The biotin is then recognized by a fluorescently labeled streptavidin-phycoerythrin compound on the array (not shown). This method is commonly used with Affymetrix gene chips. (C) Genisphere 3DNA submicro labeling, in which the 3D reagent is attached to an oligonucleotide that is complementary to the 5′ end of the primer used in cDNA synthesis, then hybridizes to it on the microarray.

dyes after the cDNA has been synthesized. Synthesis of cDNA is usually primed with polyT or with random hexamers, the latter being essential for bacteria, which do not have polyadenylated mRNAs.

Several micrograms of template RNA are required for direct incorporation protocols. This places a prohibitive constraint on experiments using biopsies, small organisms, or fractionated cell types. The linear amplification approach circumvents this problem by running off several copies of complementary RNA from each template using T7 RNA polymerase that initiates synthesis from an oligonucleotide attached to the polyT primer (Figure 4.7B). In this way, 100ng of template can be converted into micrograms of labeled aRNA, sufficient for a few hybridizations. Some protocols call for an extra step in which aRNA is reverse transcribed into more stable acDNA. As with direct labeling reactions, the dyes can be incorporated as part of the nucleotides, or attached later to an aminoallyl group. For work with Affymetrix GeneChips, biotinylated nucleotides are incorporated into the aRNA, and these are recognized after the hybridization reaction by streptavidin conjugated to fluorescent phycoerythrin.

A third labeling approach, shown in Figure 4.7C, is to use normal nucleotides throughout, but detect hybridized target through coupling of fluorophores to a substrate attached to the polyT primer. In the Genisphere 3DNA Submicro system, the sensitivity of detection can be controlled by adjusting the amount of fluorescent anti-3DNA reagent that recognizes this substrate.

Where radioactive detection is used, the same filter must be stripped and re-probed to provide second or multiple samples. Unlike glass microarrays, which are single-use, filters can be re-probed half a dozen times. However, there is a risk that washing is incomplete, or that it affects subsequent hybridizations and introduces the potential for experimental artifacts. While radioactivity is less expensive than fluorescent dyes, exposure to radiation can add up quickly, placing a working constraint on use of radioactive probes.

Hybridization is performed in special humidified chambers. Since the aim is to detect expression of specific transcripts, hybridization and washes are performed under highly stringent conditions that minimize any propensity for cross-hybridization between similar genes. Visualization of the hybridized target on a microarray can be performed by laser-induced fluorescence imaging using either a confocal detector or a CCD camera. Because fluorescence is quenched with time and exposure to light, errors at this step (due, for example, to improper setting of the laser—too much activation can saturate the signal, too little results in loss of signal) can have a profound effect on the conclusions reached. It has also been found that differential oxidation of fluorophors in the presence of high levels of atmospheric ozone can affect the ratio of signals from two dyes (Fare et al. 2003), suggesting that steps should be taken to control the laboratory environment, particularly in certain cities at particular times of the year.

Statistical Analysis of cDNA Microarray Data

Reference sample approaches. The standard approach to analyzing microarray data is to compute the ratio of fluorescence intensities, after appropriate transformation, for two samples that are competitively hybridized to the same microarray. One sample acts as a control, or "reference" sample, and is labeled with a dye (often Cy3) that has a different fluorescence spectrum from the dye (Cy5) used to label the experimental sample. In theory, three- and four-sample competitive hybridizations can be performed, but there are concerns that the fluorescence spectra of multiple dyes are nonoverlapping, so signals may be partially confounded.

Soon after the development of cDNA microarray technology, a convention emerged that twofold induction or repression of experimental samples relative to the reference sample was indicative of a meaningful change in gene expression. This convention does not reflect standard statistical definitions of significance (Cui and Churchill 2003). Nevertheless, genes that show such a response in two or more samples from a series of experiments, or a mean change above some different arbitrary threshold, are typically chosen for further analysis. This generally has the effect of selecting the top 5% or so of the clones present on the microarray. Adjustment of the fold-difference threshold results in a tightening or loosening of this fraction.

There are sound theoretical and experimental reasons for adopting ratios as the standard for comparison of gene expression (Eisen et al. 1998):

- Formulation of a ratio captures the central idea that it is a *change in relative level of expression* that is biologically interesting, emphasizing the fact that microarrays do not provide data on absolute expression levels. A corollary is that higher fluorescence intensity for one spot on the array does not necessarily mean that a particular gene is expressed at a higher level than genes that produce lesser fluorescence signals. This is because fluorescence intensity is a function of the length of the EST, the amount of label incorporated into the cDNA during reverse transcription, the efficiency of hybridization, and the concentration of DNA prepared for the particular clone, among other factors.

- Competitive hybridization removes variation among arrays from the analysis. If signal is compared from two different arrays, it is possible that a difference in fluorescence intensity is due not so much to differential gene expression as to differences between the arrays—such as the absolute amount of DNA spotted on the arrays, or local variation introduced either during slide preparation and washing or during image capture. As described below, such variation can be accounted for by replication, but where experiments involve large numbers of time points or other treatments, it remains common for investigators to perform just one or two hybridizations per treatment.

However, there are also pitfalls in analyzing simple ratios, and data must be transformed to remove obvious biases (see Figure 4.8 and Exercise 4.2).

The nature of the reference sample can also impact the analyses, since if transcript abundance is too high or too low relative to experimental samples, the power to discriminate differential expression is reduced. For this reason, many authors choose to use a pooled aliquot of all of their experimental samples as their reference, which should average representation of each transcript class. Alternatively, a standard reference that can be prepared easily by different groups facilitates more direct comparison of different experiments. A third option is to label organismal DNA or a pool of oligonucleotides uniformly representing each gene on the microarray.

Data normalization. All microarray experiments must be "normalized" to ensure that experimental biases are removed. The most prevalent bias is an overall difference in mean sample fluorescence intensity, either because inevitably slightly more mRNA is used as the template for one of the samples, or a difference in sensitivity or power is associated with detection of the fluorescence signal for one of the dyes. For this reason, raw ratios are usually centered to produce a mean ratio of one, after first log-transforming them on the base 2 scale (Figure 4.8). This procedure produces symmetry of relative increases and decreases in expression about zero: twofold induction (a ratio of 2.0) or repression (a ratio of 0.5) transform to values of +1 or –1, respectively. It also tends to ensure that the frequencies of the log ratios are closer to a normal distribution than are raw ratios, and this facilitates statistical analysis.

Recognizing that the logarithm of a ratio is equivalent to the difference in the logarithms (that is, $\log[A/B] = \log[A] - \log[B]$), it is also permissible to directly normalize each channel separately. Subtraction of the mean log fluorescence intensity for the channel from each value transforms the meas-

Red	Green	Difference	Ratio (G/R)	Log$_2$ Ratio	Centered R
16,500	15,104	−1,396	0.915	−0.128	−0.048
357	158	−199	0.443	−1.175	−1.095
8,250	8,025	−225	0.973	−0.039	0.040
978	836	−142	0.855	−0.226	−0.146
65	89	24	1.369	0.453	0.533
684	1,368	529	2.000	1.000	1.080
13,772	11,209	−2,563	0.814	−0.297	−0.217
856	731	−125	0.854	−0.228	−0.148

Figure 4.8 Simple normalization of microarray data.   The difference between the raw fluorescence intensities is a meaningless number. Computing ratios (green/red) allows immediate visualization of which genes are higher in the red channel than the green channel, and logarithmic transformation of this measure on the base 2 scale results in symmetric distribution of values around zero. Finally, normalization by subtraction of the mean log ratio adjusts for the fact that the red channel was generally more intense than the green channel.

urements such that the abundance of each transcript is represented essentially as a fold increase or decrease relative to the sample mean, namely as a relative fluorescence intensity. Some authors choose to take the further step of ensuring that the variance of each channel is also constant, meaning that the dispersion about the mean is the same for all hybridizations. This is done by dividing the relative fluorescent intensities by their standard deviation so that the log transformed values have a mean of zero and a variance of one. This preserves the rank order of expression and removes any systematic biases in the efficiency of labeling for classes of treatment, but may remove some true biological signal.

A different approach to normalization is to regress either the ratios or raw values onto a standard curve made of a set of control probes. These can be a series of spiked-in transcripts covering the full range of expression levels, from a different organism that do not hybridize to any of the probes of interest. Alternatively, the user can define a set of probes that they expect not to be differentially expressed across treatments, or choose a set of probes that empirically have invariant expression levels in the experiment. There is no single correct procedure for data normalization, and in practice it is a good idea to use several methods and compare the results.

EXERCISE 4.2 *Calculate which genes are differentially exposed*

Data for the following 10 genes are obtained from a microarray, where each number represents the fluorescence intensity for the Cy3 or Cy5 channel after background subtraction. Calculate which genes are at least twofold different in their abundance on this array using two different approaches: (a) by formulating the Cy3:Cy5 ratio, and (b) by calculating the difference in the log base 2 transformed values. In both cases, make sure that you adjust for any overall difference in intensity for the two dyes, and comment on whether this adjustment affects your conclusions.

	Cy3	Cy5
Gene 1	1,378	1,678
Gene 2	331	797
Gene 3	1,045	1,001
Gene 4	3,663	2,557
Gene 5	259	133
Gene 6	1,882	2,264
Gene 7	604	738
Gene 8	2,611	2,671
Gene 9	528	1,092
Gene 10	458	593

(Continued on next page)

EXERCISE 4.2 (continued)

ANSWER

	Ratios method				Log transformation method				
	Cy3	Cy5	Ratio	Adj. ratio	log_2 (Cy3)	log_2 (Cy5)	Adj. Cy3	Adj. Cy5	Differ.
Gene 1	1,378	1,687	0.817	0.856	10.43	10.72	0.62	0.71	−0.09
Gene 2	331	797	0.415	0.435	8.37	9.64	−1.43	−0.37	−1.07
Gene 3	1,045	1,001	1.044	1.094	10.03	9.97	0.23	−0.04	0.26
Gene 4	3,663	2,557	1.433	1.502	11.84	11.32	2.03	1.31	0.72
Gene 5	259	133	1.947	2.041	8.02	7.06	−1.79	−2.95	1.16
Gene 6	1,882	2,264	0.831	0.871	10.88	11.15	1.08	1.14	−0.06
Gene 7	604	738	0.818	0.858	9.24	9.53	−0.57	−0.48	−0.09
Gene 8	2,611	2,671	0.978	1.025	11.35	11.38	1.55	1.38	0.17
Gene 9	528	1,092	0.484	0.507	9.04	10.09	−0.76	0.09	−0.85
Gene 10	458	593	0.772	0.810	8.84	9.21	−0.96	−0.79	−0.17

Using the ratios method, without adjustment for overall dye effects, genes 2 and 9 appear to have Cy3/Cy5 < 0.5, suggesting that they are differentially regulated. No genes have Cy3/Cy5 > 2. However, the average ratio is 0.95, indicating that overall fluorescence is generally 5% greater in the Cy5 channel. One way to adjust for this is to divide the individual ratios by the average ratio, which results in the adjusted ratio column. This confirms that gene 2 is underexpressed in Cy3, but gene 9 is not, whereas gene 5 may be overexpressed.

Using the log transformation method, you get very similar results. Base 2 is used because it is intuitive to think of a twofold change as less than −1 or greater than 1. The adjusted Cy3 and Cy5 columns indicate the difference between the log_2 fluorescence intensity and the mean log_2 intensity for the respective dye, and hence express the relative fluorescence intensity, relative to the sample mean. The difference between these values gives the final column on the right-hand side, indicating that genes 2 and 5 may be differentially expressed by twofold or more. By contrast, if you just subtract the raw log_2 values, you will see that gene 9 appears to be underexpressed in Cy3, but gene 5 appears to be slightly less than twofold overexpressed.

Most imaging software provides options for dealing with fluctuation in signal intensity across a spot, as well as for subtraction of the mean or median local background fluorescence intensity, as described in Box 4.1. More complex transformations have been proposed to deal with systemic artifacts that for unknown reasons tend to bias ratios as a function of the overall intensity of fluorescence in both channels (Quackenbush 2002). A popular one is the Loess procedure that reduces any relationship between the ratio and intensity on each array due to print-tip effects or differences in the efficiencies with which the dyes may label low-abundance transcripts.

BOX 4.1 Microarray Image Processing

After labeled probe is hybridized to a microarray, several steps must be taken in order to transform the fluorescence intensity associated with each probe into a measure of transcript abundance. Most of these steps are automated by software provided with commercial scanners, but because each program performs the transformations in a slightly different manner, and since repeated measures from the same image can give slightly different results, it is worth considering just how these images are processed. There are essentially four steps: (1) image acquisition, (2) spot location, (3) computation of spot intensities, and (4) data reporting.

The raw image of a microarray scan is usually a 16-bit TIFF file that is a digital record of the intensity of fluorescence associated with each pixel in the array, represented as a number between 0 and 65,536 (i.e., 2^{16}). Higher resolution can be achieved by decreasing the pixel size (for example, 0.1 mm^2 results in approximately 130 pixels covering a spot 150 µm in diameter) or by storing the data in 32-bit format. However, the files are extremely memory-hungry (a normal scan usually requires 40 Mb of disk space), and the sources of experimental error are greater than those associated with image resolution.

The image is usually captured after first performing a pre-scan, both to confirm that the hybridization worked and to estimate the appropriate gain on the laser in order to capture as much information as possible without saturating the signal. If the gain is set too high, all high-intensity spots will converge on the same upper value, leading either to loss of data if both channels (dyes) are similarly affected, or unwanted bias if only one channel is saturated. If the gain is set too low, information at the low end of the scale is lost in the background.

Because dyes quench with time, and possibly at different rates, it is not a good idea to repeatedly scan the same array. Red-green color images of spots are actually false-color representations of underlying digital values; single channels are normally visualized in black and white, as shown in the figure here.

Once the image has been captured, the individual spots must be located. This is most simply achieved by laying a grid over the image that places a square or circle around each spot. In ScanAlyze, which is free software for Microsoft Windows available from http://rana.lbl.gov/EisenSoftware.htm (Eisen et al. 1998), the grid is produced by specifying the number of rows and columns and the spacing of the centers of each spot (which will be the same as the spacing used by the arraying robot). Subsequently, a circle with the same diameter as the average spot is drawn around the grid centers. For example, an 8×12 grid with circles 150 µm in diameter spaced at 200 µm will overlay each of 96 spots, with

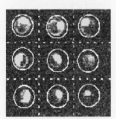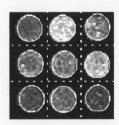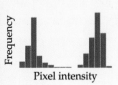

Visualization of single-channel microarrays. The microarray on the left has poor-quality spots; it is difficult to pinpoint the appropriate background or spot intensity values. In the array on the right, however, the uniformity of the spots makes this task much more obvious, and repeated measures are more likely to be the same.

(*Continued on next page*)

BOX 4.1 (*continued*)

50 µm between the edges of adjacent circles. Since there are always imperfections in the spacing of spots of perhaps up to 10 µm, the spots must be re-centered by deforming the grid so as to maximize the coverage of the spots by the circles. This is done semiautomatically by most software, focusing on subsets of the total microarray, using visual confirmation that each round of centering (which takes less than a second) improves the fit. At the same time, flaws on the microarray due to dust specks, coverslip movement, and blotches of unwashed dye can be flagged to exclude the underlying data. The whole process can take an experienced analyst half an hour or more per microarray.

Next the spot intensities are calculated. The simplest approach is to compute the mean intensity for each pixel *within* the circle surrounding a spot, and subtract from this number the mean intensity of the *background* pixels immediately surrounding the spot. Since background intensity can be strongly affected by dust specks that increase the signal, some users prefer to subtract the median background pixel intensity.

The spots produced by capillary transfer often have a donut shape, or are otherwise uneven in intensity, due to spread of the DNA solution to the perimeter of the spot. For this reason, a more accurate way of reading spot intensity is to draw a histogram of pixel intensities throughout the square grid around a spot. As shown in the right-hand figure, the distribution will usu-

ally be bimodal, with one peak associated with the background and the other with the desired hybridization signal. Subtracting the background from the signal peak values may give the most robust measure of fluorescence. The variance in pixel intensity also supplies a measure of spot quality that can be used to flag data for exclusion.

Data is usually reported as a tab-delimited text file in columns that associate the particular measures of spot intensity with row and column identifiers and spot quality values. Most software supports data normalization to remove overall biases associated with the amount of cDNA and quality of the labeling reaction, but these manipulations can also be performed with familiar software, such as Microsoft Excel or any number of statistical packages.

A final concern is to align spot numbers with clone identity. Again, the process is usually automated, but is not necessarily trivial since the spotting process results in different juxtaposition of clones on the array than those in adjacent wells of the microtiter plates.

Further linkage of the data to genome databases that allow users to call up information on the function and sequence of interesting clones requires merging the output of the data analysis with relational databases. As described later in this chapter, protocols built on XML-based languages that will allow storage and retrieval of microarray data from public databases have been developed.

It is generally assumed that microarrays provide accurate readouts of gene expression over three orders of magnitude. This is approximately the range of pixel intensities that scanners can resolve above background. Most scanners produce TIFF output files, with pixel intensities ranging from 0 to 65,536 (2^{16}, the range of 16-bit data) and background values typically between 50 and 200. While direct measures of transcript levels support the supposition that transcription is regulated within such a narrow range for the majority of genes, departures from linearity of the abundance-fluorescence intensity relationship are not accounted for routinely by microarray analysis.

Significance testing. Rather than simply adopting a twofold cutoff, most gene expression studies now require an assessment of the statistical significance of the difference between samples. One approach is simply to assess which genes fall outside the expected extremes of a binomial distribution of ratios, but more robust methods are based on formulating a Student's *t*-test for each gene. As described in Box 4.2 and illustrated in Figure 4.9, the significance of the mean difference in fluorescence intensity (or ratio to a reference sample) between a pair of samples is judged relative to the variation within each sample. **Significance Analysis of Microarrays** (SAM) software available at http://www-stat.stanford.edu/~tibs/SAM (Tusher et al. 2001) builds in some modifications to standard *t*-tests specifically for microarray data and also allows the user to assess correlations with covariates such as clinical status.

Where there are multiple factors, or multiple levels of each factor, in the experimental design, analysis of variance (ANOVA) methods similar to those adopted by quantitative and agricultural geneticists more than 50 years ago can be used (Kerr et al. 2000; Wolfinger et al. 2001). Formulation of *F*-ratios is more efficient than comparison of a series of pairwise *t*-tests, and correlations between variables are accounted for by fitting a combined statistical model. ANOVA is a robust procedure for partitioning sources of variation—for example, testing whether or not the variation in gene expression is less within a defined subset of the data than it is in the total data set. Statistical testing requires moderate levels of replication (between 4 and 10 replicates of each treatment), but the extra effort and expense of this can

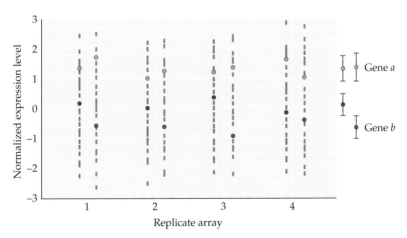

Figure 4.9 Analysis of variance (ANOVA) for gene expression data. Each column of points shows the normalized level of expression of a set of genes relative to the sample mean. Contrasting the relative expression levels for two genes across four replicate arrays shows that there is no difference between the red and green labeled samples for gene *a*, whereas gene *b* is less strongly expressed relative to the sample mean in the green samples.

BOX 4.2 Basic Statistical Methods

Most scientific experiments and surveys in some way aim to compare numerical values between two or more groups of interest. A molecular biologist may want to compare the expression level of a gene under two different cellular environments, or a human geneticist may want to know if five different human populations have the same average cholesterol level. The field of statistics allows for such comparisons to be made in a meaningful way.

Suppose two different laboratories carry out the same experiment: both labs measure the average weight of two strains of white mice. Lab A weighs four mice from each strain and finds that the average weight in strain 1 is 18.0 grams, while the average of the four mice from strain 2 is 21.0 grams. Lab B weighs 125 mice from each strain and finds averages of 20.2 grams for strain 1 and 20.1 grams for strain 2. The results from Lab A suggest that the strain 1 mice weigh an average of 3 grams less than the strain 2 mice. In contrast, the data from Lab B imply that the strain 1 mice weigh an average of 0.1 grams more than strain 2 mice. How should one evaluate such data?

Clearly, there are differences in the amounts of information in the two experiments. Lab B weighed many more mice than did Lab A, so intuitively we place more value on the result from Lab B. The statistical methods of confidence intervals and significance tests allow for quantitative answers to the question, "Do the average weights of strains A and B differ, and if so, by how much?"

Estimating Unknown Means: Confidence Intervals

Many experiments ask questions about the values of unknown mean values. In the mouse example, there are actually two unknown means, one for each strain of mouse. When we collect *sample* data from a *population*, we most likely will compute the sample mean $\overline{X} = \sum x_i / n$ as an estimate of the true population mean. When reporting such a result, one should also recognize that if the experiment were to be repeated, a different value of $\overline{X}$ would be found. Thus, it is important to provide an indication of the variability of the estimate. A *confidence interval* computed from sample data provides a range of values likely to contain the true population mean, and simultaneously accounts for the effects of sample size and individual-to-individual variation. When estimating a population mean, the formula for a confidence interval is:

$$\overline{X} \pm t_{n-1} s / \sqrt{n}$$

The value of $\overline{X}$ is the single best *point estimate* of the unknown mean. The remainder of the formula creates a symmetric interval around $\overline{X}$ to account for individual variation (the sample standard deviation, s), sample size (n) and the desired level of confidence (t_{n-1}). The calculation is illustrated below using the first column of data from Table 1. The confidence interval is interpreted as a range of values having a 95% chance of containing the true (population) mean value. For the sample calculation, there is a 95% chance that the mean weight of strain 1 mice falls between 17.37 and 22.63.

Table 1

Strain 1	Strain 2	Strain 3
21	23	24
18	19	21
23	21	25
20	20	22
18	21	24

For example, compute the confidence interval for the mean weights of the 5 mice in Strain 1 that are shown in the first column of data.

$$\overline{X} = \frac{1}{5}(21 + 18 + 23 + 20 + 18) = \frac{1}{5}(100) = 20$$

$$s^2 = \frac{\sum x_i^2 - \frac{1}{n}\left(\sum x_i\right)^2}{n-1}$$

$$= \frac{(21^2 + \cdots + 18^2) - \frac{1}{5}(100)^2}{4}$$

$$= \frac{2018 - 2000}{4} = 4.5$$

$$s = \sqrt{s^2} = \sqrt{4.5} = 2.12$$

The value of t_4 is found by using a table of the t distribution with 4 degrees of freedom (df). For a 95% confidence interval, the value is 2.776. The final confidence interval becomes:

$$20 \pm 2.776 \times 2.12 / \sqrt{5}$$
$$= 20 \pm 2.63 = (17.37, 22.63)$$

Evaluating Hypotheses about Means: Significance Tests

The fundamental steps of the scientific method include developing hypotheses, designing and carrying out an experiment to test that hypothesis, and then either rejecting or accepting the hypothesis based on the experimental data. A *significance test* (or *hypothesis test*) is a formalization of these steps for use with numerical data. We will consider a test of the *null hypothesis* that the mean of a population is equal to μ_0, a value motivated by the topic being studied. The significance test proceeds first by calculating a *test statistic, t*. The test statistic is then used to calculate a *p-value*, which is used to decide whether the null hypothesis should be rejected (small p-values, often 0.05 or less) or accepted (larger p-values, often 0.05 or larger). The formula for the test statistic for testing the hypothesis that the mean of a population is equal to μ_0 is:

$$t = \frac{\overline{X} - \mu_0}{s/\sqrt{n}}$$

The p-value can be approximated using a table of the t distribution with $n - 1$ degrees of freedom (df). For positive values of t, the desired value is the probability of a value of t greater than the one observed. The calculation is illustrated using the male data from Strain 1 in Table 1. Suppose Strain 1 has been maintained in a different lab for many generations where it is known to have an average weight of 18 grams, and you want to ensure that the mean has not changed. In this setting, the null hypothesis would be that the mean weight of the male mice in Strain 1 is 18 grams, and the computed p-value tells us the probability of getting a sample mean of 20 or higher if the true population mean really is 18 grams.

$$\begin{array}{ll} \text{Strain 1} \\ \overline{X}_1 = 20 & t = \dfrac{20 - 18}{2.12/\sqrt{5}} = 2.11 \\ s_1 = 2.12 \end{array}$$

Using a table of the t distribution with 4 df, the p-value associated with this test statistic is found to be between 0.05 and 0.1. There is some evidence that the mean has increased from 18 grams, but since the p-value is greater than 0.05, the evidence is not strong enough to reject the null hypothesis of mean 18 grams. That is to say, if we were to sample a new set of five individuals from this same population of mice, it would not be surprising for them to have an average weight of 20 grams or higher.

Comparing Three or More Unknown Means: One-way ANOVA

It might be the case that a scientist has three or more populations represented in the data and wants to know if there are any differences among the means of those populations. To address such a question, he or she would take a sample from each of the populations and apply a technique known as *analysis of variance*, more commonly called ANOVA. ANOVA leads to significance tests of the null hypothesis (which here states that there are no differences among the population means). This approach is much more efficient than performing three separate t-tests comparing each pair of means, and only gets more useful as the number of populations increases.

The calculations for ANOVA are simple, but a bit tedious. Define X_{ij} to be the jth observation from the sample taken from the ith of P populations. Furthermore, let $X_{i.}$ be the sum of all n_i observations in the sample from population i, and let $X_{..}$ and $\overline{X}$ be the grand sum of all N observations from all samples and the sample mean of those N values, respectively. With these values defined, we first compute the *total sum of squares*,

$$SSTot = \sum_{i=1}^{P}\sum_{j=1}^{n_i}\left(X_{ij} - \overline{X}\right)^2 = \sum_{i=1}^{P}\sum_{j=1}^{n_i} X_{ij}^2 - \frac{1}{N}X_{..}^2$$

The left-hand portion of the equation indicates that the total sum of squares is simply the sum of squared deviations of individual values from the overall mean. The right-hand portion is an algebraically equivalent expression that is much faster to calculate. Next, compute the *treatment sum of squares*,

$$SSTrt = \sum_{i=1}^{P}\sum_{j=1}^{n_i}\left(\overline{X}_i - \overline{X}\right)^2 = \sum_{i=1}^{P}\frac{X_{i.}^2}{n_i} - \frac{1}{N}X_{..}^2$$

(Continued on next page)

BOX 4.2   *(continued)*

Table 2

Source	df	SS	MS	F
Treatments	$dfT = p - 1$	SSTrt	$MSTrt = SST/dfT$	$F = MSTrt/MSE$
Error	$dfE = i - p$	SSE	$MSE = SSE/dfE$	
Total	$i - 1$	SSTot		

From the left-hand equation we see that the treatment sum of squares is the sum of squared deviations of the P sample means from the overall mean. Again, the right-hand portion is a faster computational form.

Finally, we must compute the *error sum of squares*,

$$SSE = SSTot - SSTrt$$

The three sums of squares, along with some other useful values, are usually arranged in an ANOVA table in the form of Table 2.

The value of F serves as a test statistic for the null hypothesis that all populations have equal means. A table of the F distribution, with dfT numerator degrees of freedom, and dfE denominator degrees of freedom, is necessary to calculate a p-value. Large values of F suggest that the means are not all equal to one another.

We illustrate this below using:

Strain 1	Strain 2	Strain 3	
$\bar{X}_1 = 20$	$\bar{X}_2 = 20.8$	$\bar{X}_3 = 23.2$	$\bar{X} = 21.3$
$X_{1.} = 100$	$X_{2.} = 104$	$X_{3.} = 116$	$X_{..} = 320$

Referring to a table of the F distribution with $dfT = 2$ and $dfE = 12$, and an F-value of 4.43, we find that the p-value for the test is between 0.05 and 0.10. There is some, but not especially strong, evidence that the three strains of mice do not all have equal means. Subsequently, so-called *post-hoc* tests can be used to assess which subsets of the strains are different from the others. In this case there is some evidence that strain 3 differs from strains 1 and 2.

Following slight modifications, both pair-wise t-tests and ANOVA are useful for assessing significance in microarray analyses. There is some debate as to the most appropriate ways to implement the methods and interpret the output from tests on tens of thousands of genes. ANOVA can also be extended into more complex two- or multi-factorial models. These procedures allow the investigator to assess the significance of interaction effects. In the introductory example above, we might want to know whether the difference in conclusions from the two laboratories is itself significant, namely whether there is a laboratory by strain inter-action effect: perhaps strain 1 does not grow so well in Lab A for some reason. Combin-ing all of the data from many microarrays of different types provides increased statistical power for hypothesis testing, but also raises computational and statistical issues that are the subject of ongoing research.

$$SSTot = \left(21^2 + 18^2 + 23^2 + \cdots + 22^2 + 24^2\right) - \frac{1}{15}(320)^2 = 6892 - 6826.67 = 65.33$$

$$SSTrt = \left(\frac{100^2}{5} + \frac{104^2}{5} + \frac{116^2}{5}\right) - 6826.67 = 6854.4 - 6826.67 = 27.73$$

$$SSE = SSTot - SSTrt = 65.33 - 27.73 = 37.60$$

$$MSTrt = SSTrt / 3 - 1 = 27.73 / 2 = 13.87$$

$$MSE = SSE / 15 - 3 = 37.60 / 12 = 3.13$$

$$F = \frac{MSTrt}{MSE} = \frac{13.87}{3.13} = 4.43$$

be counteracted by the elimination of the reference sample. As the sample size of an experiment increases, the power of the statistical tests increases.

Analysis of variance of microarray data proceeds in two steps. First, the raw fluorescence data is log-transformed on the base-2 scale, and the arrays and dye channels within each array are normalized with respect to one another. Tests for overall differences among samples (perhaps due to subtle differences in the amount of mRNA, or proportion of total RNA that is polyadenylated) can also be performed at this step. Removing the dye and array effects leaves normalized expression levels of each EST clone relative to the sample mean.

Subsequently, a second model is fit for the normalized expression levels associated with each individual gene. Essentially, the question is posed as

---

EXERCISE 4.3 *Evaluate the significance of the following gene expression differences*

Five replicate microarrays were performed contrasting an experimental treatment with a reference sample. Three of the genes gave the following results (after adjustments for dye effects). For which genes is there strong evidence for elevated gene expression in the experimental sample?

	Gene 1		Gene 2		Gene 3	
	Exp.	Ref.	Exp.	Ref.	Exp.	Ref.
Array 1	606	287	1,519	1,007	771	586
Array 2	441	198	3,783	1,166	738	502
Array 3	702	366	1,496	1,562	699	493
Array 4	597	255	3,472	1,773	800	625
Array 5	888	402	2,374	2,029	854	718

ANSWER: *The average ratios of experimental:reference are 2.16, 1.77, and 1.33 for genes 1, 2, and 3, respectively. At first glance, these values suggest that gene 1 and perhaps gene 2 have elevated expression. However, the variance of the expression ratios for gene 2 is much higher than that of the other two genes: note that in the second replicate array the level for the experimental sample was more than three times the reference sample, but in the third replicate the measurement for the reference sample actually had the higher value. We can test for gene expression differences using the t-test described in Box 4.2. Begin by computing the experimental:reference ratio for each replicate array. The null hypothesis of interest is that the mean of the ratios is equal to 1 (i.e., no change in expression level). The t-statistics for each of the three genes are 7.3, 0.8, and 3.0, respectively, each with 4 degrees of freedom, leading to p-values less than 0.05 for genes 1 and 3, but not for 2. Thus, we conclude that the level of expression is increased only in genes 1 and 3.*

to whether the replicate samples within a treatment are more like one another than those in the other treatments, irrespective of uncontrolled differences among the arrays due to spotting effects. Furthermore, where multiple different treatments (e.g., sex, drug, and cell type) are being contrasted in the same experiment, it is possible to ask whether in addition to the individual treatment effects, interactions between treatments affect the gene expression levels.

Statistical power is a major issue with all microarray experiments. Given the variation from experiment to experiment, at least 10 microarrays must be performed to detect a change of expression of 1.2-fold with confidence. Such changes may be among the most significant biological effects (for example, if they involve the level of expression of key regulatory kinases). A volcano plot of significance against magnitude of effect such as that shown in Figure 4.10 may assist in choosing genes for more detailed analysis. On this particular plot, genes in the lower left and right sectors (C) represent poten-

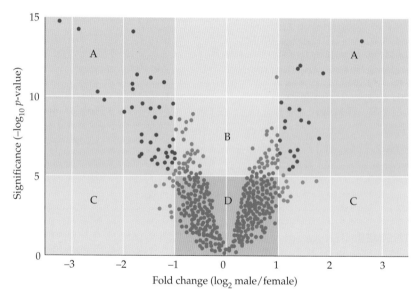

Figure 4.10 Volcano plot of significance against effect.   Based on 24 replicate arrays contrasting male against female adult *Drosophila*, the *x*-axis shows the difference in normalized log-transformed expression level between the two treatments, and the *y*-axis the significance of the difference as the negative logarithm of the *p*-value. On this scale, more significant effects (smaller *p*-values) are at the top of the plot, and the nominal $\alpha = 0.05$ takes a value of 1.3. If fold change is used as the measure of significance, genes are selected that lie to the left or right of the two vertical lines (−1 and 1), which represent the cutoff for twofold difference in expression. If significance is chosen, genes are selected that lie above a horizontal line that represents a chosen significance threshold, for example $p = 0.00001$ ($-\log p = 5$). The two regions marked A represent genes with a large-fold change and high significance; region B indicates genes with high significance but only a small difference; genes in the two C regions show large but insignificant differences; and region D genes do not differ by either criteria. (After Jin et al. 2001.)

tial false positives, showing a large difference that is not significant. Those in the top center (B) are potential false negatives if fold change is the criterion for acceptance, since the effect is relatively small but is highly significant.

Bioconductor. Freely distributed software for statistical analysis of microarray data is available from a number of sources (Draghici 2003), including the Bioconductor Project (http://www.bioconductor.org). Software from this site is written in the open-source statistical programming language R, and implements procedures that are similar to those that can be found in commercial products such as SAS/JMP Genomics. Implementation requires more knowledge of statistics and computer science than most molecular biologists possess, but the basic skills can be learned quickly through collaboration. Various commercial packages for microarray analysis also offer statistical algorithms, but these tend to be more directed toward identifying the genes that have extreme values in a single hybridization, and toward data mining procedures as discussed in the next section.

A particularly useful program in Bioconductor is *qvalue*, a program for assessing the false discovery rate (FDR) in microarray experiments (Storey and Tibshirani 2003). FDR has emerged as a useful procedure for conservative selection of genes based on significance testing. This is because thousands of contrasts are performed, and significance thresholds must be adjusted to reflect this fact. In a set of 10,000 probes, approximately 500 should exceed the nominal 0.05 p-value by chance, and appear as false positives, so a much more stringent p-value cutoff is advised. Traditionally, the multiple comparison problem is addressed by Bonferroni adjustment, namely by dividing the test-wide significance level by the number of tests to arrive an experiment-wide significance level. With 10,000 tests, this is usually too stringent and results in exclusion of many false negatives. The q-value is a measure of the proportion of genes at any significance level that are expected false positives: a q-value of 0.05 implies that only 5% of the genes at that level are likely to be false positives. Intuitively, if in a set of 10,000 probes, 5 are expected to be positive at $p = 0.005$, but 100 are observed significant at this level, then 95 of these are probably true positives. The use of FDR allows the investigator to select a set of genes with a level of confidence that is most appropriate for their purposes.

Given a list of significant genes, it is usually useful to know whether particular families of gene are under- or overrepresented relative to all of the genes on the microarray. For example, if 10% of the probes represent mitochondrial proteins, but 20% of the differentially expressed genes are of this type, this difference in percentage may itself be significant. A useful Web-based tool for assessing this question is the EASE software (Expression Analysis Systematic Explorer; http://david.abcc.ncifcrf.gov/ease/ease.jsp), which converts lists of accession numbers for a wide range of organisms into gene ontology classes, and then tests for the significance of changes in abundance of GO categories (Hosack et al. 2003). Numerous similar tools are available from the GO Consortium Web site (http://www.geneontology.org/GO.tools.annotation.shtml) and elsewhere.

Microarray Data Mining

In attempting to make biological sense of microarray data, it is useful to convert strings of hundreds of thousands of numbers into a format that the human brain can process. Invariably, this entails graphical representation, either in the form of line drawings or color-coding, that places genes into clusters with similar expression profiles (Eisen et al. 1998). Clustering implies co-regulation, which in turn may imply that the genes are involved in a similar biological process. Consequently, in addition to describing how individual genes respond to certain treatments, microarray analysis describes the level of coordinate regulation of gene expression on the genome-wide scale. Since the clustering process groups unknown genes with annotated genes, it can lead to the formulation of hypotheses concerning the possible function of the unknown genes.

Researchers recognized early on that color-coding provides a direct (though not particularly precise) means to immediately identify co-regulated genes. Raw fluorescence intensities are transformed into false-color representations according to the convention that relatively high ratios of expression of experimental-to-reference sample are coded red and low ratios are coded green, with the brightness of the color proportional to the magnitude of the differential expression (Figure 4.11). A ratio of 1 is black.

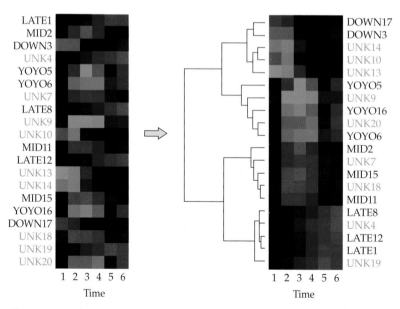

Figure 4.11 Hierarchical clustering of gene expression. An initially disordered set of gene expression profiles (left) can be converted into an immediately intelligible set of clusters by hierarchical clustering and rendering of the profiles in color, as in this hypothetical TreeView representation of a time-series with 20 genes (right). The observation that the genes of the DOWN, YOYO, MID, and LATE classes cluster together suggests that the unknown (UNK) genes may have functions of the respective groups in which they cluster.

(Because most colorblind people cannot contrast red and green, some authors are now using yellow and blue, respectively.) Alignment of all the genes on an array one above the other with the experimental contrasts across the figure allows the researcher to see patterns, especially if some sort of clustering algorithm has been used to filter the sample.

As described in Box 4.3, clustering can be performed in a supervised or unsupervised manner. A common initial approach is to use correlation coefficients to sort the genes that share the most similarity in expression profile across a range of conditions into their own clusters. Such **hierarchical clustering** is an excellent method for identifying groups of co-expressed genes, and is incorporated into several programs available through Bioconductor, as well as in most commercial packages.

BOX 4.3 Clustering Methods

The rapid advances in technology for simultaneously measuring levels of gene expression at many loci have presented a number of challenging data analysis problems for bioinformaticians. One of the most common tasks is the identification of "clusters" of genes that share an **expression profile.** Experiments of this sort collect expression data from G genes using E experiments. A typical example is measuring expression from many genes at a number of time points (e.g., the expression data from $G = 2000$ genes at $E = 8$ time intervals). Intuitively, the object is to identify groups of genes that appear to undergo coordinated changes in expression level, either positive or negative, as indicated in Figure 4.12.

Fortunately, methods for defining such clusters have been around for a number of years in different contexts. Thus, while new approaches for clustering that take explicit account of the nuances of gene expression data are being created, most widely used algorithms are simple modifications of traditional statistical methodologies. Most approaches fall into one of two categories. *Bottom-up* clustering methods begin with each gene in its own cluster. Clusters are then recursively clustered based on similarities, creating a hierarchical, treelike organization. (Indeed, many of the same methods are used for both gene expression clustering and phylogeny reconstruction.) *Top-down*

methods begin by selecting a predetermined number of clusters. Genes are then assigned to those clusters to minimize variation within clusters and maximize variation between them.

We will demonstrate a variation of the popular clustering algorithm proposed by Eisen et al. (1998). Like other bottom-up methods, it consists of three distinct steps: (1) construct a matrix of similarity measures between all pairs of genes; (2) recursively cluster the genes into a treelike hierarchy; and (3) determine the boundaries between individual clusters.

We denote the (normalized) measurement of expression for gene g in experiment e as x_{ge}. For each of the $[G(G-1)]/2$ pairs of genes i and j, compute the correlation coefficient r_{ij} between the E experimental measurements for the two genes:

$$r_{ij} = \frac{1}{E} \sum_{e=1}^{E} \left(\frac{x_{ie} - \overline{x}_i}{s_i} \right) \left(\frac{x_{je} - \overline{x}_j}{s_j} \right)$$

where $\overline{x}_i$ is the average expression level for gene i over the E experimental conditions and s_i is the standard deviation of those same E measurements. Genes with similar expression profiles will have values of r_{ij} near unity.

With the matrix of correlations in hand, we proceed to the clustering portion of the

(Continued on next page)

BOX 4.3 *(continued)*

algorithm. The process is easily described in the following recursion:

1. Find the pair of clusters with the highest correlation and combine the pair into a single cluster.
2. Update the correlation matrix using the average values of the newly combined clusters.
3. Repeat steps 1 and 2 $G - 1$ times until all genes have been clustered.

This algorithm is simple enough to provide a numerical example. Suppose the initial correlation matrix for $G = 5$ genes is:

	2	3	4	5
1	0.3	0.2	0.8	0.1
2		0.9	0.1	0.8
3			0.2	0.7
4				0.1

The highest correlation is the 0.9 observed between genes 2 and 3, so we combine them into a cluster and recompute the correlation matrix:

	23	4	5
1	0.25	0.8	0.1
23		0.15	0.75
4			0.1

Note that the "correlation" between the clusters 23 and 4, for example, is the average of the correlations going into the new cluster, $r_{23,4} = \frac{1}{2}(r_{24} + r_{34}) = \frac{1}{2}(0.1 + 0.2) = 1.5$. We continue this process, clustering 1 with 4, then 23 with 5. The resulting hierarchy takes the form:

The final step in the clustering process is to determine the boundaries of individual clusters. In the above example, do 2 and 3 form a cluster to the exclusion of 5, or is there a single 235 cluster? Eisen at al. (1998) made no formal attempt to address this question, relying instead on defining clusters based on shared biological function. For instance, if genes 2, 3, and 5 were all heat-shock proteins, it might be sensible to include all of them in a single cluster. More recently, authors such as Hastie et al. (2000) and Kerr and Churchill (2001) have used metrics based on principles of ANOVA or the bootstrap procedures used in phylogenetic analysis to define well-supported clusters.

Hierarchical clustering can be performed both on the genes and the treatments, allowing detection of patterns in two dimensions. In cases where the treatments represent a series of drugs or mutations or tissue types rather than a temporal sequence, this two-dimensionality can be an extremely powerful mechanism for identifying similarities in genome-wide responses. It is important to recognize that there are actually numerous different types of hierarchical clustering algorithms that may or may not give the same tree structure, and that there is often no solid statistical support for the clustering produced by these methods. Also, ambiguities can arise as a result of the somewhat arbitrary orientation of branches stemming from each node.

A more sophisticated approach to clustering is to specify the number of clusters that are desired in advance, and then force the data to conform to this structure. Self-organizing maps (SOMs) and *k*-means clusters are similar procedures that automate this task (Tamayo et al. 1999). In *k*-means clus-

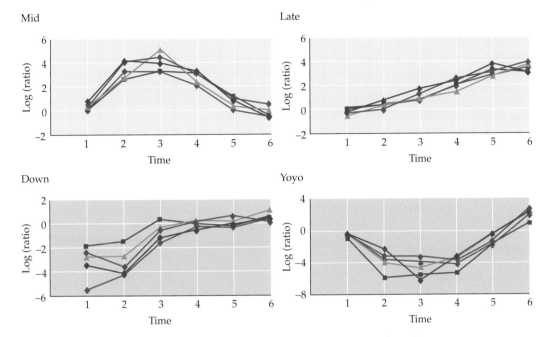

Figure 4.12 Profile plots. The hypothetical data set in Figure 4.11 is re-plotted, showing the actual ratios at each time point. In this case, *k*-means clustering recognizes the same clusters as hierarchical linkage clustering.

tering, all genes are initially assigned at random to one of *k* clusters. The mean value for each treatment in each cluster is then computed, and each gene is reassigned to the cluster to which it shows the closest similarity (Figure 4.12). This procedure is reiterated until a stable structure is achieved. Self-organizing maps are also assembled by an iterative procedure that moves the centroid of each cluster in each step. The choice of the number of clusters can be made according to biological criteria, or may depend on a post-hoc comparison of the results of analyses run with different numbers of clusters. Since the objective is to identify coordinately regulated genes, the precise statistical significance of clusters is less important than failing to see a potentially significant relationship. False-positive groupings will presumably be invalidated by subsequent analyses—for example, if co-expressed genes are later found to be expressed in completely different cell types under circumstances that make no biological sense.

An alternative to clustering methods that rely on pairwise comparisons of values within each treatment is to ask how expression of each gene relates to the major sources of variation in the total sample. In **principal component analysis** (**PCA**, also known as **singular value decomposition**), the major axes of variation among the treatments are identified and each gene is assigned a value representing how that component of the variation contributes to its profile of expression (Holter et al. 2000). These axes will typically contain contributions from multiple treatments, so the clustering is sensitive to interaction as well as to primary effects. For example, in a study contrasting the effects of sex, cancer type, and chemotherapeutic agent, it may be that an interaction between sex and cancer type is the major source of gene expression variation, followed by a major effect of chemotherapy. Plotting the principal component scores of all genes on these two axes will identify clusters of genes that respond in like manner to the major sources of variation. An implication of this approach is that it does not confine genes to a single cluster, since different genes may cluster on different pairs of axes.

Cluster analysis is a useful procedure for seeing trends in data, but there are no well-established procedures for assessing the significance of individual clusters. In fact, there is no reason why genes cannot belong to multiple clusters, reflecting differential regulation in different tissues and under different circumstances. In the final analysis, however, the biological meaning of clusters is assessed by their predictive power and their capacity for generating hypotheses. These features in turn are a function of the ability to align cluster data with gene ontology and other genomic annotation, including linkage to literature databases (Jenssen et al. 2001).

EXERCISE 4.4 *Perform a cluster analysis on gene expression profiles*

Suppose that while performing a microarray analysis of gene expression at four successive time points, you are particularly interested in whether any of the following five genes are co-regulated. By estimating the correlation between each profile, draw a plausible cluster diagram indicating which genes are most closely co-regulated, and comment on the pattern you observe.

	T1	T2	T3	T4	T5
Gene A	10.2	13.6	15.1	14.9	12.1
Gene B	13.7	13.5	12.9	13.3	13.8
Gene C	8.6	11.2	12.7	12.8	12.5
Gene D	13.3	13.3	11.1	9.6	9.6
Gene E	12.2	14.5	11.1	11.2	13.7

ANSWER: *Use software such as the "Correlation Function Data Analysis" tool in Microsoft Excel (or compute the correlations by hand; see Box 4.3) to show that the correlation between each expression profile is as follows:*

	Gene B	Gene C	Gene D	Gene E
Gene A	−0.82	0.82	−0.43	−0.36
Gene B	—	−0.49	0.14	0.65
Gene C	—	—	−0.80	−0.18
Gene D	—	—	—	0.34

This matrix shows that genes A and C are most alike, and that B and E also have somewhat similar profiles. By contrast, genes A and B are strongly negatively correlated, as are genes C and D. Consequently, a plausible clustering of the profiles looks like:

However, the strong negative correlation between A and B, and between C and D, might imply that the two major clusters of genes are actually regulated by some of the same factors, just with opposite effects. If you plot the expression values of each gene across the time course, you will see that whereas A and C increase from T1 to T3, B and D tend to decrease. Clustering by the absolute value of the correlation coefficients results in quite a different picture of the relationships among the genes.

ChIP Chips and Gene Regulation

In addition to finding differentially regulated genes, microarrays can be used to help dissect the mechanisms by which differential transcription is achieved. There are two aspects to this problem: identifying the genes that are targets for specific transcription factors, and finding the transcription factor binding sites in the regulatory regions of those genes. The two pursuits can reinforce one another, and are most successful when combined with experimental manipulation to confirm the interactions.

An indirect approach to identifying target genes is to compare gene expression between wild-type and mutant organisms that either have a deletion of a transcription factor gene, or ectopically express it under control of a tissue-specific or inducible promoter. Differential regulation need not imply that the gene regulation is direct, since an intermediate factor may lie in the regulatory pathway.

A more direct approach is to isolate fragments of DNA that bind to the transcription factor, and determine which gene(s) those fragments regulate. The genomic technique of **chromatin immunoprecipitation microarrays (ChIP chips**; Lieb 2003) has been developed for this purpose, as diagrammed in Figure 4.13A. Protein complexes are cross-linked to chromatin by light chemical treatment, and then an antibody specific for the transcription factor of interest is used to purify the protein-DNA complexes by affinity chromatography (see Figure 5.5). The DNA is then liberated, and competitively hybridized with whole genomic DNA to a DNA microarray containing probes for intergenic DNA. Fragments that are bound to the transcription factor will have elevated hybridization intensity on the microarray. A variation on this approach, known as the ChIP Seq assay, is to directly

Figure 4.13 Chromatin immunopre-cipitation and regulatory pathways.
(A) ChIP chips involve capture of DNA crosslinked to a transcription factor protein, which is then hybridized to a genomic DNA microarray. Fragments that produce a stronger signal in the immunoprecipitated fraction than whole gDNA identify likely targets of the transcription factor. (B) Comprehensive assembly of lists of targets of each of the 106 known yeast transcription factors allowed computational reconstitution of regulatory pathways in yeast. These networks adopt five common forms, as shown, with an estimate of the number of examples of each.

(A)

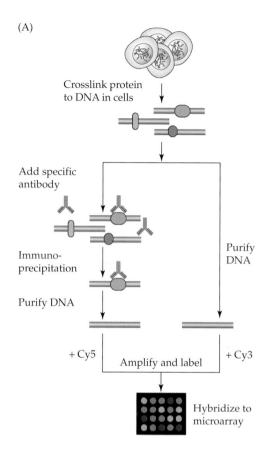

(B)

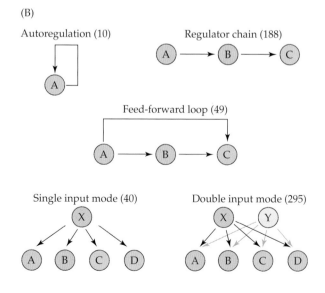

sequence the immunoprecipitated sequences using one of the next generation parallel re-sequencing methods (Robertson et al. 2007).

ChIP chip has been used to systematically explore the design of regulatory pathways in budding yeast, *Saccharomyces cerevisiae*. Target genes for each of the 100 known transcription factors were determined, and ordered into the five types of network shown in Figure 4.13B. These include simple two- or three-step pathways in which gene A regulates gene B which regulates gene C. Feedback and feedforward loops are, respectively, cases where the last gene in a pathway represses and activates the first gene. Branching networks are as common as linear pathways and loops.

Binding sites tend to be less than 12 bp in length—too short to detect by standard local alignment search algorithms—so considerable attention is being given to development of new bioinformatic approaches to binding site detection. The Gibbs sampling approach is described in some detail in Box 4.4. Other methods include regression of motif counts on transcript abundance (Bussemaker et al. 2001), and searching for clusters of biochemically characterized binding sites upstream of genes. Such procedures generally result in good success in finding novel target genes, but remain subject to high rates of detection of false targets and failure to detect known targets. Incorporation of empirical and biophysical data into search algorithms in software such as Ahab (http://gaspard.bio.nyu.edu/Ahab.html) is likely to increase the performance of regulatory motif discovery (Segal et al. 2008).

DNA Applications of Microarrays

Microarrays can also be used to study differences in gene content in the genomic DNA of individuals and species. The principle is the same as cDNA microarrays, except that the probes are derived from any portion of the genome (including regulatory, intronic, and intergenic sequences), and the target is labeled genomic DNA instead of an RNA derivative. Applications contrasting individuals include detection of copy number variation due to large insertion-deletion polymorphisms and segmental duplications, as well as loss of heterozygosity and gene amplification in tumor samples (Emanuel and Saitta 2007). Array comparative genome hybridization (CGH) has also seen wide application at the species level, for example to compare the gene content of related microbial species and to detect the abundance of various species in complex microbial samples from soil, thermal vents, intestinal flora, and other niches.

Two broad classes of array are used for mammalian applications (Carter 2007). **Large insert arrays** generated from BAC or fosmid clones have high signal-to-noise ratio and provide resolution to tens of kilobases, so long as repeat sequences are masked. **Oligonucleotide tiling arrays** provide much higher resolution, potentially detecting indels a kilobase or less in size, but may suffer from high background hybridization. Several companies have developed dedicated arrays that guarantee an even spacing of oligonucleotide probes every two kilobases across the human genome, but for other

BOX 4.4 Motif Detection in Promoter Sequences

Given a set of co-regulated genes, it is generally assumed that transcription of all or most of the genes is likely to be regulated by a common transcription factor or factors. Many of these factors act by binding to specific DNA elements upstream of the promoters. Consequently, there ought to be conserved regulatory DNA sequences in these promoter regions. In some cases, such elements have been identified by biochemical and/or molecular genetic procedures (for example gel-shift assays) in which the transcription factor protein is shown to bind to the DNA sequence. In general, however, most regulatory sequences remain to be identified, and bioinformatic strategies are being employed to help find them. There are two major obstacles: *statistical* and *biological* constraints.

The statistical problem is that regulatory elements tend to be short (between 6 and 10 bases in length, reflecting the length of helical groove that the proteins recognize through specific hydrogen bonding) and somewhat variable. The expected frequency of any n-base sequence is one in every 4^n nucleotides, which for a 7-base element is once every 16 kb. Thus, even in a microbial genome, every short sequence will be represented thousands of times, so the problem of identifying elements that are overrepresented in the promoters of just a few percent of the genes (in each case in just a handful of copies) is acute. Adding to this the possibility that one or two of the bases diverge from the canonical motif only exacerbates the problem.

The biological constraint is that transcriptional regulation is complex. The same pattern of transcription can be generated by different transcription factors, so not all genes with the same profile will share a particular motif. Also, the same transcription factor can be involved in the generation of a wide variety of expression profiles, sometimes because the response it elicits is concentration-dependent (the same factor can be an activator at some concentrations, a repressor at others), sometimes because of variation in the spacing between the motif and the pro-

moter, and often as a function of the co-factors it associates with on the DNA. Thus, regulatory element function is heavily context-dependent, and while it may be possible to identify conserved core elements, there is no one-to-one correspondence between the presence of that core and transcriptional output.

Importantly, the binding sites that confer important biological responses may often be the ones with the lowest affinity for the transcription site (which can be more sensitive to slight changes in protein concentration), and hence the ones that diverge most from the consensus sequence. Furthermore, in higher eukaryotes, regulatory sequences tend to be dispersed over tens and even hundreds of kilobases, and can be positioned downstream of, within, or upstream of genes. This greatly increases the amount of space that must be searched in order to find these elements.

One common statistical strategy for identifying motifs in unaligned promoter sequences is based on a method called **Gibbs sampling.** The idea is to iterate randomly through (ungapped) alignments of the promoter regions, searching for alignments that result in the identification of blocks of conserved residues of some pre-specified length w. In essence, the result is an optimal local multiple sequence alignment.

Discussion of the exact details of the Gibbs sampling method require statistical concepts beyond the scope of this text. However, the flavor of the method is easily demonstrated by example. Suppose the following sequences are promoter regions of a cluster of co-regulated genes (clearly, the sequences are shorter than what we would find in practice):

ACCGTGGTGT

TGGCACAAGC

GCCGATAGTC

AGTGGCGAAC

CCTGTGGTCA (sequence Z)

Initialize the iteration by selecting a random starting sequence (the last one, for our example) and designating it as sequence Z. The idea of each step in the iteration is to use the remaining sequences to find the location of the shared regulatory motif in Z. Pick a random initial alignment for the remaining sequences and starting point for the motif of width $w = 4$.

<pre>
            ****
        ACCGTGGTGT
    TGGCACAAGC
          GCCGATAGTC
    AGTGGCGAAC
</pre>

For each position i, where $i = 1..w$, in the alignment within the motif, tabulate the nucleotide frequencies q_{ij}:

	1	2	3	4
A	0.25	0.25	0.00	0.50
C	0.25	0.50	0.00	0.25
G	0.50	0.00	1.00	0.25
T	0.00	0.25	0.00	0.00

Compute the corresponding nucleotide frequencies p_j for the pool of sites outside the pattern:

$p_A = 5/24 = 0.21$ $\qquad$ $p_C = 6/24 = 0.25$

$p_G = 7/24 = 0.29$ $\qquad$ $p_T = 6/24 = 0.25$

Select a starting point, x, for the motif in sequence Z:

<pre>
        ****
    CCTGTGGTCA
</pre>

Calculate the probability of the pattern using the values from the profile (q's):

$Q(x) = 0.5 \times 0.25 \times 1.0 \times 0.25 = 0.00625$

and also the probability of the pattern using the "background" values from sequence regions outside the profile (p's):

$P(x) = 7/24 \times 6/24 \times 7/24 \times 7/24 = 0.006203$

The ratio of the two probabilities, $R(x) = Q(x)/P(x) = 1.01$, is indicative of how likely it is that sequence Z has an example of the motif beginning at position x, so we select as the location of the motif the position with largest $R(x)$. The current sequence Z is

added to the alignment, with a new sequence designated as Z for the next cycle of the iteration. Placement of the motif in sequence Z becomes more and more refined each iteration, and the complete sequence alignment (i.e., the placement of the motif's starting point in each sequence) eventually converges.

The idea driving Gibbs sampling methods is that the better the description of the profile probabilities (q's), the more accurately the position of the motif in sequence Z can be identified. In its early stages, the locations of the motifs are chosen essentially at random. Frequencies in the p's and q's are similar, and, consequently, values of $R(x)$ hover near 1. As the iterations continue, however, some of the motif locations are placed correctly by chance, leading to more pronounced differences between the profile frequencies in q and the background frequencies in p, higher ratio values, and better placement of the motif in that iteration's sequence Z.

For the example data, the alignment eventually converges to:

<pre>
        ****
    ACCGTGGTGT
        TGGCACAAGC
    GCCGATAGTC
        AGTGGCGAAC
        CCTGTGGTCA
</pre>

with q's:

	1	2	3	4
A	0.00	0.20	0.00	0.00
C	0.00	0.00	0.00	0.40
G	0.00	0.80	1.00	0.00
T	1.00	0.00	0.00	0.60

and p's:

$p_A = 9/30 = 0.30$ $\qquad$ $p_C = 11/30 = 0.37$

$p_G = 8/30 = 0.27$ $\qquad$ $p_T = 2/30 = 0.07$

For further details, including methods for selecting the optimal value of the motif width w, see Lawrence et al. (1993) and Roth et al. (1998); the latter describes the popular AlignACE software.

(Continued on next page)

BOX 4.4 (*continued*)

Iteration proceeds by dropping one sequence from the analysis and recomputing the weight matrix and alignment. At the end of the process, a motif is identified that is over-represented in the sample, so long as some pre-set likelihood cut-off threshold is exceeded. Subsequently, further criteria might be specified, such as the location and orientation of the motif relative to the promoter, whether it is palindromic, and how specific it is relative to other sequences in the genome that were not included in the initial collection. Cross-species contrasts might also be performed on the supposition that functional elements are likely to be conserved, and the availability of complete genome sequences will greatly facilitate such contrasts (McGuire et al. 2000). A useful program for performing this type of search online is the AlignACE program at http://atlas.med.harvard.edu/cgi-bin/alignace.pl.

Other approaches to motif recognition have been considered. A hidden Markov model (HMM) approach is implemented in the MEME program described in Bailey and Gribskov (1998) and available at http://meme.sdsc.edu/meme/website. An alternative to searching for motifs in a cluster of genes is to ask whether a previously identified motif has a different frequency among clusters. Jakt et al. (2001) applied this approach to characterize and refine motifs present in several types of transcription profile observed during the cell cycle of yeast. A list of all of the matches and their respective likelihood scores in the complete set of promoters is assembled, and the number of matches in each cluster at several specificity levels is compared with the number expected in a sample of the same size, computed from a hypergeometric distribution. This approach shows not just which clusters are enriched for a particular motif, but also whether the enrichment is only for perfect matches or includes more divergent motifs.

A different approach again is to avoid clustering, and instead just regress the presence of every possible combination of nucleotide sequences up to 7 or 8 bases, on the relative gene expression level difference between a single time point or condition and a reference point (Bussemaker et al. 2001). That is to say, a linear model is set up for each gene in which the relative expression level is fit as a function of the number of times a particular precise motif is detected in the promoter region. Multiple regression can also be applied to fit the effects of several motifs simultaneously, and to test for synergistic or antagonistic interactions between them. Applied to a single time point of synchronized yeast cells, the method identified 11 motifs involved in initial cell cycle progression, including the previously identified stress response, SFF, MCB, and M3a/b elements, as well as two apparently new motifs. Undoubtedly, a range of novel approaches to association of gene expression and regulatory motif finding await development.

---

species expression arrays can also be used for CGH. Similarly, SNP genotyping arrays have proved useful for the detection of microdeletions and copy number variants (CNV) (Stranger et al. 2007a).

Conversely, short oligonucleotide arrays can also be used to detect SNPs, facilitating rapid linkage mapping of mutations (Borevitz et al. 2003). SNPs that happen to lie near the center of a 25-mer oligonucleotide will tend to change the hybridization intensity when comparing two strains or individuals. An array with hundreds of thousands of probes will typically detect several thousand "single-feature polymorphisms" (SFPs). After pooling

DNA from progeny with similar phenotypes in a procedure known as **bulked segregant analysis (BSA)**, it is possible to resolve large effect QTL or Mendelian mutations to within several hundred kilobases using fewer than ten gene chips, which is both less expensive and less time-consuming than traditional methods based on single-marker genotyping of hundreds of individuals.

Parallel Analysis of Gene Expression: RNA Sequencing

Serial Analysis of Gene Expression

Serial analysis of gene expression, commonly known as **SAGE**, is a method for determining the absolute abundance of every transcript expressed in a population of cells. The SAGE protocol is based on the serial sequencing of 10-bp (short SAGE) or 17-bp (long SAGE) tags that are unique to each and every gene (Velculescu et al. 1995). These gene-specific tags are produced by an elegant series of molecular biological manipulations and then concatenated for automated sequencing, with each sequencing reaction generating up to 50 high-quality tags. At least 50,000 tags are required per sample to approach saturation, the point where each expressed gene is represented at least twice in the tag collection. The advantage of SAGE is that the method is unbiased by factors such as reference samples, hybridization artifacts, or clone representation, and gives what is believed to be an accurate reading of the true number of transcripts per cell.

The procedure used to generate tags is diagrammed in Figure 4.14. Starting with less than a microgram of polyA mRNA, a double-stranded cDNA library is constructed using biotinylated polydT oligonucleotides to prime the reverse transcription reaction. This cDNA is then cleaved with an "anchor enzyme"—typically NlaIII, a four-cutter restriction enzyme that cleaves DNA an average of once every 400 bases. Streptavidin-coated magnetic beads are then used to purify the 3′ end of each cDNA from the anchoring site to the polyA tail, which binds to the streptavidin through the biotin group. An adapter oligonucleotide is then directionally ligated to the anchoring sites. This adapter contains a recognition site for a type II restriction endonuclease, BsmFI, which liberates the tags adjacent to the anchoring site by cleaving 15 bp into the cDNA fragment. The adapter-tag fragments are then ligated tail-to-tail to form **ditags**, which are amplified by PCR using primers complementary to the 5′ end of the adapter. These PCR products are then purified by extraction from an acrylamide gel, and again cleaved with the anchoring enzyme NlaIII, which liberates the adapters and leaves 30-bp ditags. These ditags are purified once more, then ligated end-to-end under controlled conditions that produce concatenated chains from a few hundred base pairs to a few kilobases in length. The approximately 1-kb fraction is excised from a gel one last time, cloned into a sequencing vector, and several thousand clones are sequenced directly.

Figure 4.14 Principle of SAGE. In serial analysis of gene expression, the absolute abundance of every transcript expressed in a cell population is determined by serial sequencing of 15-bp, gene-specific tags produced by molecular manipulations and concatenated for automated sequencing. The text describes the method in detail.

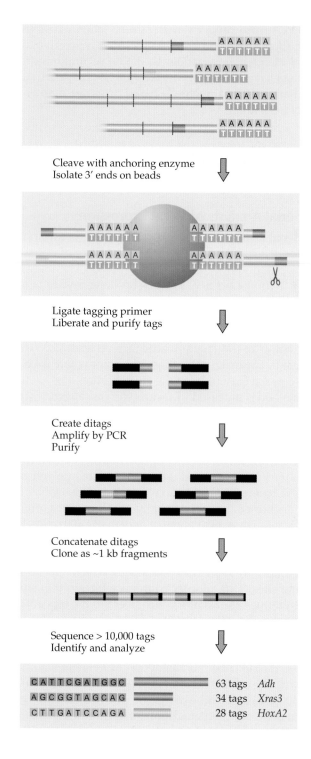

Cleave with anchoring enzyme
Isolate 3′ ends on beads

Ligate tagging primer
Liberate and purify tags

Create ditags
Amplify by PCR
Purify

Concatenate ditags
Clone as ~1 kb fragments

Sequence > 10,000 tags
Identify and analyze

CATTCGATGGC	63 tags	*Adh*
AGCGGTAGCAG	34 tags	*Xras3*
CTTGATCCAGA	28 tags	*HoxA2*

(A)

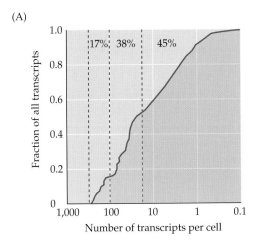

Figure 4.15 SAGE analysis of yeast and colorectal cancer transcriptomes. (A) Almost 50% of yeast transcripts are expressed at less than 10 copies per cell, whereas 17% of the total mRNA is due to a small number of transcripts at greater than 100 copies per cell. (B) A comparison of over 60,000 tags each from both normal colon epithelium and from a colorectal tumor sample revealed 83 candidate genes that are either up- or downregulated at least tenfold in the tumor. (A after Velculescu et al. 1997; B after Zhang et al. 1997.)

(B)

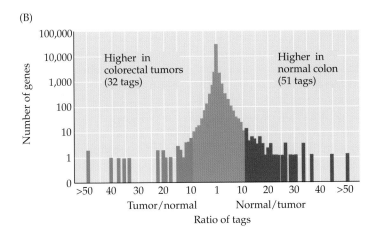

Tag identification is automated, and for model organisms the gene corresponding to each tag can be identified immediately and unambiguously. Since each tag commences with the sequence of the anchoring site, SAGE analytical software identifies tags with the correct length and spacing and filters out artifacts due to end-filling, cloning, and PCR errors. A list of each unique tag and its abundance in the population is assembled, and where possible the tags are annotated with whole genome and/or EST information.

Transcript abundance measured by SAGE can either be expressed in relative terms or converted to an estimate of the number of transcripts per cell (Figure 4.15). Typically, across all eukaryotic cell types, fewer than 100 transcripts account for 20% of the total mRNA population, each being present in between 100 and 1000 copies per cell. These include transcripts that encode ribosomal proteins and other core elements of the transcriptional and translational machinery, histones, and some taxon-specific genes such as Rubisco in plants and polyadenylated mitochondrial transcripts in

Drosophila. A further 30% of the transcriptome comprises several hundred intermediate-frequency transcripts with between 10 and 100 transcripts per cell. These include housekeeping enzymes, cytoskeletal components, and some unusually abundant cell type-specific proteins.

The remaining half of the transcriptome is made up of tens of thousands of low-abundance transcripts, some of which may be expressed at less than one copy per cell and many of which are tissue-specific or induced only under particular conditions. Thus most of the transcripts in a cell population contribute less than 0.01% of the total mRNA. If one-third of a higher eukaryotic genome is expressed in a particular tissue sample, then somewhere in the neighborhood of 10,000 different tags should be detectable. Taking into account that half of the transcriptome is relatively abundant, at least 50,000 tags must be sequenced to approach saturation.

For comparative purposes, SAGE is most useful for identifying genes that show large increases in one or more samples, or are completely missing from others. Because even a single tag is good evidence that a gene is expressed in a tissue, SAGE also performs better than microarray analysis with respect to determining which genes are expressed at low levels, since signals of similar intensity to true low-abundance signals often cloud microarray analysis. However, failure to observe a tag is poor evidence that a gene is not transcribed, even if the collection is approaching saturation (two or more tag sequences) for most transcripts.

Quantitative comparison of SAGE samples is not always easy to interpret. A tag present in four copies in one sample of 50,000 tags and two copies in another may actually be twofold induced in the first sample; but such a difference is also expected to arise by random sampling. Even the contrast of 20 tags to 10 is not obviously significant given the large number of comparisons that are performed, though it is at least suggestive. Web-based tools for performing online comparisons of tag abundance across experiments, as well as facilitating download of complete data sets, are well developed and accessible through SAGEnet (http://www.sagenet.org).

RNA-Seq

Mid-2008 saw the publication of the first studies demonstrating the potential of **RNA-Seq**—the direct sequencing of fragments of cDNA—for characterization of transcriptomes. Nagalakshmi et al. (2008) and Wilhelm et al. (2008) both surveyed yeast strains under a variety of conditions, while Lister et al. (2008) demonstrated several applications in *Arabidopsis.* The basic RNA-Seq approach is shown in Figure 4.16, but optimal protocols are still being established. The polyA fraction of cellular RNA is isolated and fragmented into 200 base sequences that are used to prime random cDNA synthesis, resulting in **short quantitative random RNA libraries (SQRLs)**. These libraries become the template for one of the next-generation sequencing platforms described in Chapter 2 (currently 454 Life Sciences, Illumina Genome Analyzer (Solexa), or ABI SOLiD), with the goal of generating at least 10 million short sequence reads for each RNA sample. These short sequences are then mapped back to the reference genome of the species and aligned with predicted and annotated exon sequences in order to arrive at a digital estimate of the abundance of each transcript.

For mammalian genomes, it appears that upwards of 40 reads per gene are required to be confident that a gene is expressed and to estimate its abun-

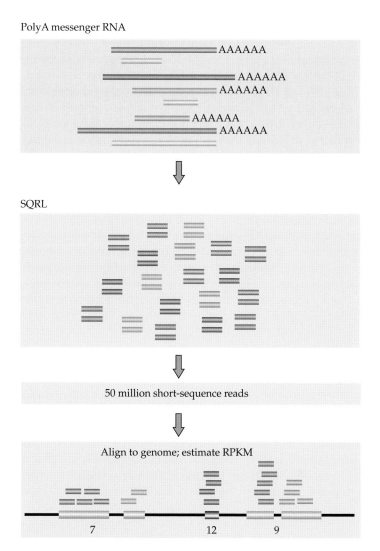

Figure 4.16 RNA-Seq Analysis. Starting with a preparation of messenger RNA (the polyA fraction), a short quantitative random RNA library (SQRL) is generated by fractionation and conversion to cDNA using one of a number of strategies to overcome potential biases. These fragments are then read by a next-generation sequencing apparatus, and the tens or hundreds of millions of short reads are aligned to the whole-genome sequence. Finally, comparison with predicted exons allows estimation of transcript abundance along the gene in RPKM units. In this example, there are fewer tags for the middle gene, but the tags derive from less exon sequence (darker shading) and so the RPKM is higher (12 versus 7 and 9).

dance. Mortazavi et al. (2008) introduced "reads per kilobase of predicted exon per million total reads" (RPKM) as a scalable measure of abundance. One transcript in a typical human or mouse cell is expected to yield approximately 3 RPKM, and only a few hundred transcripts produce more than 1,000 RPKM. Cloonan et al. (2008) adopted an approach to SQRL production that retains the directionality of the transcript, providing an extra quality control step for distinguishing gene expression from background contamination. The vast majority of reads correspond to exons, as expected, and they also conclude that in excess of a gigabase of sequence is needed to robustly detect all transcripts in a cell.

All of these studies demonstrate levels of complexity to the transcriptome that cannot be discerned by microarray or SAGE analyses. Since the short reads often cover exon-exon junctions, they provide an accurate snapshot of alternative splicing, and since they are not biased by probe selection, they document alternate 5' and 3' ends of transcripts. By obtaining multiple reads per sequence, it is possible to detect sequence polymorphisms that might contribute to diseases or traits, and to directly observe allele-specific transcription. Modifications of the protocol facilitate specific applications such as quantification of micro RNA abundance or documentation of the distribution of methylation of genomic DNA. Approximately one quarter of the short reads correspond to repetitive sequences or to sequences that are common to duplicate genes, but with appropriate bioinformatic procedures it is possible to filter out these effects and observe clear differences in gene expression that are difficult to distinguish with other methods.

Single-Gene Analyses

One of the primary objectives of all comparative gene expression studies is to identify a small subset of genes, further study of which may be illuminating with respect to understanding the biological basis of differences in the samples under study. Whether a change in gene expression causes or merely correlates with a response, the first step in further study is to use a different experimental procedure to confirm that the gene really *is* differentially expressed. Several methods for studying expression one gene at a time have been developed since the early days of molecular biology; two of these are described in this section.

Northern Blots

Northern blots are the simplest procedure used to determine whether a gene is expressed in a sample, but being semiquantitative they really only confirm twofold or greater differences in gene expression. PolyA mRNA is isolated from the tissue of interest and the transcripts are separated according to length on a denaturing agarose gel. The mRNA molecules are then transferred to a nylon or nitrocellulose filter by capillary or electrolytic blotting, fixed in place by ultraviolet cross-linking, and hybridized to a labeled short DNA probe synthesized from a cloned fragment of the gene of interest. The label can be radioactive, chemiluminescent, or histochemical, and will produce a band on the blot wherever hybridization occurs, in proportion to the amount of mRNA in the sample. If the transcript is alternatively spliced, multiple bands corresponding to the different length transcripts will be detected. (In fact, a common way to identify the location of exons in a gene, short of sequencing full-length cDNAs, remains the probing of northern blots with a series of fragments isolated from different portions of the genomic DNA covering a locus.)

In order to be quantitative, the intensity of signal from the gene being probed must be compared with an internal control for the amount and quality of the mRNA in different lanes on the blot. "Housekeeping" genes (such as ribosomal proteins or particular actin subunits that are thought not to fluctuate significantly in expression level across treatments) are used to reprobe the blot. The specific gene signal is then normalized to the control signal. In cases where a single transcript is produced or information about alternative splicing is not of interest, the gel electrophoresis step can be omitted, in which case mRNA samples are blotted directly onto the filter in an array of small dots or slots. Such dot/slot blots enable simultaneous comparison of expression levels from tens or even hundreds of samples and only take hours rather than a couple of days to perform.

Quantitative PCR

The now-standard method for quantifying individual gene expression is **quantitative reverse-transcription PCR**, also known as **Q-PCR** or **Q-RT-PCR**. Standard PCR (the polymerase chain reaction) is not generally quantitative because the end product is observed after the bulk of the product has been synthesized, at a point where the rate of synthesis of new molecules has reached a plateau. Consequently, small differences in the amount of target at the start of the reaction are masked. As with the measurement of biochemical reaction rates, quantitative measures of nucleic acid polymerization must be made during the linear phase of the reaction. The number of cycles required to attain the linear phase of increase in product biosynthesis is regarded as a quantitative indicator of the amount of template in the RNA population. To be truly quantitative, in Q-PCR the reaction rate is compared with that of a similar template (ideally, one with the same base composition, such as the same gene with a small internal deletion) that is spiked into a dilution series. This establishes the sensitivity of the reaction to the number of molecules of template.

Real-time measurement of product accumulation provides the most sensitive assay; otherwise, multiple reactions must be set up in parallel, with replicates, and stopped at chosen times. Commercial machines for Q-PCR utilize a transparent reaction cuvette or capillary exposed to a fluorescence detector. The detector measures the signal from a dye that only fluoresces when intercalated with double-stranded nucleic acid. Readings are taken just prior to the denaturation step between $80°$ and $85°$, since primer-dimers and nonspecific products denature below this temperature and contribute minimally to the fluorescence signal.

The Q-PCR reaction is usually primed with cDNA synthesized by one cycle of reverse transcription, following which gene-specific primers are used to amplify the locus of interest. With each cycle, the fluorescence measures the number of molecules of double-stranded DNA that have been synthesized. If a reaction is primed with 10 molecules of the template, the rate of observable product accumulation will be delayed relative to one primed with 1,000 molecules of template because it takes at least half a dozen cycles to produce sufficient specific product to begin seeing a signal (Figure 4.17).

An alternative to using an intercalating dye is Applied Biosystem's Taq-Man assay. This method relies on the property of **fluorescence resonance energy transfer (FRET)**, in which the fluorescence from one dye is transferred to and thereby quenched by a nearby dye that emits at a different

(A)

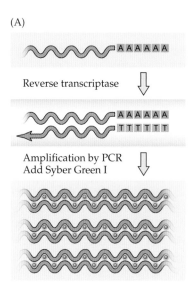

Figure 4.17 Quantitative RT-PCR.
(A) mRNA is reverse transcribed and the transcript of interest is amplified using a gene-specific 5′ primer. (B) In the presence of a fluorescent dye such as Syber Green I, amplification can be monitored in real time. (C) The crossing point of the linear phase of the reaction with background fluorescence is used to establish a standard curve from which the relative amount of unknown product can be determined. (After Rasmussen et al. 1998.)

(B)

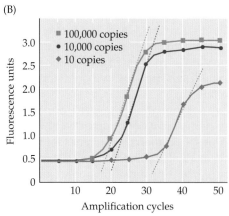

(C)

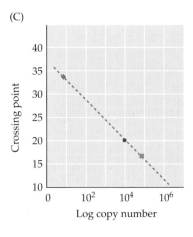

wavelength. The two dyes are incorporated at either end of an internal primer that hybridizes to the template cDNA. As the PCR reaction proceeds, a 5′-to-3′ exonuclease activity of the polymerase digests this FRET primer, freeing the reporter dye into solution so that its fluorescence emission is no longer quenched. Thus, the observed fluorescence intensity increases as more PCR product is made, allowing quantification of transcript levels.

Properties of Transcriptomes

In this section, we describe some applications of gene expression profiling on a genome-wide basis. Literature that includes microarray data is increasing at an exponential rate as the method becomes incorporated as a tool in genetic dissection, so this section is not, nor is it intended to be, a comprehensive survey. The intention is merely to illustrate how the procedure is being used to investigate a wide range of biological processes.

Microbial Transcriptomics

Microarrrays have been used extensively both for gene expression profiling and comparative genomics in a wide range of microbes. In bacteria as diverse as *E. coli*, extremophiles, and industrial fermenters, transcription profiling has been employed to study the organization of biochemical pathways, adaptive responses to antibiotics, and the mechanistic basis of phenomena such as biofilm formation. Hybridization of bacterial DNA to microarrays is being employed to characterize strain diversity and the species content of environmental samples, and as a rapid detection technology for infectious diseases and food-borne pathogens. Similarly, fungi of various sorts and protozoan parasites have received much attention aimed toward dissection of cellular processes and identification of target genes for use in immunization and drug development.

Many of the first cDNA and oligonucleotide microarray studies were done using the budding yeast *Saccharomyces cerevisiae*, starting with an analysis of transcriptional changes that occur during the haploid cell cycle (Spellman et al. 1998). One of the advantages of yeast as an experimental system is that the early completion of the genome sequence quickly led to the annotation of the entire complement of just over 6,200 genes, which is small enough to allow complete representation of the open reading frames (ORFs) on a single microarray. Comparison of gene expression profiles after synchronization of cell growth resulted in the identification of hundreds of genes that are transcribed during, and presumably required for, procession through each stage of mitosis.

Similarly, sporulation has been studied in great detail in *S. cerevisiae* (Figure 4.18; Chu et al. 1998). In this yeast equivalent of meiosis, diploid cells are converted to haploid gametes in response to starvation on nitrogen-deficient medium. Decades of genetic and molecular analysis had already identified 50 or so genes that were induced in four temporal waves during sporulation. cDNA microarray analysis identified more than 1,000 genes that

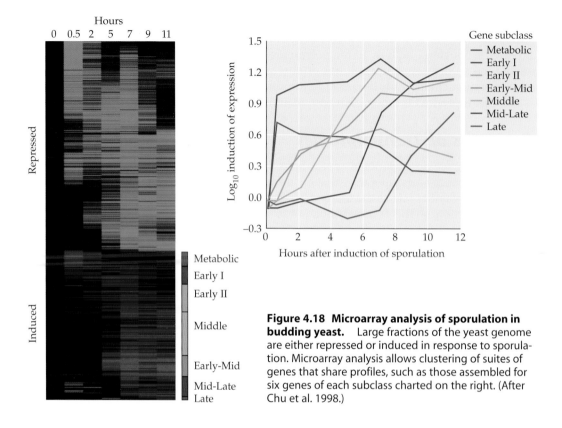

Figure 4.18 Microarray analysis of sporulation in budding yeast.   Large fractions of the yeast genome are either repressed or induced in response to sporulation. Microarray analysis allows clustering of suites of genes that share profiles, such as those assembled for six genes of each subclass charted on the right. (After Chu et al. 1998.)

showed a more than threefold change for at least one of six time points, or an average 2.2-fold change across the entire time course. In addition, global analysis allowed the early phase to be subdivided into four more subtle groupings, including a set of 52 immediate but transiently induced metabolic genes that are likely to be responding to nitrogen starvation rather than initiation of sporulation per se. Intriguingly, only a relatively small fraction of cell-cycle regulated genes are regulated in a similar fashion in the fission yeast *Schizosaccharomyces pombe*, implying considerable divergence in all but the core functions that coordinate meiosis in divergent yeasts (Rustici et al. 2004).

A very different application of cDNA microarrays is in experimental evolution. Yeast cells cultured in glucose-limiting media adapt by reducing their dependence on glucose fermentation and switching to oxidative phosphorylation as a more efficient means of generating energy. Three replicate chemostat experiments starting from the same isogenic culture were allowed to evolve for over 250 generations, in which time half a dozen selective sweeps led to profound metabolic adaptation (Ferea et al. 1999). Remarkably, gene expression profiling demonstrated that 3% ($^{184}\!/_{6,124}$) of all genes showed an average twofold change in gene expression across all three repli-

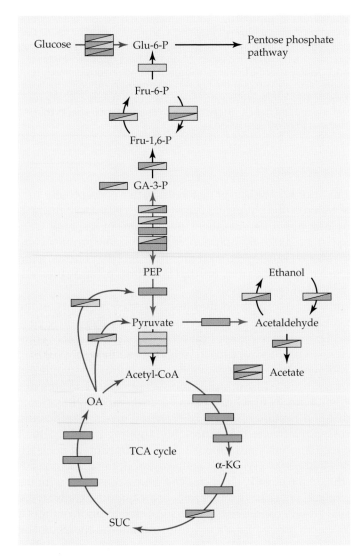

Figure 4.19 Gene expression changes in yeast. In this diagram of yeast central metabolism, the boxes represent transcripts encoding enzymes that are repressed (green) or induced (red) at least twofold, either after a shift to glucose-limiting media (the diauxic shift; colored triangles in the upper left half of boxes) or after 250 generations of adaptive evolution (colored triangles in lower right of boxes). Both conditions lead to reduced glucose fermentation and increased oxidative phosphorylation through the TCA cycle. (After DeRisi et al. 1997; Ferea et al. 1999.)

cate evolved strains, in most cases including similar responses in at least two of the strains. This corresponded to about a third of all genes that showed at least a twofold change in one strain, indicating that despite the small number of adaptive mutations responsible for the response to glucose starvation, hundreds of genes responded in a coordinate manner.

About half of the genes that changed in at least two populations were previously characterized, and most of these have roles in respiration, fermentation, and metabolite transport (Figure 4.19). Furthermore, many of the genes showed parallel changes as a result of physiological adaptation to glucose depletion (the "diauxic shift"), indicating that long-term evolution occurred in a somewhat predictable direction that mirrored expecta-

Figure 4.20 The compendium approach.   This technique is used to identify genes ▶ and treatments that act on similar pathways. Profiling a large number of mutations or treatment conditions identifies clusters of genes that are co-regulated across a range of conditions; for example, here genes in clusters A1 and A2 have opposite effects, while those in B and C display specific features of interest. Simultaneous hierarchical clustering by treatment (across the top of the figure) also groups conditions that lead to similar overall transcription profiles, allowing researchers to generate hypotheses as to the functions of the mutant genes, drugs, or other environmental agents that led to these perturbations in gene expression.

tions derived from prior knowledge of metabolism. In a conceptually similar study of adaptation of the pathogenic yeast *Candida albicans* to fluconazole exposure, at least two different global mechanisms of response were observed (Cowen et al. 2002). Similar evolved expression profiles were observed in drug-resistant isolates sampled directly from HIV patients, highlighting how microarray analyses may be used to address pressing biomedical problem of antiobiotic resistance.

Numerous studies have examined the effect of gene knockouts on the yeast transcriptome, leading to insights into the targets of regulatory gene function. A particularly impressive application of this strategy is the "compendium of expression profiles" approach (schematized in Figure 4.20), in which over 300 different strains and conditions were contrasted (Hughes et al. 2000). The data set included 11 inducible transgenes, 13 drug treatments, and 276 deletion mutants, 69 of which removed unclassified ORFs. Some 20–30% of all deletion mutants had more than 100 genes significantly induced or repressed relative to the control, highlighting the ubiquity of coordinate gene regulation in yeast. Interestingly, the fraction of unclassified ORF deletions that had a large effect on the transcriptome was considerably less than that of the already characterized genes, consistent with the idea that these "orphan" genes have minor roles or are important only under extreme or unusual growth conditions. Hierarchical classification of the 300 samples and 6,000 genes proved to be a powerful mechanism for clustering genes with unknown or poorly studied functions into processes shared by other genes, such as cell wall biosynthesis, steroid metabolism, mitochondrial function, mating, and protein synthesis.

Parasite gene expression profiling is exemplified by analysis of the causative agent of malaria, *Plasmodium falciparum*. The genome of this protozooan contains 5,400 predicted genes, most of which have no known function. Strikingly, 60% of the genes are expressed while the organism infects human red blood cells, in a tightly orchestrated program that follows the functional needs of the parasite (Bozdech et al. 2004). Following invasion, the organism upregulates protein synthesis and metabolic activity, and thereafter waves of genes related to DNA replication, growth, plastid formation, the next cycle of invasion, and motility are turned on. One application of this type of analysis is in the identification of candidate vaccine targets expressed on the cell surface.

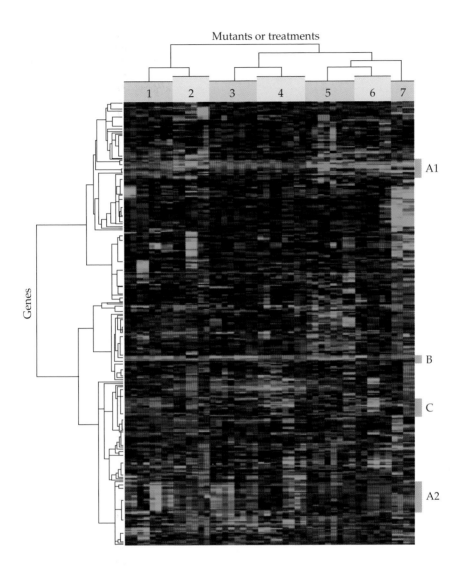

Cancer and Clinical Applications

Both SAGE and microarray technology have been employed extensively to characterize the human transcriptome in diverse cell types. Estimates from SAGE analysis of the number of genes expressed in any given cell type fall in the range of 15,000 to 25,000 (Table 4.1). A "minimal" transcriptome consisting of at least 1,000 genes expressed in all cell types has also been suggested. The greatest variety of gene expression is apparently found in brain tissue, perhaps reflecting the complexity of neuronal types. Aside from assembling atlases of gene expression, numerous studies have begun to characterize differences in gene expression in cancer cells as well as differences associated with other human diseases.

TABLE 4.1 *Estimates of Transcript Diversity in Human Tissues*

Tissue[a]	Total SAGE tags	Number of unique tags
Colon epithelium	98,089	12,941
Keratinocytes	83,835	12,598
Breast epithelium	107,632	13,429
Lung epithelium	111,848	11,636
Melanocytes	110,631	14,824
Prostate	98,010	9,786
Monocytes	66,673	9,504
Kidney epithelium	103,836	15,094
Chondrocytes	88,875	11,628
Cardiomyocytes	77,374	9,449
Brain	202,448	23,580

[a]Each tissue was represented by at least two libraries.

Source: Data from Velculescu et al. (1999).

Microarray analyses of cancer have been carried out with the following objectives:

- Enhanced classification of cancer types, including identifying cell-type of origin.
- Characterization of expression profiles that may help predict therapeutic response.
- Clustering genes in order to generate hypotheses concerning their mode of action in carcinogenesis.
- Identification of novel gene targets for chemotherapy.

The usefulness of microarrays for cancer classification has been demonstrated both through studies of cancer cell lines and cancer biopsies. One early microarray study of 9,700 cDNAs in 60 cancer cell lines of diverse origin clearly demonstrated that samples of similar cancer types—for example, neuroblastomas, melanomas, leukemias, and colon and ovarian cancers (though not necessarily breast and lung carcinomas)—tend to share gene expression profiles that in part reflect differences retained from their tissue of origin (Figure 4.21; Ross et al. 2000). Hierarchical clustering identified groups of genes that provide signature profiles for each cancer subtype and hence identify markers that may prove useful in clinical diagnosis, and suggest functions for previously uncharacterized genes.

Biopsy gene expression profiles can also be clustered relative to one another and to normal tissues to subclassify cancer types. Breast cancers in particular are not always readily classified using classical histological mark-

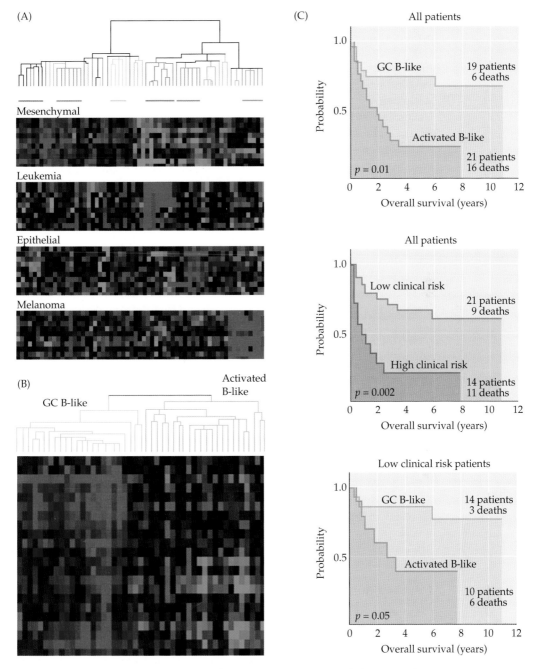

Figure 4.21 Molecular pharmacology of cancers. (A) Hierarchical clustering of biopsy expression profiles reveals that different tumors can be identified on the basis of type-specific expression profiles. (B) Similarly, clustering of distinct cancer types, such as diffuse large B-cell lymphomas (DLBCL), uncovers the existence of novel molecular subtypes (GC B-like and Activated B-like) that may be predictive of survival probability, as indicated by the standard Kaplan-Meier plots shown in (B,C). (A after Ross et al. 2000; B,C after Alizadeh et al. 2000.)

ers, but genome-wide comparisons have the power to resolve the likely cell-type of origin (Perou et al. 2000). While leukemias can be broadly classified by histology, until the advent of microarrays no markers were available that predicted either mortality or therapeutic response. Expression profiles have now been identified that cluster leukemias into groups that correlate with long-term prognosis, as shown in Figure 4.21C (Alizadeh et al. 2000). Metastatic and solid tumors can be distinguished on the basis of transcription of an extensive set of genes that regulate passage through the cell cycle, and ongoing studies focus on the quantitative correlation of specific genes with proliferation rates. Similarly, molecular signatures in breast cancers may predict metastatic potential and responsiveness to hormone or chemotherapy, particularly if coupled with genotype information on the status of susceptibility loci such as *BRCA1* (van't Veer et al. 2002). However, there are numerous issues to be resolved concerning the power and repeatability of predictors based on microarrays, and the standards required for regulatory implementation are different from those used during exploratory research (Petricoin et al. 2002).

Microarrays are also being used in the context of infectious disease, the immune system, and disorders from asthma to autoimmunity. For example, profiling lesions due to infection with leprosy revealed how *Mycobacterium* affects the cytokine balance in T-cells in a manner that correlates with the manifestation of disease (Bleharski et al. 2003). Applications range from finding biomarkers for infection and exposure to toxins or drugs, to learning how infection with one agent may affect tolerance to another, assessing contamination in blood products, and monitoring the effects of systemic parasite infection. Comparison of peripheral blood expression profiles of 13 adults from each of two different ethnic groups in Burkina Faso, West Africa, demonstrated downregulation of genes indicative of T-regulatory cell production among the Fulani people, who are relatively resistant to malaria. Subsequent depletion of this cell population from blood samples of the Mossi group enhanced their proliferative response to *Plasmodium* antigens (Torcia et al. 2007), demonstrating how microarray studies can be combined with experimental manipulation to address the mechanistic basis of human disease susceptibility.

Development, Physiology, and Behavior

Gene expression profiling has been adopted by biologists working with model systems to address a wide range of questions in development, physiology, and behavior (Reinke and White 2002). One immediate application is in annotating gene expression as development proceeds in *Drosophila*, nematodes, and plants. The life cycle of a fruit fly consists of four distinct phases: *embryogenesis*, in which the basic features of the body pattern are established; a series of larval *instars* in which the precursor imaginal tissues of the future adult fly form; the *pupa*, in which the larva is dissolved and the adult assembled; and the *adult*. One comprehensive study of 66 sequential time points, summarized in Figure 4.22, revealed that 86% of the 4028 genes

(A)

(B)

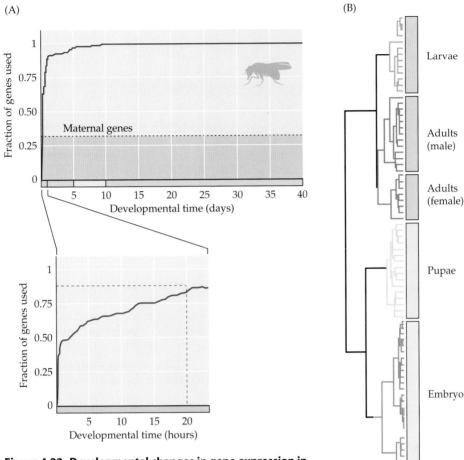

Figure 4.22 Developmental changes in gene expression in _Drosophila_. (A) Over 80% of all genes are transcribed during embryo-genesis, starting with 30% of genes as maternal transcripts, followed by a gradual accumulation of usage over the next 24 hours. (B) Clustering of profiles across the complete developmental time-course reveals that transcription in adults is globally closer to that observed in larvae than in pupae and embryos. (After Arbeitman et al. 2002.)

studied change in expression during development. Remarkably, many genes are expressed in two waves: embryonic resembles pupal expression, and larval expression resembles adult. In some striking cases, such as myogenesis, microarrays reveal how analysis of RNA from whole organisms can detect transcriptional events in cellular subpopulations (White et al. 1999). It has even been possible to detect induction of a gene in a single pair of cells in the brain.

With this baseline, it is now possible to identify early and late transcriptional targets of single mutations and so to assemble pathways and networks of gene expression. This approach supplements classical genetic and cell biological studies, providing a genome-wide perspective that can suggest hypotheses which can be tested by directed studies of specific genes. As well as providing insight into organ specification, sex determination, embryonic and appendage patterning, and hormonal regulation of metamorphosis, microarrays have been used to dissect physiological responses to aging, diet (notably, caloric restriction), and sexual behavior.

Microarrays have also been used to identify genes associated with wood biosynthesis in forest trees grown on soils of varying quality; to characterize responses to viral and nematode pathogenesis; and to dissect nodule induction by symbiotic nitrogen-fixing bacteria. Dairy farmers are interested in characterizing resistance and susceptibility to mastitis, and sheep farmers would like to know more about the properties of wool follicles. In each of these cases, simple expression profile comparisons are just the first step in analysis, but careful experimental design can be employed to hone the tool for an extraordinarily wide range of applications.

Studies oriented toward nervous system function also cover a diversity of questions, including psychogenomics (the genomic basis of psychological disorders such as Alzheimer's disease and epilepsy), neuropharmacogenomics (the effect of drugs on specific parts of the brain), and behavioral genomics (the molecular basis of behavioral differences). For example, Whitfield et al. (2003) generated a molecular profile of transcript abundance in the brains of honey bees that successfully predicted whether 57 out of 60 bees were foragers or nurses. Since adult bees tend to transition from hive work to foraging with age, they used principal components analysis to compare the contributions of the different factors to overall expression variance, and found that behavior has twice the impact as age on transcription in the brain.

Evolutionary and Ecological Functional Genomics

A new discipline of evolutionary and ecological functional genomics is emerging, building heavily on comparative genomics and transcription profiling (Feder and Mitchell-Olds, 2002). Some applications, such as studying interactions between biotic and abiotic factors, for example plant-bacterial symbiosis or insect dessication resistance, focus on finding crucial genes and biochemical or cellular pathways. Many others require complex experimental designs involving multiple levels of several factors—perhaps young and old males and females from a dozen different ecotypes. In these cases, the statistical tools used to analyze data tend to be more directed toward precise estimation of variance components, and avoidance of false negatives at the expense of occasional false positives. Much recent focus has been on integration of population and quantitative genetic methods with gene expression profiling.

Comparative approaches promise to shed much light on the relationship between transcriptional divergence, genetic divergence, and phenotypic divergence. Among closely related species, it is not unusual to see significant differences in the transcript levels of as many as 20% of the genes expressed in a tissue. A landmark comparison of expression profiles in humans, chimps, and an orangutan (Enard et al. 2002) suggests that more transcripts have evolved in the liver than the brain, but that there tends to be greater divergence between humans and other primates in those genes that are different. Nevertheless, scaling by the amount of variation among individuals within a species, most divergence at the transcriptional level is consistent with genetic drift, and stabilizing selection may be more predominant than directional selection (Figure 4.23).

Gene expression variation among individuals within species is often likely to highlight genes that may contribute to quantitative traits. In a study contrasting the muscle-tissue genes from a northern and a southern population of *Fundulus* fish, more divergence was seen between latitudes than between two sibling species in the southern region, implicating these genes

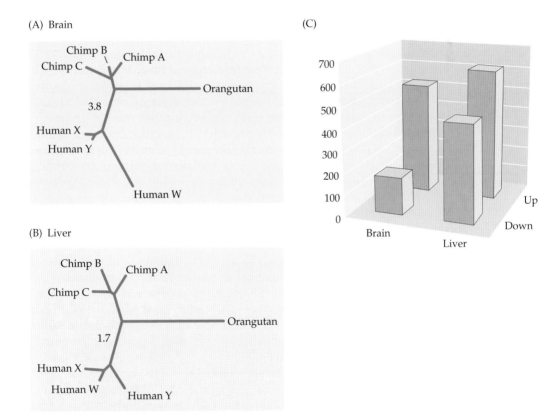

Figure 4.23 Divergence in gene expression in primates. When the relationship between the transcriptome in the brain (A) or liver (B) of individual primates is drawn according to the average distance between transcript abundance, it is observed that the branch leading to humans is twice as long in the brain as it is in the liver. This suggests an accelerated rate of transcriptional divergence in the brain, although counts of the total number of differentially expressed genes suggests greater divergence in the liver (C). (After Enard et al. 2002; Hsieh et al. 2003.)

Figure 4.24 Expression quantitative trait loci (eQTL). (A) If two parents with divergent levels of transcript abundance for a gene (red, high; green, low) are crossed, it is generally expected that their F_1 progeny will have an intermediate level of transcription (yellow). Abundance will vary in F_2 individuals, and may be associated with the genotype at a number of loci sampled throughout the genome. For example, the first locus shows a dominant effect of the A allele, since both AA and AC genotypes have relatively low transcription in a sample of 5 individuals, whereas CC have high abundance. (B) Mapping of such markers onto the genome often indicates that the expression QTL maps to the same location of the gene that encodes the transcript, suggesting that a *cis*-acting regulatory polymorphism affects the transcript abundance. (After Schadt et al. 2003; Yvert et al. 2003.)

in adaptive divergence (Oleksiak et al. 2002). In the progeny of crosses between phenotypically divergent parents, simultaneous expression profiling and tracking of genetic markers (**genetical genomic analysis**; de Koning and Haley 2005) can be used both to map genes whose expression correlates with a trait of interest, as well as genes that regulate transcript abundance. The latter are called **expression quantitative trait loci** (**eQTL**) and they have been reported in relation to obesity in mice, lignin content in eukalyptus, and variation in natural isolates of yeast. Some early conclusions from these studies are that eQTL often co-localize with the genes they regulate (implying regulatory polymorphism), much of the transcriptome is affected by relatively large-effect eQTL, and the same eQTL may regulate many target genes, whether or not they encode transcription factors (Figure 4.24).

Genotypes can also be associated with transcript abundance in populations of unrelated individuals, including humans. The majority of human transcripts show genetic influence on levels of abundance, often in the same range as the heritability of visible phenotypes. Using stringent criteria, Stranger et al. (2007a,b) found significant associations between SNP or CNV polymorphisms and 1,500 transcripts in human lymphoblastoid cell lines, most of which involve regulatory SNPs in the genes themselves. Up to one quarter of all expressed transcripts showed evidence for divergence in expression between the three HapMap population groups (Yoruban, European, and Asian), implying that regulatory variation is pervasive in humans, and presumably most species.

The combination of gene expression profiling with whole-genome genotyping is being used to study how genetic and environmental factors combine to affect disease susceptibility. An analysis of peripheral blood gene expression in over 1,000 Icelanders found only a modest correlation between transcript abundance and clinical measures of obesity, whereas parallel analysis of gene expression in adipose tissue of over 600 Icelanders found that two-thirds of all transcripts are predictive of body mass and waist-to-hip ratio (Emilsson et al. 2008). Over 3,000 transcript abundance measures in adipose tissue were associated with *cis*-regulatory polymorphisms,

(A)

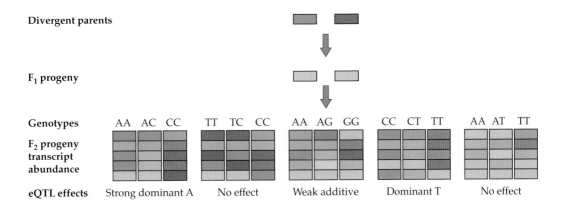

Divergent parents

F₁ progeny

Genotypes	AA AC CC	TT TC CC	AA AG GG	CC CT TT	AA AT TT

F₂ progeny
transcript
abundance

| eQTL effects | Strong dominant A | No effect | Weak additive | Dominant T | No effect |

(B)

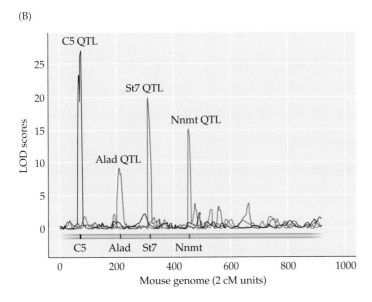

slightly more than in the blood. One module of 886 co-regulated genes that are significantly associated with body mass defines a macrophage-enriched metabolic network that is also upregulated in obese mice, implicating inflammation in the etiology of obesity. Further analyses suggest an enriched association of promoter polymorphisms in these genes with obesity, showing how expression profiling can add significant value to genetic association studies.

Many more applications are sure to emerge in the next few years. Combined with protein expression profiling as described in the next chapter, biologists' perspective on the genetic basis of variation, divergence, and pathology is itself evolving rapidly.

Gene Expression Databases

Gene expression profiling has been criticized for its focus on data generation rather than hypothesis testing. However, practitioners point out that, first, hypotheses tend to flow from large data sets and, second, whole-genome approaches often engender a shift in perspective that can spur new research directions. The successful use of microarrays thus depends on making the data available to as many potential users as possible, which in turn requires the establishment of Internet-based databases. There are three features of database establishment that need to be addressed: importing data in a standard format into the database, facilitating queries, and generating novel data-mining tools.

A consortium of users known as the Microarray Gene Expression Data Society (MGED; http://www.mged.org) have agreed on community standards and proposed a set of **m**inimal **i**nformation for the **a**nnotation of **m**icroarray **e**xperiments (**MIAME**). The minimal features of an experiment that are required to annotate an experiment are:

- Definition of the clones represented on the array (the "platform").
- Precise description of the samples and controls (reference samples).
- Characterization of the hybridization and labeling protocols.
- Researchers' names and institutional addresses.
- Data values associated with each point on the array.

Different software associated with scanners output data in different formats, and use different criteria for calculating spot and background intensities, so users are asked to reformat their output to a standard set of criteria. The alternative of depositing TIFF files with the raw data and allowing centralized extraction of intensity values is unrealistic due to the extremely large volumes of data that would have to be stored. Nevertheless, individual users may opt to store their own data locally in a format that allows others to visualize actual spot profiles.

Ideally, databases would also allow direct contrast of experiments conducted with each of the three established methodologies (Aach et al. 2000). The major obstacle here is that SAGE reports the number of tags observed in a sample, short oligonucleotide arrays report estimates of expression level derived by subtraction of the mismatch control, and cDNA microarrays only report relative expression levels. Differences among platforms—namely, the set of clones on the array—do not necessarily present novel computational problems, but do demand relational or object-oriented database management. A further concern is quality control, which entails assuring that data meets minimal standards of consistency using parameters such as approximate linearity of the relationship between the two dye signals.

The simplest form of query is just to call up raw ratio measures from single experiments. More sophisticated protocols allow comparisons across experiments. Typically, a query entails the three steps shown in Figure 4.25:

Select expression data set	Example
Type of data	Microarray, SAGE, etc.
Organism	*Drosophila melanogaster, Mus musculus*
Tissue	Embryo, heart
Treatment	25∞C, nicotine ingestion

Select measures	Example
Specific data set	Laboratory; platform; samples
Items	Intensity; normalized expression; ratio
Subset of ORFs	Gene families; search by ontology

Select query terms	Example
QC filters	At least 4 replicates
Fold change	>2
Significance of difference	$p < 0.0001$
Clustering algorithm	Hierarchical, *k*-means, etc.

Figure 4.27 Flow diagram of microarray database queries.

1. Selection of a set of arrays or treatments to contrast.
2. Selection of a set of genes of interest.
3. Definition of the criteria by which a change in expression is regarded as meaningful and/or significant.

Tools for undertaking this kind of data collection have been developed as part of the cancer genome anatomy project, initially for comparison of EST clusters in different cancer types, and for online comparison of SAGE data from human data sets. Sophisticated data-mining tools that use statistically robust approaches and facilitate online contrasts of data from different experimental groups, and potentially systems and even organisms, are under development. In the meantime, the Expression Connection of the *Saccharomyces* Genome Database (see pp. 54–57) represents the state-of-the-art of how gene expression can be profiled over the Internet, and the Gene Expression Omnibus (GEO) at the NCBI and ArrayExpress at the European Bioinformatics Institute have become the central repositories for expression data.

Summary

1. Gene expression profiling can be performed with microarray technologies that allow thousands of transcripts to be assayed simultaneously. cDNA microarrays consist of PCR-amplified EST fragments arrayed on a glass microscope slide; GeneChips consist of a series of

25-mer oligonucleotides synthesized directly on a silicon chip; and long oligonucleotides can be spotted or synthesized in situ on glass slides or beads.

2. Most cDNA microarray applications use two different fluorescent dyes so that expression is measured as the ratio of transcript abundance in the experimental versus a reference sample.

3. Statistical procedures are used to compute the significance of changes in gene expression based on variance among replicates and fold change. Experiment-wide thresholds are evaluated according to an acceptable false discovery rate (FDR) criterion.

4. A variety of clustering methods have been adopted to identify groups of co-regulated genes, including hierarchical clustering, self-organizing maps, and principal component analysis.

5. Similarity of expression profiles is often regarded as evidence for similarity of function of a cluster of genes, allowing putative annotation of the function of unknown genes. Though not definitive, this type of analysis is at least sufficient to generate hypotheses that can be tested by more traditional molecular biological approaches.

6. Investigation of the mechanisms responsible for co-regulation of sets of genes is facilitated by detailed monitoring of temporal expression profiles in mutant backgrounds. Bioinformatic methods can be used to identify shared upstream regulatory sequences that may lead to the isolation of transcription factors that mediate particular expression profiles.

7. Serial analysis of gene expression (SAGE) is an alternative method for characterizing transcript abundance in cells. It relies on sequencing large numbers of gene-specific tags. High-volume sequencing of cDNA may also be used to estimate the abundance of transcripts in an unbiased manner.

8. Northern blots and quantitative reverse transcription-PCR (Q-RT-PCR) can be used to confirm expression differences detected on microarrays.

9. The complete set of microbial ORFs can be represented on a single microarray, facilitating a wide range of studies including characterization of the cell cycle, response to nutrient starvation or other environmental changes, and adaptation in long-term culture.

10. Microarray analysis has revealed that cancer cells tend to retain transcriptional features of the normal cell from which they derive, allowing molecular phenotyping of cancers. In addition, gene expression profiles may be diagnostic of the metastatic potential of tumors and thus may play a role in the design of treatment protocols.

11. Applications of microarrays in human disease research range from characterization of transcriptional changes in diseased tissue to toxicology, immunology, and pharmacology. Characterization of clusters of induced or repressed genes may generate hypotheses as to the etiology of disease.

12. As costs decrease and the availability of array technology expands, gene expression profiling will be increasingly important in the developmental biology of model organisms, as well as the ecology and evolutionary biology of a wide range of parasitic and infective species.

13. Genetical genomics refers to the joint application of gene expression profiling and genome-wide genotyping to study the molecular basis of gene expression variation.

Discussion Questions

1. Compare and contrast the advantages and disadvantages of cDNA and oligonucleotide-based gene expression profiling. What might the effect of cross-hybridization be with each method?

2. Why did the majority of early studies of transcriptional differences adopt fold-change as the criterion for detecting changes in gene expression, rather than a measure of statistical significance?

3. Discuss the potential pitfalls in the "guilt-by-association" approach to assigning probable gene function, in which similarity of expression profile is used to infer similarity of function.

4. In what ways can transcriptional profiling contribute to evolutionary and ecological genomics?

5. Is gene expression profiling suitable for providing clinical diagnoses that a patient or his/her physician may use to guide treatments?

Web Site Exercises

The Web site linked to this book at http://www.sinauer.com/genomics provides exercises in various techniques described in this chapter.

1. Analyze two 4,000-gene microarray experiments using ScanAlyze.

2. Compare a variety of methods for computing ratios and normalizing the microarray data.

3. Perform an analysis of variance (ANOVA) of a three-treatment experiment.

4. Compare several methods for clustering the data in Exercise 4.3.

Literature Cited

Aach, J., W. Rindone and G. Church. 2000. Systematic management and analysis of yeast gene expression data. *Genome Res.* 10: 431–445.

Alizadeh, A. A. et al. 2000. Distinct types of diffuse large B-cell lymphoma identified by gene expression profiling. *Nature* 403: 503–511.

Arbeitman, M. N. et al. 2002. Gene expression during the life cycle of *Drosophila melanogaster*. *Science* 297: 2270–2275.

Bailey, T. L. and M. Gribskov. 1998. Methods and statistics for combining motif match scores. *J. Comp. Biol.* 5: 211–221.

Bleharski, J. R. et al. 2003. Use of genetic profiling in leprosy to discriminate clinical forms of the disease. *Science* 301: 1527–1530.

Borevitz, J. O. et al. 2003. Large-scale identification of single-feature polymorphisms in complex genomes. *Genome Res.* 13: 513–523.

Bozdech, Z., M. Llinás, B. L. Pulliam, E. D. Wong, J. Zhu and J. L. DeRisi. 2004. The transcriptome of the intrerythrocytic developmental cycle of *Plasmodium falciparum*. *PLoS Biology* 1: 85–100.

Bussemaker, H. J., H. Li and E. Siggia. 2001. Regulatory element detection using correlation with expression. *Nat. Genet.* 27: 167–174.

Carter, N. 2007. Methods and strategies for analyzing copy number variation using DNA microarrays. *Nat. Genet.* 39 (Suppl.): S16–S21.

Chu, S. et al. 1998. The transcriptional program of sporulation in budding yeast. *Science* 282: 699–705.

Churchill, G. A. 2002. Fundamentals of experimental design for cDNA microarrays. *Nat. Genet.* 32 (Suppl): 490–495.

Cloonan, N. et al. 2008. Stem cell transcriptome profiling via massive-scale mRNA sequencing. *Nature Methods* 5: 613–619.

Cowen, L. E., A. Nantel, M. Whiteway, D. Thomas, D. Tessier, L. M. Cohn and J. B. Anderson. 2002. Population genomics of drug resistance in *Candida albicans*. *Proc. Natl. Acad. Sci. (USA)* 99: 9284–9289.

Cui, X. and G. A. Churchill. 2003. Statistical tests for differential expression in cDNA microarray experiments. *Genome Biol.* 4: 210.

de Koning, D. J. and C. S. Haley. 2005. Genetical genomics in humans and model organisms. *Trends Genet.* 21: 377–381.

DeRisi, J., V. Iyer and P. O. Brown. 1997. Exploring the metabolic and genetic control of gene expression on a genomic scale. *Science* 278: 680–686.

Draghici, S. 2003. *Data Analysis Tools for DNA Microarrays.* Chapman and Hall/CRC Press, Princeton, NJ.

Eisen, M. B., P. Spellman, P. Brown and D. Botstein. 1998. Cluster analysis and display of genome-wide expression patterns. *Proc. Natl. Acad. Sci. (USA)* 95: 14863–14868.

Emanuel, B. S. and S. C. Saitta. 2007. From microscopes to microarrays: Dissecting recurrent chromosomal rearrangements. *Nat. Rev. Genet.* 8: 869–883.

Emilsson, V. et al. 2008. Genetics of gene expression and its effect on disease. *Nature* 452: 423–428.

Enard, W., et al. 2002. Intra- and interspecific variation in primate gene expression patterns. *Science* 296: 340–343.

Fare, T. L. et al. 2003. Effects of atmospheric ozone on microarray data quality. *Anal Chem.* 75: 4672–4675.

Feder, M. E. and T. Mitchell-Olds. 2003. Evolutionary and ecological functional genomics. *Nat. Rev. Genet.* 4: 651–657.

Ferea, T., D. Botstein, P. O. Brown and F. Rosenzweig. 1999. Systematic changes in gene expression patterns following adaptive evolution in yeast. *Proc. Natl. Acad. Sci. (USA)* 96: 9721–9726.

Gautier, L., L. Cope, B. M. Bolstad and R. A. Irizarry. 2004. affy- analysis of Affymetrix GeneChip data at the probe level. *Bioinformatics* 20: 307–315.

Gilad, Y., S. A. Rifkin, P. Bertone, M. Gerstein and K. P. White 2005. Multi-species microarrays reveal the effect of sequence divergence on gene expression profiles. *Genome Res.* 15: 674–680.

Gunderson, K. L. et al. 2004. Decoding randomly ordered DNA arrays. *Genome Res.* 14: 870–877.

Hastie, T. et al. 2000. "Gene shaving" as a method for identifying distinct sets of genes with similar expression patterns. *Genome Biol.* 1: research003.1–003.21

Holter, N., M. Mitra, A. Maritan, M. Cieplak, J. Banavar and N. Federoff. 2000. Fundamental patterns underlying gene expression profiles: Simplicity from complexity. *Proc. Natl. Acad. Sci. (USA)* 97: 8409–8414.

Hosack, D. A., G. Dennis Jr., B. T Sherman, H. C. Lane and R. A. Lempicki. 2003. Identifying biological themes within lists of genes with EASE. *Genome Biology* 4: R60

Hsieh, W.-P., T-M. Chu, R. D. Wolfinger and G. Gibson. 2003. Mixed-model reanalysis of primate data suggests tissue and species biases in oligonucleotide-based gene expression profiles. *Genetics* 165: 747–757.

Hughes, T. R. et al. 2000. Functional discovery via a compendium of expression profiles. *Cell* 102: 109–126.

Hughes, T. R. et al. 2001. Expression profiling using microarrays fabricated by an ink-jet oligonucleotide synthesizer. *Nat. Biotechnol.* 19: 342–347.

Irizarry, R. A., B. Hobbs, F. Collin, Y. Beazer-Barclay, K. Antonellis, U. Scherf and T. P. Speed. 2003. Exploration, normalization, and summaries of high density oligonucleotide array probe level data. *Biostatistics* 4: 249–264.

Jakt, L. M., L. Cao, K. Cheah and D. Smith. 2001. Assessing clusters and motifs from gene expression data. *Genome Res.* 11: 112–123.

Jenssen, T. -K., A. Lægreid, J. Komorowski and E. Hovig. 2001. A literature network of human genes for high-throughput analysis of gene expression. *Nat. Genet.* 28: 21–28.

Jin, W., R. Riley, R. Wolfinger, K. White, G. Passador-Gurgel and G. Gibson. 2001. Contributions of sex, genotype, and age to transcriptional variance in *Drosophila*. *Nat. Genet.* 29: 389–395.

Kerr, M. K., M. Martin and G. Churchill. 2000. Analysis of variance for gene expression microarray data. *J. Comput. Biol.* 7: 819–837.

Kerr, M. K. and G. Churchill. 2001. Statistical design and the analysis of gene expression microarray data. *Genet. Res.* 77: 123–128.

Lawrence, C. E., S. Altschul, M. Boguski, J. Liu, A. Neuwald and J. Wootton. 1993. Detecting subtle sequence signals: A Gibbs sampling strategy for multiple alignment. *Science* 262: 208–214

Lieb, J. D. 2003. Genome-wide mapping of protein-DNA interactions by chromatin immunoprecipitation and DNA microarray hybridization. *Methods Mol. Biol.* 224: 99–109.

Lipshutz, R. J., S. Fodor, T. Gingeras, and D. Lockhart. 1999. High-density synthetic oligonucleotide arrays. *Nat. Genet.* 21 (Suppl.): S20–S24.

Lister, R, R. O'Malley, J. Tonti-Filippini, B. Gregory, C. Berry, A. Millar and J. R. Ecker. 2008. Highly integrated single-base resolution maps of the epigenome in *Arabidopsis*. *Cell* 133: 523–536.

Lockhart, D. J. et al. 1996. Expression monitoring by hybridization to high-density oligonucleotide arrays. *Nat. Biotechnol.* 14: 1675–1680.

McGall, G., J. Labadie, P. Brock, G. Wallraff, T. Nguyen and W. Hinsberg. 1996. Light-directed synthesis of high-density oligonucleotide arrays using semiconductor photoresists. *Proc. Natl. Acad. Sci. (USA)* 93: 13555–13560.

McGuire, A. M., J. Hughes, and G. M. Church. 2000. Conservation of DNA regulatory motifs and discovery of new motifs in microbial genomes. *Nat. Genet.* 10: 744–757.

Mortazavi , A., B. Williams, K. McCue, L. Schaeffer and B. Wold. 2008. Mapping and quantifying mammalian transcriptomes by RNA-Seq. *Nature Methods* 5: 621–628.

Nagalakshmi, U., Z. Wang, K. Waern, C. Shou, D. Raha, M. Gerstein and M. Snyder. 2008. The transcriptional landscape of the yeast genome defined by RNA sequencing. *Science* 320: 1344–1349.

Oleksiak, M. F., G. A. Churchill, and D. L. Crawford. 2002. Variation in gene expression within and among natural populations. *Nat Genet.* 32: 261–266.

Perou, C. M. et al. 2000. Molecular portraits of human breast tumors. *Nature* 406: 747–752.

Petricoin, E. F. III. et al. 2002. Medical applications of microarray technologies: A regulatory science perspective. *Nat. Genet.* 32 (Suppl.): S474–S479.

Quackenbush, J. 2002. Microarray data normalization and transformation. *Nat. Genet.* 32 (Suppl.): S496–S501.

Rasmussen, R., T. Morrison, M. Herrmann and C. Wittwer. 1998. Quantitative PCR by continuous fluorescence monitoring of a double strand DNA specific binding dye. *Biochemica* 2: 8–11.

Reinke, V. and K. P. White. 2002. Developmental genomics approaches in model organisms. *Annu. Rev. Genomics Hum. Genet.* 3: 153–178.

Robertson, G. et al. 2007. Genome-wide profiles of STAT1 DNA association using chromatin immunoprecipitation and massively parallel sequencing. *Nature Methods* 4: 651–657.

Ross, D. T. et al. 2000. Systematic variation in gene expression patterns in human cancer cell lines. *Nat. Genet.* 24: 227–235.

Roth, F. P., J. Hughes, P. Estep and G. M. Church. 1998. Finding DNA regulatory motifs within unaligned noncoding sequences clustered by whole-genome mRNA quantitation. *Nat. Biotechnol.* 16: 939–945.

Rouillard, J-M., C. J. Herbert and M. Zuker. 2002. OligoArray: Genome-scale oligonucleotide design for microarrays. *Bioinformatics* 18: 486–487.

Rustici, G. et al. 2004. Periodic gene expression program of the fission yeast cell cycle. *Nat. Genet.* 36: 809–817.

Schadt, E. E. et al. 2003. Genetics of gene expression surveyed in maize, mice and man. *Nature* 422: 297–302.

Schena, M. et al. 1995. Quantitative monitoring of gene expression patterns with a cDNA microarray. *Science* 270: 467–470.

Segal, E., T. Raveh-Sadka, M. Schroeder, U. Unnerstall and U. Gaul. 2008. Predicting expression patterns from regulatory sequence in *Drosophila* segmentation. *Nature* 451: 535–540.

Spellman, P. et al. 1998. Comprehensive identification of cell cycle-regulated genes of the yeast *Saccharomyces cerevisiae* by microarray hybridization. *Mol. Biol. Cell* 9: 3273–3297.

Storey, J. D. and R. Tibshirani. 2003. Statistical significance for genome-wide studies. *Proc. Natl. Acad. Sci. (USA)* 100: 9440–9445.

Stranger, B. E. et al. 2007a. Relative impact of nucleotide and copy number variation on gene expression phenotypes. *Science* 315: 848–853.

Stranger, B. E. et al. 2007b. Population genomics of human gene expression. *Nat. Genet.* 39: 1217–1224.

Tamayo, P. et al. 1999. Interpreting patterns of gene expression with self-organizing maps: Methods and application to hematopoietic differentiation. *Proc. Natl. Acad. Sci. (USA)* 96: 2907–2912.

Torcia, M. G., et al. 2007. Functional deficit of T regulatory cells in Fulani, an ethnic group with low susceptibility to *Plasmodium falciparum*

malaria. *Proc. Natl. Acad. Sci. (USA)* 105: 646–651.

Tusher, V. G., R. Tibshirani and G. Chu. 2001. Significance analysis of microarray data applied to the ionizing radiation response. *Proc. Natl. Acad. Sci. (USA)* 98: 5116–5121.

van't Veer, L. J., et al. 2002. Gene expression profiling predicts clinical outcome of breast cancer. *Nature* 415: 530–536.

Velculescu, V. E., L. Zhang, B. Vogelstein and K. Kinzler. 1995. Serial analysis of gene expression. *Science* 270: 484–487.

Velculescu, V. E. et al. 1997. Characterization of the yeast transcriptome. *Cell* 88: 243–251.

Velculescu, V. E. et al. 1999. Analysis of human transcriptomes. *Nat. Genet.* 23: 387–388.

White, K. P., S. Rifkin, P. Hurban and D. S. Hogness. 1999. Microarray analysis of *Drosophila* development during metamorphosis. *Science* 286: 2179–2184.

Whitfield, C. W., A.-M. Cziko and G. E. Robinson. 2003. Gene expression profiles in the brain predict behavior in individual honey bees. *Science* 302: 296–299.

Wilhelm, B. T., S. Marguerat, S. Watt, F. Schubert, V. Wood, I. Goodhead, C. Penkett, I. Rogers and J. Bähler. 2008. Dynamic repertoire of a eukaryotic transcriptome surveyed at single-nucleotide resolution. *Nature* 453: 1239–1243.

Wolfinger, R., G. Gibson, E. Wolfinger, L. Bennett, H. Hamadeh, P. Bushel, C. Afshari and R. S. Paules. 2001. Assessing gene significance from cDNA microarray gene expression data via mixed models. *J. Comput. Biol.* 8: 625–637.

Yvert, G. et al. *Trans*-acting regulatory variation in *Saccharomyces cerevisiae* and the role of transcription factors. *Nat. Genet.* 35: 57–64.

Zhang, L. et al. 1997. Gene expression profiles in normal and cancer cells. *Science* 276: 1268–1272.

5 Proteomics and Functional Genomics

Proteomics may be defined loosely as the study of the structure and expression of proteins, and of the interactions between proteins. Our discussion of proteomics begins with a description of how proteins are annotated computationally, then moves on to a survey of how data on the expression and identity of proteins in cells is obtained. We also describe several methods that are used to study the structure of proteins.

Functional genomics, which here refers to documentation of the functions of large numbers of genes using mutational and recombinant molecular biological approaches, is included in this chapter to emphasize that it shares with proteomics the common goal of ascertaining biological function.

Functional Proteomics

Protein Annotation

The major database for protein sequence and function is the **UniProt Universal Protein Resource** (http://www.pir.uniprot.org; Apweiler et al. 2004). This site has been assembled from three contributing resources: the European Bioinformatics Institute (EBI), the Swiss Institute of Bioinformatics' Expert Protein Analysis System (ExPASy), and Georgetown University's Protein Information Resource (PIR). It consists of three components: a comprehensive repository of individual protein sequences (UniParc), a collection of related sequences with 100%, 90%, or 50% sequence identity (the UniRef clusters), and a "knowledgebase" (**UniProtKB**) that is the major portal for accessing curated information about any protein in the database. UniProtKB is in turn divided into two sections, one termed SwissProt that consists of entries that have been manually edited for content and quality, and the other called TrEMBL that consists of computationally generated records only. Each data-

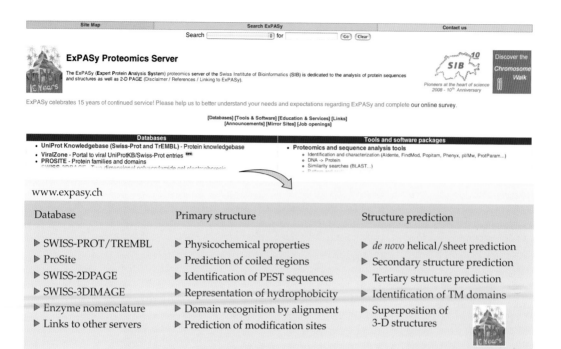

Figure 5.1 Proteomics tools on ExPASy. The layout of ExPASy, the Expert Protein Analysis System Web site, is documented at **www.expasy.org/sitemap.html**.

base can be searched either by text (for example sequence ID, gene ontology, or sequence properties) or by BLAST similarity alignment.

A standard UniProtKB entry includes the amino acid sequence, protein name or description, taxonomic data, citations, and cross-references to more than a dozen other genomic databases with nucleotide sequence, gene expression, and functional data. Typically, it also links to numerous other types of information, including known amino acid sequence variants, structural features of the predicted protein, gene ontology information, and known or predicted types of protein modification. The ExPASy site (http://ca.expasy.org; Figure 5.1) further provides access to a wide range of tools for protein analysis and links to Web sites that facilitate analysis of proteomic data as discussed below.

Proteins can be regarded as strings of domains, with each domain having its own structure, function, and evolutionary history. The standard database of protein domains extracted from the Swiss-Prot database is **Prosite**, which defines domains according to a pattern-recognition algorithm. In essence, a set of residues commonly found in a cluster of related proteins is identified and used as the basis for comparison with other protein sequences. For example, I-G-[GA]-G-M-[LF]-[SA]-x-P-x(3)-[SA]-G-x(2)-F is the consensus signature for aromatic amino acid permeases. All proteins with this domain have a sequence resembling "isoleucine then glycine then glycine or alanine then glycine then methionine then leucine or phenylalanine, and so on (with x representing any amino acid). Such consensus signatures are not always obvious, so a more general approach to domain recognition is to generate a **weight matrix** (also known as a **profile**) of position-specific amino acid weights and gap penalties, as outlined in Box 5.1.

The options for exploring protein similarities are bewildering. Since different pattern and profile search methods are not guaranteed to yield the same results, it is generally a good idea to compare multiple domain search methods. All new protein sequences deposited in a protein database are compared with the complete list of profiles in the various databases by way of hidden Markov model algorithms. For example, the Prosite database represents a collection of protein sequences that are structurally related to one of over 1,400 known functional protein domains. Similar approaches are used to represent the Pfam, TIGRFAM, and Panther families of proteins and protein domains. Other domain-finding methods linked to the UniProt annotation of a protein include SMART, PRINTS, and ProDom, which may give alternate perspectives on the structure and function of the sequence. The TRANSFAC database brings together information on transcription factors, while SENTRA serves a similar role for prokaryotic signal transduction proteins.

In addition to assigning each gene product to a family of similar proteins, protein annotations aim to provide information on the primary, secondary, and tertiary structures of proteins. As shown in Figure 5.2, primary structure

(A) Primary

```
MKVLLRLICFIALLISSLEADKCKEREEKIILVSSANEIDVRPCPLNPNEHKGTITWYKD
DSKTPVSTEQASRIHQHKEKLWFVPAKVEDSGHYYCVVRNSSYCLRIKISAKFVENEPNL
CYNAQAIFKQKLPVAGDGGLVCPYMEFFKNENNELPKLQWYKDCKPLLLDNIHFSGVKDR
LIVMNVAEKHRGNYTCHASYTYLGKQYPITRVIEFITLEENKPTRPVIVSPANETMEVDL
GSQIQLICNVTGQLSDIAYWKWNGSVIDEDDPVLGEDYYSVENPANKRRSTLITVLNISE
IESRFYKHPFTCFAKNTHGIDAAYIQLIYPVTNFQKHMIGICVTLTVIIVCSVFIYKIFK
IDIVLWYRDSCYDFLPIKASDGKTYDAYILYPKTVGEGSTSDCDIFVFKVLPEVLEKQCG
YKLFIYGRDDYVGEDIVEVINENVKKSRRLIIILVRETSGFSWLGGSSEEQIAMYNALVQ
DGIKVVLLELEKIQDYEKMPESIKFIKQKHGAIRWSGDFTQGPQSAKTRFWKNVRYHMPV
QRRSPSSKHQLLSPATKEKLQREAHVPLG
```

(B) Secondary

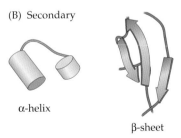

α-helix

β-sheet

(C) Tertiary

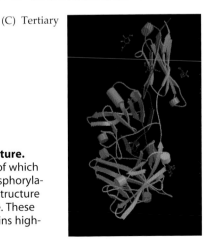

Figure 5.2 Primary, secondary, and tertiary protein structure. The primary structure is the sequence of amino acids, some of which can be identified as likely sites for modifications such as phosphorylation (shown in color for S, T, and Y residues). The secondary structure is the local folding of alpha helix and beta sheet, for example. These fold into tertiary structure, and very often into distinct domains highlighted in blue, green, and red.

EXERCISE 5.1 *Structural annotation of a protein*

Using the conceptual translation of the gene that you identified in Exercise 1.2, perform the following steps designed to annotate structural features of the protein.

Access the UniProt file for the protein and note the following:

a. Find the Ensembl translation ID and UniProt accession number of the protein.

b. List the start and endpoints of the Prosite and Pfam domains.

c. What is the predicted isoelectric point, charge, and mass of the peptide?

d. What is the Prosite Consensus Pattern for this protein?

e. If you can, find an image of the 3D structure of the predicted protein.

ANSWER: *From the UniProtKB home page (http://www.pir.uniprot.org/ database/knowledgebase.shtml), type "IL13" in the search box and add a second search box for "Human." On the page that comes up, IL13 HUMAN may not be the first entry, but it should be listed as UniProt ID P35225. Scroll down to the "Database Cross-References" and you will find that the Ensembl entry is ENSG00000169194. This links you to the gene page, but the column on the left provides a "Protein Information" link that brings up the Ensembl ProtView. The protein ID is ENSP00000304915. Toward the bottom of the page, the "Interleukin" domain is listed as residues 43–70 for Prosite and 15–57 for Pfam. Beside the peptide sequence, various statistics are listed, including isoelectric point (8.23), charge (+4), and molecular weight (15,788). Find the cross-reference to the Prosite view, entry number PS00838, which documents a 25–28 amino acid consensus pattern for the protein as*

[LI]-x-E-[LIVM](2)-x(4,5)-[LIVM]-[TL]-x(5,7)-C-x(4)-[IVA]-x-[DNS]- [LIVMA]

Numerous links in the PDB section at the bottom of the page show ribbon structures of the protein similar to the one shown here.

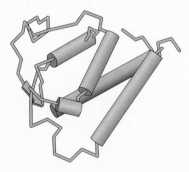

refers to the sequence and modifications of the amino acids. Secondary structure is the local folding into alpha helix, beta sheet, or coiled coils. Tertiary structure is the folding of these elements into domains, and domains can fold into quaternary conformations and multiprotein complexes.

It should be appreciated that much of the information associated with a given protein annotation is generated by computer prediction and/or comparison with similar proteins that have been studied in other organisms. Even the primary protein sequence is often a best-guess in the sense that the "start" methionine is usually assumed to be encoded by the first AUG identified by the gene-finding algorithm or by mapping of the 5′ end of a transcript. Furthermore, alternative splice variants are not accurately described by predictive algorithms or EST sequence surveys, and can be missed even after extensive molecular characterization of the gene. It is also generally assumed that posttranscriptional editing has not occurred and that rare amino acids have not been inserted, even though examples of both these phenomena have been described in a wide variety of organisms. (One prominent example is RNA editing of the vertebrate glutamate receptors; see Bass 2002.)

Annotation of the primary sequence often includes the predicted sites of posttranslational protein modification. Identifying these sites is a particularly acute problem for eukaryotic proteomes, since protein modification is ubiquitous and subcellular structure is so rich. Software is available online to suggest likely sites for phosphorylation by a variety of kinases; for glycosylation, acetylation, methylation, and myristoylation; for peptide cleavage (including ubiquitin conjugation sites that target proteins for degradation); and for targeting of proteins to intracellular compartments such as the mitochondrion, lysosome, chloroplast, or nucleus. These tools have variable accuracy. Methods exist for confirming that each type of modification or intracellular targeting actually occurs, but they rely on the prior generation of monoclonal antibodies against the protein and/or the purification of native protein, both of which are time-consuming endeavors. Yet because the activity of a protein is so often modulated by chemical modifications, annotating these effects is just as important as annotating the distribution of protein expression. High-throughput methods for detecting posttranslational modification are described in Kannicht 2002.

Protein annotations often link to tools that enable graphical portrayal of the predicted or described protein secondary structure and other protein features, such as those shown in Figure 5.3 for a *Drosophila* octopamine receptor. Numerous algorithms have been developed for predicting the distribution of alpha helix, beta sheet, or coiled coils, based on properties such as the distribution of charge and bulk of amino acid side chains. Each of the programs listed on the ExPASy proteomics tools site adopt distinct heuristics that incorporate alignments of similar sequences to improve predictions. If random assignment of protein secondary structure yields a score of 33 on an accuracy scale from 0 to 100, predictive algorithms will tend to increase the score only to 75%. Hydrophobicity plots are at the core of good methods for predicting membrane-spanning domains; their performance accu-

BOX 5.1 Hidden Markov Models in Domain Profiling

One of the more active areas of current investigation is the identification of **sequence motifs** from genomic data. This task spans a wide variety of specific applications. Examples include finding tRNA genes, identifying α-helix regions in proteins, and detecting transcription factor binding sites in the upstream regions of genes. Many of these seemingly disparate tasks can be approached through the use of **profile hidden Markov models.** We introduced hidden Markov models in Box 2.3; here we build on those basic concepts to describe profile HMMs.

Profile HMM methods begin with the construction of a probabilistic description of the variation in a motif of interest—a protein secondary-structure element, for example, or a DNA binding site. Typically, the parameters and structure of the model are obtained by training, using an empirical data set of examples of the motif in question. For example, a profile HMM for a DNA binding site is constructed by aligning many examples of that binding site and extracting characteristic properties.

Profile HMMs are a probabilistic extension of the notion of a **consensus sequence,** as illustrated in this simple example:

	1	2	3	4	5	6	7
A	A	C	T	–	–	T	A
B	A	C	T	C	–	T	G
C	A	T	T	–	–	T	T
D	T	C	T	C	T	T	G
E	A	T	T	C	–	G	G

One possible consensus sequence for this multiple alignment is:

[AT][CT][T][CT]*[GT][AGT]

This consensus sequence successfully captures much of the information in the alignment. It indicates the main columns of the motif, identifies the potential locations of insertion/deletion variation, and describes potential variation at sites in the motif.

Consider the two sequences ACTCTG and TTTTGA. Both match the consensus sequence, but it is clear after inspecting the entire alignment that the second sequence does not appear to be an example of the motif. The alignment contains more information than the consensus sequence, and it is that information that profile HMMs try to capture and exploit.

A simple profile HMM describing this alignment is shown in Figure A. Let us consider how the HMM in this figure generates sequences. The sequence TCTTA can be generated if the model emits a T from main state 1 (with probability 0.2), moves to main state 2 (with probability 1), emits a C (probability 0.6), moves to main state 3 (probability 1), emits a T (probability 1), moves to main state 4 (probability 0.4), emits a T

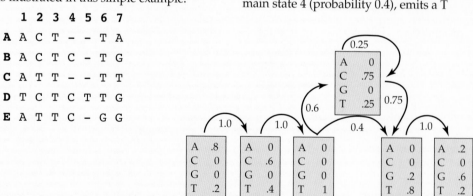

Figure A Simple profile HMM.   The model contains five *main states* (along the bottom row) that are represented in every example of the motif. Insertion/deletion variation is handled, in this case, by a single *insert state* (the top row). The *emission probabilities* (beside each base) match the empirical frequencies of bases in each column of the alignment used for training. The *transition probabilities* (curved arrows) reflect the empirical locations of indels, which are observed only between main states 3 and 4.

(probability 0.8), moves to main state 5 (probability 1), and emits the final A (probability 0.2). Thus, the probability of this HMM generating this sequence is:

$$\Pr(TCTTA) = 0.2 \times 1 \times 0.6 \times 1 \times 1 \times 0.4 \times 0.8 \times 1 \times 0.2 = 0.00768$$

The presence of the insert state allows an arbitrary number of bases to be inserted between main states 3 and 4. Thus, the probability of the sequence TCTCCTA is:

$$\Pr(TCTCCTA) = 0.2 \times 1 \times 0.6 \times 1 \times 1 \times 0.6 \times 0.75 \times 0.25 \times 0.75 \times 0.75 \times 0.8 \times 1 \times 0.2 = 0.001215$$

The movement from main state 3 to the insert has probability 0.6; once there it emits the C with probability 0.75; and it remains in the insert state with probability 0.25. We see that this is much greater than the probability of this sequence occurring in random sequence with equal frequencies of each base ($\Pr = 4^7 = 0.00006$).

Finally, consider the sequence AGTTG. Although it appears to match the alignment fairly well, it has probability 0 under this HMM because G has never been observed at the second position. When training, it is generally undesirable to have 0 emission probabilities, to avoid such situations. The use of **pseudocounts** insures that all emission probabilities are nonzero.

The current model only allows for indel variation between main states 3 and 4. This assumption may or may not be biologically

reasonable. In order to allow insertions at any location in the sequence, in more general HMMs every main state is linked to an insert state. Furthermore, to allow examples of the motif that are missing one or more main states, delete states are added to the HMM. The general form of a profile HMM is shown in Figure B.

Main states can be thought of as the core of the motif. Insert states, as their name implies, allow for the insertion of residues beyond the motif core, while delete states provide a mechanism for the (typically rare) circumstance where a main state is missing from a representative of the motif being modeled. Training and determination of the final form of the model are explained in more detail by Sonnhammer et al. (1998).

The profile HMM has numerous uses. Given a specific sequence, one can ask whether or not it is an example of a motif by comparing its probability of being generated by the profile HMM with the probability that the sequence would be observed in a random stretch of DNA. Extending this notion, the PFAM database consists of profile HMMs constructed for many protein families. A query sequence can be compared to each of the HMMs in the database to see if it is representative of any of those families. Profile HMMs can also be used to probe genomic sequences and identify likely examples of the motif. Finally, the transition and emission probabilities within a profile HMM help us understand the molecular biology of the motif being modeled.

Figure B General form of a profile HMM.   Every main state is linked to an insert state and delete states are added.

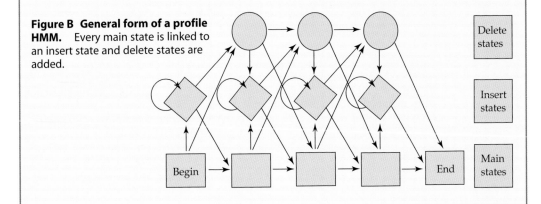

(A) Alpha-helical content prediction

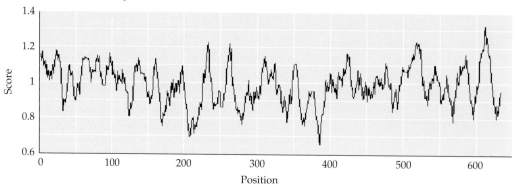

(B) Transmembrane domain prediction

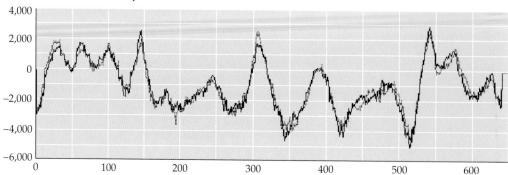

(C) Phosphorylation site prediction

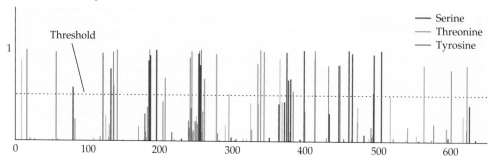

Figure 5.3 Protein structural profiles. Profiles of the *Drosophila* octopamine receptor OAMB protein, as determined using software linked to ExPASy. (A) Prediction of α-helical content, using the Chou-Fassman algorithm with ProtScale. (B) Prediction of location of the seven transmembrane domains typically found in G-protein linked receptors, using TMpred. (C) Prediction of probable serine, threonine, and tyrosine phosphorylation sites, using NetPhos.

racy is increased by including the presence of charge bursts on either side of the membrane. Gene annotations also provide direct links to representation of tertiary structure in the form of ribbon or cylinder diagrams, described later in this chapter.

Protein Separation and 2D-PAGE

In order to characterize the proteome of a cell type, it is necessary first to separate the proteins from one another and then to determine the identity and, if possible, the relative quantity of each of thousands of different proteins. Protein separation has conventionally been achieved using **two-dimensional polyacrylamide gel electrophoresis**, or **2D-PAGE** (Figure 5.4). Separation according to charge is followed by separation according to mass. While this procedure does not guarantee separation of all proteins, it is nevertheless possible to identify several thousand different spots on a carefully prepared gel. The proteins are detected most simply by staining the gel with either silver stains or fluorescent dyes such as SYPRO Ruby; for some applications, a radioactive group can be incorporated to increase sensitivity.

Separation according to charge, or more technically **isoelectric point (pI)**, originally relied on carrier molecules known as ampholytes to establish a pH gradient in a column of acrylamide gel during electrophoresis. This procedure is not sufficiently reproducible for side-by-side comparison of gels prepared from two different tissue samples, so has largely been replaced by immobilized pH gradients (IPGs) that are built into commercially supplied gel slices. These IPGs are now being produced to cover overlapping narrow pH ranges, with the result that greater resolution can be achieved by running sub-samples under a series of different conditions that are optimized for a particular comparison.

Separation according to molecular mass is performed by running samples out of the isoelectric focusing gel into an SDS-PAGE slab gel. The SDS detergent both masks any charge differences and denatures the proteins so that they migrate through the pores in the gel according to size of the protein, much like electrophoretic separation of DNA molecules.

2D-PAGE is not quantitative over the complete range of protein concentrations, which are known to cover several orders of magnitude. There are at least three sources of error: non-stoichiometric extraction of proteins from various cellular constituents; failure of proteins to absorb into or migrate out of the isoelectric focusing gel; and non-linear responses of the dye used for detection. Highly charged and low-abundance proteins tend to be underrepresented due to poor extraction, though buffers with special non-ionic detergents and other reagents can improve efficiency. Similarly, proteins sequestered in organelles and membranes or bound up in nuclear or extracellular matrix are not necessarily represented in proportion to their abundance, while differences in abundance between treatments may merely reflect altered cellular conditions that affect the efficiency of extraction. Very large proteins may not migrate into the SDS gel, and very small ones may migrate off it under conditions that separate the majority of cellular con-

Figure 5.4 Two-dimensional polyacrylamide gel electrophoresis. (A) In 2D PAGE, protein extracts are applied to the center of an isoelectric focusing strip and allowed to diffuse along the ionic gradient to equilibrium. The strip is then applied to an SDS gel, where electrophoresis in the second dimension separates proteins into "spots" according to molecular mass. The size of each spot is proportional to the amount of protein. (B) A partially annotated human lymphoma 2D gel from the ExPASy database.

(A)

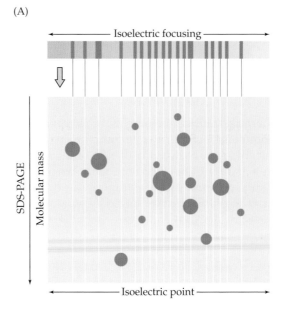

(B)

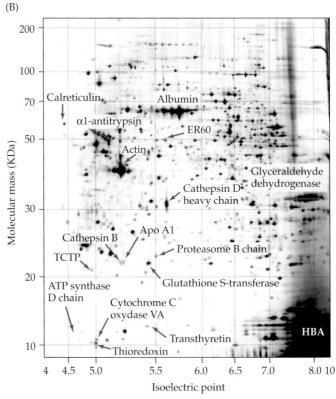

stituents. Fluorescent dyes reportedly have greater linear range than silver stains for monitoring of concentration differences, but may not accurately report concentration differences over more than two orders of magnitude.

Web sites are available that allow individual researchers to download images and documentation of characteristic 2D-gels from a variety of tissues in a variety of species under well defined conditions. For example, the SWISS-2DPAGE database (http://ca.expasy.org/ch2d) offers several options for searching for proteins on gels, or for querying gels to identify proteins (see Figure 5.4B). Side-by-side comparison of 2DPAGE gels is not straightforward due to variation in migration rates and local distortions, but can be facilitated by software that distorts images so that they can be overlaid. The Melanie package (http://ca.expasy.org/melanie), developed by the Swiss Institute of Bioinformatics, provides statistical data on the probability that a particular spot corresponds to one that has already been annotated on a reference gel.

Semi-quantitative estimation of protein concentrations can be made by integration of pixel intensities over the surface area of a spot, allowing crude calculation of relative abundance between treatments after appropriate normalization. Comparisons between gels are only suitable for contrasts of very similar tissues, as variation in posttranslational modification can cause subtle shifts in migration of the same protein prepared from different tissues. However, new two-color fluorescence detection methods have been introduced that allow direct comparison of protein abundance from two conditions that are separated on the same gel (Unlü et al. 1997; Minden 2007). This approach is also known as **2D-difference gel electrophoresis** (**DIGE**).

Several other methods available for protein separation offer distinct advantages, both for thorough characterization of a proteome and for isolation of native protein complexes. Standard chromatographic methods similar to those classically used for protein purification can be employed to separate subfractions of proteins prior to 2D-PAGE. This step allows more concentrated samples to be loaded, and also increases the resolution of low-abundance proteins.

Affinity chromatography, illustrated in Figure 5.5, is a proven technique for purifying groups of proteins that form physical complexes within a cell. The idea is to reversibly link one of the components of the protein complex to a chromatographic matrix, then wash the cellular constituents over the column under gentle conditions that allow normal protein-protein interactions to occur. Once the bulk of the cellular proteins have washed through, the remaining proteins can be eluted with a different buffer, and then characterized by a variety of methods.

One method of transient crosslinking is to use a monoclonal antibody directed against the known component. Another is to use recombinant DNA technology to tag the "bait" protein with a peptide fragment such as polyhistidine, maltose binding protein (MBP), or glutathione S-transferase (GST) that will bind to commercially supplied columns. The sequences that encode these tags are built into cDNA cloning vectors so that when expression of the protein is induced in bacterial cells, the tag is incorporated at either the

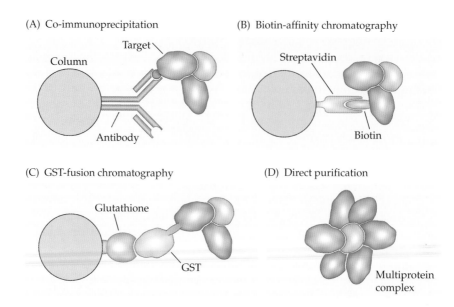

Figure 5.5 Affinity chromatography.   A variety of methods exist for purifying proteins that interact with a target protein. (A) In co-immunoprecipitation, an antibody tethered to the chromatographic column matrix binds to the target protein, which in turn binds loosely to interacting partners. (B) Biotin can be chemically crosslinked to the target protein, which will then bind to streptavidin on the column. (C) GST fusion proteins synthesized using recombinant DNA methods consist of glutathione S-transferase, which binds to glutathione on the column, translated in frame with the target protein. (D) Large macromolecular complexes can be purified directly on sedimentation gradients or by other chromatographic methods.

N-terminal or C-terminal end where it is unlikely to interfere with protein folding. A library of cDNAs in such vectors is known as an **expression library**. Combinations of two or more tags together on the same protein can increase efficiency and purity of purifications, and is particularly useful where thousands of different proteins are being purified. For nucleic acid-binding complexes, affinity chromatography has been performed with biotinylated RNA or DNA probes that bind to streptavidin matrices. Examples of this technology include description of up to 100 proteins that constitute molecular machines such as the spliceosome, nuclear pore complex, and spindle pole body.

Mass Spectrometry

Protein identities are routinely determined using a combination of peptide sequencing and mass spectrometric (MS) methods that achieved high-throughput scale in the late 1990s (Chalmers and Gaskill 2000). The basis of this approach, shown in Figure 5.6, is that each protein can be identified by the overlap between a set of identified and predicted signature peptides.

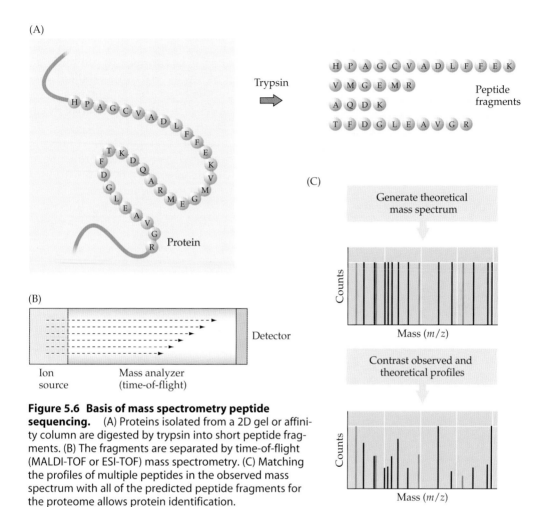

Figure 5.6 Basis of mass spectrometry peptide sequencing. (A) Proteins isolated from a 2D gel or affinity column are digested by trypsin into short peptide fragments. (B) The fragments are separated by time-of-flight (MALDI-TOF or ESI-TOF) mass spectrometry. (C) Matching the profiles of multiple peptides in the observed mass spectrum with all of the predicted peptide fragments for the proteome allows protein identification.

Rather than directly determining the amino acid sequence by chemical means, peptide fragments generated by trypsin digestion of whole proteins are separated according to their mass-to-charge ratio (m/z). Each m/z peak corresponds to a peptide 5–20 amino acids in length, whose precise mass is a function of the actual sequence of amino acids. Any given peak may correspond to dozens of possible peptides and so does not uniquely identify a protein; but several peaks derived from one or even several proteins provide a statistically supported identification of the protein (Fenyö 2000).

A **mass spectrophotometer** consists of three units. The first is an *ionization device* that moves individual peptide fragments from the solid phase into a gaseous ion phase after extraction from a 2D gel or elution from a chromatography column. Two common ionization units are the matrix-assisted laser desorption ionization (MALDI) and electrospray ionization (ESI) devices. The second unit is a *separation chamber*, in which the ions move through a vacuum according to their charge and mass and are separated on the basis of time-of-flight (TOF). The third device is a *detector* with the sensitivity and resolving power to separate peaks for over 10,000 separate molecular species, with a mass accuracy as low as 20 parts per million. Just a few picograms of fragments, corresponding to several million molecules, can be detected with resolution of differences of less than one-tenth of a Dalton in relative mass of molecules 2,000 Da in size.

Protein identities are computed automatically from peptide m/z spectra, by comparison with a database of the predicted spectra in the proteome of the organism under study. These databases are assembled by *in silico* trypsin digestion of conceptually translated EST or cDNA sequences, and/or predicted protein coding regions in genomic DNA. Ambiguities due to an incomplete database, to overlap of the spectra of two or more possible pep-

(A)

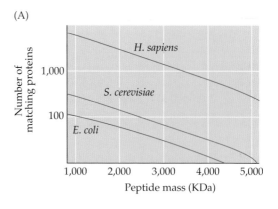

(B)

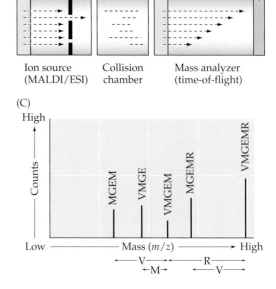

Figure 5.7 Peptide sequencing by tandem mass spectrometry.   (A) Any peptide mass corresponds to multiple fragments from different proteins in the proteome, with the number of possible matches increasing with size of the proteome. This plot shows the predicted number of fragments of each size for *E. coli*, yeast, and human. (B) The problem of aligning peptide fragments with proteins is simplified by partial sequencing of fragments using tandem mass spectrometry. (C) The peaks of MS/MS profiles are separated by characteristic widths that correspond to the amino acid that is knocked off either end of the peptide sequence in the collision chamber. (A data from Fenyö 2000.)

(C)

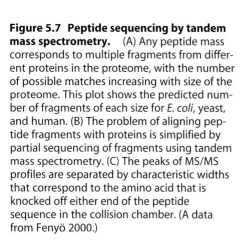

tides, and to differences in posttranslational modification can be resolved by obtaining actual peptide sequence data using the **tandem mass spectrometry** (**MS/MS**) technique (Figure 5.7). In this procedure, before allowing a species of ion to strike the detector, it is shunted into a collision chamber in which nitrogen or argon gas molecules at low pressure collide with the ions. This causes the peptide backbone to break, liberating subfragments that are then separated by time-of-flight and detected. Because the subfragments are derived from a single peptide, the distances between the peaks define the mass differences of the subfragments, and hence identify the constituent amino acids sequentially from both ends of the peptide.

EXERCISE 5.2 *Identification of a protein on the basis of a mass spectrometry profile*

Given the protein sequence:

MALWMRGFFYTPKPGAGSLQPLRALEGSLQKGIVEQCCTKSICSR

compute the number of tryptic digestion products and their lengths in amino acids. Given the approximate amino acid masses listed below, also compute the expected approximate masses of the different peptides, and determine whether the protein is likely to be present in MS profile (A) or (B) shown in the figure.

Ala	*A*	*89*	*Arg*	*R*	*174*	*Asp*	*D*	*133*	*Pro*	*P*	*115*
Gly	*G*	*75*	*Lys*	*K*	*146*	*Glu*	*E*	*147*	*Phe*	*F*	*165*
Met	*M*	*149*	*Gln*	*Q*	*146*	*Cys*	*C*	*121*	*Val*	*V*	*117*
Leu	*L*	*131*	*Asn*	*N*	*132*	*Ser*	*S*	*105*	*Trp*	*W*	*204*
Ile	*I*	*131*	*His*	*H*	*155*	*Thr*	*T*	*119*	*Tyr*	*Y*	*181*

(A) (B)

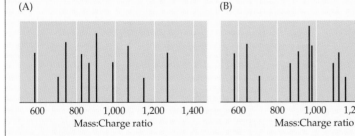

Mass:Charge ratio Mass:Charge ratio

ANSWER: *Trypsin cleaves at arginine and lysine residues, so the predicted fragments are:*

MALWMR	*(889)*	*ALEGSLQK*	*(970)*
GFFYTPK	*(966)*	*GIVEQCCTK*	*(1,123)*
PGAGSLQPLR	*(1,156)*	*SICSR*	*(636)*

Profile B includes peaks corresponding to each of these fragment sizes, so is most likely to correspond to the protein.

In order to characterize the entire proteome of a cell population, **liquid chromatography** can be used in combination with tandem mass spectrometry (**LC/MS/MS**), obviating the need for prior electrophoretic separation of spots. The combination of partial sequence data with peptide masses and appropriate analytical software is sufficient to allow protein identification in complex mixtures of tens of proteins that emerge from each fraction of the chromatograph. Using the more complex two-dimensional cation exchange/reversed phase liquid chromatography approach known as Multidimensional Protein Identification Technology (MudPIT), Washburn et al. (2001) were able to detect and identify 1,484 yeast proteins, including many rare species and integral membrane proteins that are not detected with other methods. Furthermore, the identity of the individual peptides represented in the MS/MS spectra provides information on the likely folding pattern of membrane-spanning regions, which are not exposed to chemical cleavage. MudPIT provides enhanced resolution of peptide mixtures and is being used to study protein interactions as well as the protein content of organelles, viruses, molecular machines, and affinity-purified protein complexes (Liu et al. 2002).

Neither 2D-gels nor mass spectrometry are quantitative methods, since the relative intensities of spots on a gel or peaks in a spectrum provide at best qualitative representations of protein concentration. This is true of comparisons within and between samples, and relates to factors such as nonstoichiometric extraction of proteins from tissues, overlap of spots and peaks, and the nonlinearity of protein-dye response. However, techniques for quantitative proteomics do exist and they are reviewed in Panchaud et al. (2008).

One approach to quantifying protein expression between samples or treatments is to label one of the samples with a heavier stable isotope such that the MS peaks lie immediately adjacent to one another. In this case, the ratios of the heights of the peak are indicative of relative expression level. Heavy isotopes include ^{15}N and ^{2}H, which can be incorporated into the proteins either in vivo or after extraction. In vivo labeling involves growing cells in a medium supplemented with the heavy isotope, which is incorporated into normal biosynthesis. Incorporation following extraction can be performed by direct coupling to reactive amide groups on lysine residues in particular.

Isotope-coded affinity tag (ICAT) reagents crosslink to cysteine residues on proteins, and include a biotin group that allows purification of labeled proteins. The commercially supplied reagents carry eight light (hydrogen) or heavy (deuterium) atoms substituted on the carbon side chain, ensuring uniform separation of labeled fragments during MS (Figure 5.8). This procedure was first applied to the characterization of differences between yeast cells grown on two different carbon sources (Gygi et al. 1999), but has since been used to study such phenomena as trypanosome development, apoptosis, protein degradation, liver disease, prostate and brain cancers, and Alzheimer's disease. Amine-specific iTRAQ reagents have also been introduced (Ross et al. 2004). These isobaric tagging adducts allow

(A)

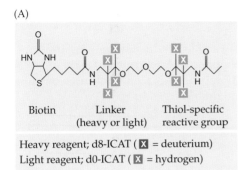

Biotin · Linker (heavy or light) · Thiol-specific reactive group

Heavy reagent; d8-ICAT (**X** = deuterium)
Light reagent; d0-ICAT (**X** = hydrogen)

(B) Mass difference from stable isotopes

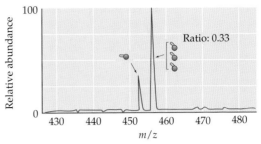

(B) Mass difference from stable isotopes

Ratio: 0.33

(C) Identify peptide by sequence information (tandem MS)

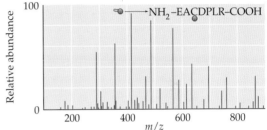

NH_2–EACDPLR–COOH

(D)

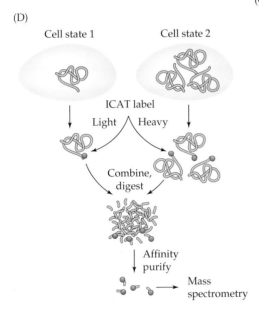

Cell state 1   Cell state 2

ICAT label

Light   Heavy

Combine, digest

Affinity purify

Mass spectrometry

Figure 5.8 Quantitative proteomics using ICAT reagents.   (A) Uniform labeling of two different protein samples is achieved using commercially available ICAT reagents that have eight light or heavy hydrogen atoms incorporated onto the carbon chain of a side-group that is crosslinked to reduced cysteine residues. (B) After tryptic digestion, ICAT-labeled fragments are purified by affinity chromatography against the biotin group on the label. (C) Mass spectrometry is used to identify differentially regulated peaks, and (D) tandem MS is subsequently performed to determine the identity of chosen differentially regulated fragments. (After Gygi et al. 2000.)

relative and absolute quantitation; they typically tag lysine residues and can be multiplexed for comparison of up to eight biological samples simultaneously.

Direct comparison of protein and mRNA levels in the same types of cell is technically difficult owing to differences in detection thresholds and array annotation, and the fact that transcriptomic and proteomic methodologies demand different expertise. It is clear that changes in transcript abundance need not imply that the corresponding proteins change, and protein levels can change without alteration of transcript levels (Celis et al. 2000; Unwin and Whetton 2006). Nevertheless, the two measures are correlated, and both approaches provide complementary windows onto the dynamic nature of gene activity.

There are several reasons why transcript and protein profiles may differ, including:

- Gene expression is regulated at several levels, including translation and degradation, which leads to uncoupling of transcript and protein levels (RNA turnover can be extremely rapid, taking only minutes in some cases; proteins can be cleaved into multiple products).

- Alternative splicing and protein modification lead to misrepresentation of the total levels of both classes of gene product when single spots are measured.

- Proteins are present over a large range of concentrations inside cells, whereas detection methods are not sensitive over more than two or at most three orders of magnitude, while low level proteins are often undetected.

These observations imply that some caution should be placed on the interpretation of the biological significance of differences in mRNA abundance detected by microarrays and chips. It should also be recognized that protein expression levels are not necessarily indicative of protein function. This is because posttranslational modification, subcellular localization, and association with small molecules and other proteins all greatly affect protein function.

Immunochemistry

A convenient way to visualize specific proteins, either in extracts or in whole cells and tissues, is through the use of **antibodies**. Antibodies are secreted immunoglobulin proteins that are a part of the adaptive immune response mounted by vertebrates against foreign agents. When a purified protein is injected into a mouse or rabbit, for example, that protein becomes an antigen, and the animal responds by generating a series of antibodies that recognize a variety of epitopes on the protein. These polyclonal antibodies can be converted into **monoclonal antibodies** (**MAbs**) by fusion of the mouse B cells with a myeloma cell line to create an immortal hybridoma that expresses a single class of immunoglobulin. MAbs recognize a single **epitope**, or short peptide, and so are highly specific for individual proteins or even modified protein isoforms. Binding of a MAb to a protein is detected indirectly: a commercially supplied secondary antibody that is conjugated to some type of label is bound to the constant region of the MAb (Figure 5.9). The label can be a radioactive tag such as ^{125}I, a fluorescent dye, or an enzyme that catalyzes a pigment-forming reaction.

The use of MAbs to detect proteins that have been separated according to mass on a gel and transferred to a nylon membrane is known as **Western blotting** (by analogy with Southern and Northern blotting for DNA and RNA hybridizations). A particularly sensitive type of detection performed in the wells of a microtitre plate is known as **enzyme-linked immunosorbant assay**, or **ELISA**, and is useful for high-throughput and semi-quantitative analysis of protein expression in cell extracts.

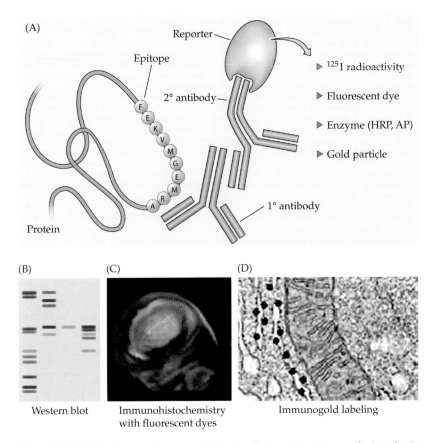

Figure 5.9 Antibodies and immunohistochemistry. Proteins are detected using primary antibodies directed against an epitope in the target protein. The primary antibody is detected by a secondary antibody conjugated to one of a variety of labels, including radioactivity, fluorescence, gold particles, and enzymes. Labeled protein can then be detected in blots (B), tissue preparations (C), and thin sections (D).

The localization of proteins to tissues or whole organisms either in wholemount or thin sections is known as **immunohistochemistry**. Subcellular localization of protein distribution can be performed using **immunogold labeling**, in which gold particles of various sizes are bound to the antibodies and visualized by electron microscopy. Co-localization of two or three different proteins in a tissue can be achieved by clever choice of primary and secondary antibodies from different animals, and is very useful for directly assessing whether the same proteins are expressed in the same cells.

Protein Microarrays

Protein microarrays are analogous to transcription profiling arrays, except that proteins are arrayed on a slide surface or on microbeads. They are gain-ing in usage as the unique problems associated with tethering folded and active proteins to an array surface are overcome (Uttamchandani et al. 2006). The two key technological innovations behind protein microarrays were (1) the development of methods for high-throughput production and purification of micromolar amounts of protein, and (2) the demonstration that proteins can be displayed on a glass surface in such a way that they retain the capacity to bind to other molecules in a native manner.

Proteins have traditionally been produced by cloning random cDNA fragments into an inducible expression vector in *E. coli*. The bacteria are plated on a filter, grown, and then lyzed to expose the expressed protein. Since eukaryotes use different translation initiation signals, the translation start site is supplied by the cloning vector and (so long as the fusion preserves the open reading frame) a nearly full length protein can be expressed. The major drawbacks of the method are that *E. coli* does not modify proteins correctly, that other proteins in the bacteria may interfere with function, and that miniaturization is not feasible. Some of these problems have been overcome by using different host cells—including yeast, lepidopteran, and mammalian cell lines—or plant protoplasts.

Several alternative methods for high-throughput protein purification can be used. One such method, called **phage display**, is to express the protein as a fusion to a phage capsid protein, so that it is expressed on the surface of readily purified phage particles. Another is to express full length cDNAs in reticulocyte lysates, which are a cell-free protein synthesis system. Perhaps the most common method is to incorporate polyhistidine or glutathione S-transferase fusions into expression libraries, as these protein fragments can be recognized by affinity chromatography.

Proteins can also be arrayed as desired either on aldehyde-coated glass slides (which form covalent crosslinks to amine residues at the N-terminus or with exposed lysine residues) or on nickel-coated slides that crosslink to polyhistidine tags. With appropriate care to maintain the hydration of samples during printing, and using the same arraying robots as are used to print cDNA microarrays, protein microarrays can be used to interrogate a variety of interactions (Figure 5.10; MacBeath and Schreiber 2000). A current trend is the development of microbead-based arrays where interactions occur in suspension or in microchambers similar to those described in Chapter 4 for transcript profiling.

Specific protein-protein interactions can most simply be detected by labeling a probe protein (for example, an antibody) with a fluorescent tag. Similarly, small molecules will bind to microarrayed proteins (a fact that provides a potent assay for the recognition of candidate drugs). Substrates for different kinase classes have been detected by allowing the kinase to transfer a radioactively labeled phosphate specifically to the target protein, demonstrating that enzymes can be induced to act on proteins on an array. Methods for determining the abundance of expressed proteins in complex extracts by binding to microarrays of printed specific antibodies, or of mixtures of antibodies to arrayed antigens, are also available (Wingren and Borrebaeck 2004).

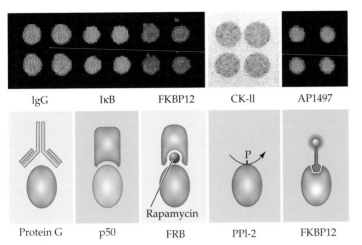

IgG IκB FKBP12 CK-ll AP1497

Protein G p50 FRB PPl-2 FKBP12

Rapamycin

Figure 5.10 Protein micro-arrays. Proof-of-principle experiments have shown that immobilized proteins can be bound to antibodies (Protein G–IgG), protein cofactors (p50–IκB), and protein/small molecule complexes (FRB–FKB12). In addition, specific phosphorylation will occur (PPI-2 by CKII), and proteins will bind dye-labeled small molecules (AP1497–FKB12). (After MacBeath and Schreiber 2000.)

The first demonstration that protein arrays can be used to identify novel protein interactions was provided by Zhu et al. (2001), who expressed and printed 5,800 yeast ORFs as GST/polyHis fusion proteins. This protein array was used to interrogate over 80% of the yeast proteome with calmodulin, and revealed 39 calmodulin-binding proteins, 33 of which were previously unknown. Interrogation with six different types of lipid identified a total of 150 phosphatidylinositol (PI) binding proteins, some of which are specific for particular PI lipids and many of which are likely to have roles in membrane signaling.

A promising approach utilizing on-chip protein synthesis from DNA templates may circumvent the inefficiencies of high throughput protein purification and crosslinking of proteins to the glass surface. Nucleic-acid programmable protein arrays (NAPPA; Figure 5.11) are built by spotting a protein expression library onto the slide, then using cell-free reticulocyte

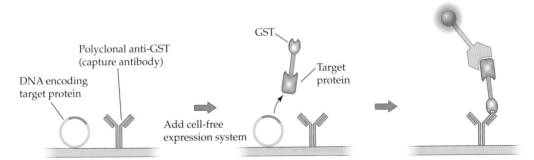

DNA encoding target protein

Polyclonal anti-GST (capture antibody)

Add cell-free expression system

GST

Target protein

Figure 5.11 Self-assembling NAPPA protein microarrays. Each spot on a nucleic-acid programmable protein microarray crosslinks a plasmid vector and anti-GST antibody to the glass slide. Rabbit reticulocyte lysate is used to transcribe and translate the protein, which is trapped by the antibody. In this way, protein is exposed on the array without any need for high-throughput purification.

lysate to transcribe and translate proteins *in situ*. These are captured locally by a monoclonal antibody against the GST epitope at the C-terminal end of each protein, most if which is left available in native conformation to interact with other proteins and reagents. Ramachandran et al. (2004) introduced the technology in a study of the pairwise protein-protein interactions among 29 components of the human DNA replication initiation complex.

Protein Interaction Maps

A dramatically different approach to determining which proteins interact with which other proteins is to allow the interactions to occur in vivo and then to detect an artificial physiological consequence of the interaction. This is the strategy behind **two-hybrid screens**. The gene encoding a protein that normally requires two physically adjacent domains to function is split into two genes. Each of these is fused to a library of cDNA fragments that produce hybrid proteins. Function is only restored if the two hybrid proteins physically interact, bringing the two functional domains of the original protein back together (Figure 5.12).

The original yeast two-hybrid (Y2H) method (Fields and Song 1989) used a transcription-activating factor, GAL4, which consists of a DNA-binding domain (BD) and an activation domain (AD). Neither domain is capable of activating transcription on its own, since the BD requires an activation sequence while the AD must be brought to the promoter of a gene through the BD. Classically, the BD is fused to a "bait" protein for which interacting partners are sought, while the AD is fused to a library of random cDNA clones—the "prey." When the two pieces are brought together in the same

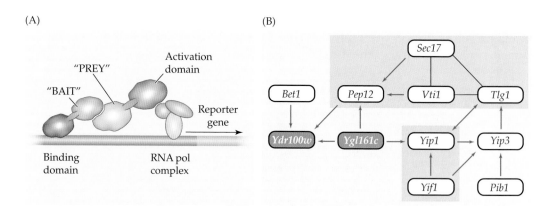

Figure 5.12 Yeast two-hybrid screens.   (A) Y2H technology works on the principle that interaction between two fusion proteins can reconstitute some biochemical function, such as transcriptional activation. (B) A protein interaction map for vesicular transport in yeast. Red arrows link interactions detected by whole genome Y2H; the black lines indicate interactions characterized by traditional methods. Darker shaded boxes show previously known interaction clusters. The two blue genes have unknown functions. (After Ito et al. 2000.)

yeast cell, any prey peptide that binds to the bait will bring the BD and AD together, restoring the capacity of the GAL4 transcription factor to induce expression of a reporter gene.

With modifications, the basic yeast two-hybrid method can be used with other systems, such as *E. coli* and mammalian cells, while alternative reporter assays have been developed (Koegl and Uetz 2007). The biggest limitation of Y2H technology is that it is prone to a high level of false positives (some baits activate transcription alone; some interactions occur by chance) and false negatives (the assay may not be sensitive enough; the fusion proteins lose their appropriate structure). Despite the fact that the interactions are induced to occur in the nucleus, Y2H has led to the successful documentation of numerous cytoplasmic and membrane-bound interactions, and is a powerful tool when appropriate controls are performed.

For genome-scale screens, Y2H has been applied using a matrix approach to search for interactions between all possible combinations of proteins in an organelle, or even an organism (Ito et al. 2000). In the matrix approach, two distinct libraries of bait and prey proteins are constructed, and these are mated together in an ordered manner. Budding yeast is ideal for this purpose, since one library can be maintained in **a** mating type haploid cells, the other in α cells, mating of which leads to co-expression of the fusion genes in **a**/α diploid cells. The matrices can be screened in a systematic manner, in which each individual bait clone is screened in replicate against the matrix of prey clones and every interaction is assayed individually. This process results in increased resolution of false negatives and positives, but is laborious and time-consuming relative to the mass-mating approach, in which whole plates (or rows and columns of plates) are crossed *en masse* (Uetz et al. 2000). For this purpose, a selectable marker such as drug resistance is used as the target for transcriptional activation rather than a visible reporter gene. Positive clones can be characterized either by sequencing the cDNA inserts, or by tracing the interaction in a more systematic set of crosses based on which clones were present in the initial pool.

EXERCISE 5.3 *Formulating a network of protein interactions*

A yeast two-hybrid experiment detects the following protein–protein interactions when the cell surface receptor genes CSR1 and CSR2 are used as bait:

> *CSR1 interacts with STK1, STK2, TFP4, CSR2, and CSR3, while CSR2 interacts with CSR1, CSR3, and STK1.*

By contrast, after purifying protein complexes associated with signal transduction kinases STK1 and STK2, the following proteins are identified by mass spectrometry:

(Continued on next page)

EXERCISE 5.3 (continued)

STK1 complexes with TFP3, TFP4, CSR2, and CSP7, while
STK2 complexes with TFP7, CSR1, CSR3, and CSP4

*Draw a network showing all possible pairwise interactions. Comment on
which interactions you feel are most robustly supported. If gene expression
profiling indicates that CSR3 is never expressed in the same tissue as CSR1
or CSR2, what conclusion might you reach about the function of the protein interaction network?*

ANSWER: *The following network displays all of the yeast two-hybrid
interactions in blue, and the affinity chromatography MS interactions in
red. The most strongly supported interactions are between CSR1 and CSR2,
STK1 and CSR2, and STK2 and CSR1, as each of these pairs of interactions
are supported by two independent data points. The microarray result
would imply that the interactions between CSR3 and the other two CSR
genes are probably false positives that occur because each of these proteins has a similar structure. A reasonable hypothesis is that CSR1 and CSR2
form heterodimers that signal through STK1, whereas STK2 signals from
either CSR1 or CSR3.*

Interactome maps involving tens of thousands of interactions between
thousands of genes have been constructed for flies and nematodes, and have
been used to support a conceptual prediction of the human protein interactome (Lehner and Fraser 2004). Although only a minority of the interactions
detected by high-throughput two-hybrid screens may be physiologically relevant, these approaches have been shown to reproduce many known complexes of interactions and to suggest interactions that have subsequently been
shown to occur upon experimental verification. By examination of similarities in the GO functional terms associated with multiple predicted interaction
partners, it is also possible to assign putative functions to unknown proteins.
Similarities in these predictions across organisms should further increase confidence in such predictions. Some of the theoretical issues surrounding visualization and comparison of interaction networks are discussed in Box 5.2.

Several other bioinformatic methods for describing protein interaction
maps have been proposed; to date most of these have been for the characterization of bacterial proteomes. One, known as the **Rosetta Stone approach**,
asks which pairs of genes in one species are found as a single gene in another

BOX 5.2 Biological Networks in Genome Science

The initial thrust in genome research was heavily focused on cataloging information: What are the sequences of genomes? What genes are present in genomes? What are the functions of genes? What are the structures of proteins? Which genes are expressed in which tissues? As the field moves into the "post-genomic" era, research is moving beyond these single dimensional questions to topics addressing higher-order interactions of biomolecules. How are metabolic pathways regulated through interactions of DNA, RNA, and protein sequences? Which proteins have physical interactions, and what localized domains enable those interactions? These types of questions require methods for describing, visualizing, and statistically analyzing interactions between molecules. Graph theory and *topological networks* provide a common mathematical framework for these endeavors.

Networks are convenient structures for displaying interactions of many types, and over the years, they have been applied successfully to problems in many fields. For example, Internet traffic flow has been studied using network methods to identify aspects of "closeness" among major hubs on the Internet (for example, are users of nytimes.com also likely to surf their way to ESPN.com?). Here we describe the basic principles of graphs and networks, briefly outline how they can be applied to answer some important problems in genome science, and present some representative results from actual data analyses.

From the computational viewpoint, networks consist of *nodes* and *edges*. A simple example is provided in Figure A. This simple network might describe a situation where transcription factor A positively induces expression of gene B, which in turn induces expression of gene C, which then acts as a negative regulator of A. The actual molecules that (potentially) interact are the nodes in the graph (A, B, and C), and the nature of their interactions are indicated by the edges. In this example, the edges are *directed* since they imply that one node

affects the other asymmetrically; in other cases, such as protein–protein interactions, the edges may be undirected and simply indicate a physical interaction. A major effort in the gene expression arena has been the application of both Boolean and Bayesian networks to extract and understand genetic pathways. In a Boolean network, interactions take the form of on/off switches. For instance, a link between A and B may take the form "If A is expressed, then B is also expressed." Bayesian networks extend this notion to allow some degree of randomness. For instance, "If A is expressed, then B is expressed 40% of the time." More quantitative variants also exist: "If A is expressed with level X, then the probability of B being expressed at level Y is 90%."

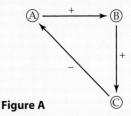

Figure A

The challenges for constructing networks are many. While it seems straight-forward to document who interacts with whom, it is less obvious how to determine if the pattern of observed interactions fits some hypothesized network structure, or if it deviates from the structure observed in another tissue or organism. Assuming that an acceptable measure of fit between observed and hypothesized network topology has been identified, the search through all possible networks is computationally demanding, as the number of potential networks grows explosively with the number of nodes. Given a network and some data, there are both theoretical and computational challenges in carrying out formal statistical procedures. Models of random networks must be developed, starting with such properties as levels of *connectivity* (on average, how

(Continued on next page)

BOX 5.2 *(continued)*

many nodes are connected to a randomly chosen node in the network?). The distribution of connectivity in biological networks often turns out to be nonrandom, in the sense that there may be a small number of nodes with an excessively large number of connections. As discussed further in Chapter 6, such observations may be very informative with regard to inferring the fate of duplicate genes, or predicting the consequences of genetic and environmental perturbations.

One of the immediate benefits of network analysis is helping molecular biologists to focus their attention on clusters of genes that have been shown to be related because of physical interactions between their protein products, shared transcriptional profiles, or genetic interactions in double mutant combinations. An example of the latter illustrates some of the practical aspects of network analysis, and is provided by Tong et al. (2004), who constructed a gene interaction network involving approximately 1,000 genes and 4,000 interactions. The phenotypes of the 132 core mutants that they studied were modified by interaction with between 1 and 146 other mutations, for an average of 34 genetic interactions, which is four times as many as the observed average number of direct protein–protein interactions. Figure B shows

sample topologies for interaction networks centered on three specific genes from their study (located in the center of each network). The interactions are represented by undirected gene linkages in the networks.

Taking this information one step further, Tong et al. formed networks showing the linkages of gene *functions* assessed according to gene ontology attributes (see Box 2.5). First, for more than ten percent of 756 GO attribute classes, genes sharing similar GO attributes are more likely to interact genetically with one another than with randomly selected genes. Reciprocally, over one quarter of all interactions were observed between genes with similar GO attributes. Figure C shows, even more interestingly, that particular classes of GO have increased likelihood of connections with one another.

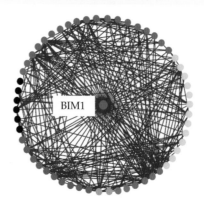

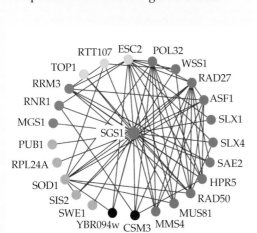

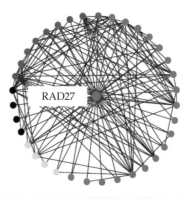

Figure B (Figure from Tong et al. 2004.)

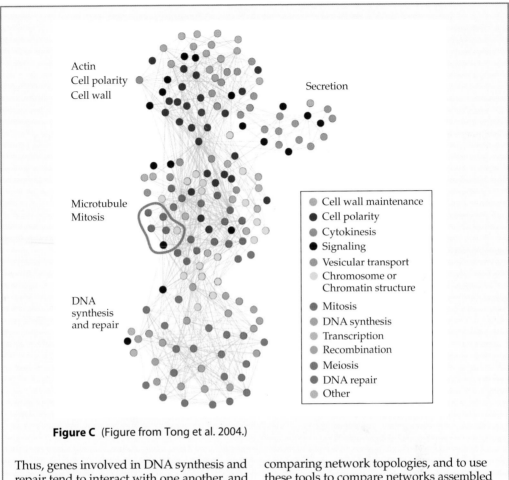

Actin
Cell polarity
Cell wall

Secretion

Microtubule
Mitosis

DNA
synthesis
and repair

- Cell wall maintenance
- Cell polarity
- Cytokinesis
- Signaling
- Vesicular transport
- Chromosome or
  Chromatin structure
- Mitosis
- DNA synthesis
- Transcription
- Recombination
- Meiosis
- DNA repair
- Other

Figure C (Figure from Tong et al. 2004.)

Thus, genes involved in DNA synthesis and repair tend to interact with one another, and also with genes in mitosis and chromosome movement, which in turn interact with genes involved in cell polarity and cell wall maintenance.

Two of the major challenges in network theory are to develop the statistical tools for comparing network topologies, and to use these tools to compare networks assembled from genetic, proteomic, transcriptional, and other databases. The unification of networks is expected to provide a more holistic picture of physiological and developmental networks than can be provided by classical genetic analysis.

species (Marcotte et al. 1999). The assumption is that two genes are unlikely to fuse to form a single protein unless they are involved in the same physiological process. This strategy has resulted in the assignment of putative functions to hundreds of previously unknown bacterial genes.

A similar approach is to ask which genes are always found in the same cistron, indicating that they are co-regulated. Identification of cistrons (stretches of DNA in microbes that are transcribed as a single mRNA but encode multiple proteins) is not trivial, but physical proximity is a good starting point for comparisons among divergent taxa. A third approach based on evolutionary comparison is to simply document the patterns of presence and absence of families of genes across a wide range of taxa (Pellegrini et al. 1999). To the extent that loss of one gene signifies that the biochemical pathway it participates in is dispensable in that organism, other genes in the pathway might also be expected to be lost. Consequently, cosegregation of genes that show variable phylogenetic distribution provides a hint that these genes may interact.

Structural Proteomics

Objectives of Structural Proteomics

Structural proteomics strives toward the prediction of the three-dimensional structure of *every* protein. Physical solution of all protein structures is not feasible, but it is thought that if high-resolution structures are obtained for a sufficiently large number of proteins, then essentially all possible protein structures will be within modeling distance of at least one solved structure (Stevens et al. 2001).

In mid 2007, the central repository for structural data, the Protein Data Bank (PDB; http://www.pdb.org), held 44,000 protein and nucleic acid structures (Berman et al. 2000). Over 5,000 protein structures are added each year, but only about 10% of these represent novel domains. It is estimated that there are between 4,000 and 10,000 distinct protein folds in nature, and that between 10,000 and 20,000 more protein structures may be sufficient to cover the full range of domain space (Burley 2000).

The concept of a protein domain is a difficult one, with slightly different meanings for biochemists, structural biologists, and evolutionary biologists. For our purposes, a **domain** is *a clearly recognizable portion of a protein that folds into a defined structure*. Most proteins are thought to resemble globular beads on a string, where the "beads" are domains, generally ranging in length from 50 to 250 residues, each of which performs a specific biochemical function. Some protein activities, however, are performed at the interface between two or more protein domains—often on two different proteins. If the two molecules are the same protein, the structure is a homodimer; otherwise it is a heterodimer (or heteromultimeric if there are multiple molecules in the protein complex). Folding of protein domains into quaternary structures and intermolecular complexes is largely beyond the purview of current structural biology, but clearly will be an important aspect of future structural proteomics.

Comprehensive structure determination requires broad sampling of protein sequence space. Traditional approaches have focused on proteins of known biological interest in humans, microbes, and model organisms. Greater sampling depth of these proteins—notably kinases, proteases, phosphodiesterases, nuclear hormone receptors, phosphatases, G-protein coupled receptors, and ion channels—is being pursued by industry, as these classes of molecule are proven targets for drugs and other pharmaceuticals. Public efforts focus on sampling breadth (for example, extending to the complete proteomes of one or more small microbial genomes); and representative sampling of novel ORFs from higher eukaryotes (vertebrates, invertebrates, and plants). The latter class includes proteins with no sequence similarity to other proteins and constitutes up to one-third of the predicted proteome of every species whose genome is sequenced.

The Protein Structure Initiative (PSI) is an NIH-sponsored collaborative effort working toward the ability to model the structure and function of essentially any protein within a decade. PSI hosts a knowledgebase at http://kb.psi-structuralgenomics.org/KB that is designed to assist in translating structural data into a better understanding of biological phenomena, as well as hastening drug discovery. Each of four large-scale high-throughput centers with automated pipelines generates hundreds of new structures every year, while a network of smaller research centers develop new technologies and tackle specific types of proteins expressed in unusual organisms and tissues, or in diseased tissue.

The utility of a new protein structure is a function of the novelty of the domain and the level of knowledge of its function (Figure 5.13). Highly refined structures of known classes of proteins with known functions are of

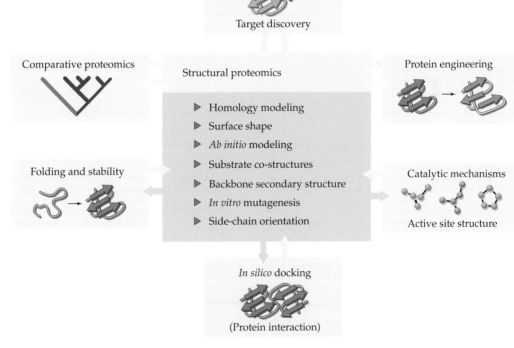

Figure 5.13 Applications of structural proteomics. As structural biologists articulate and refine the structures of increasing numbers of proteins, the knowledge is put to practical use in other areas of biology and medicine.

most relevance in biomedical research, since they support rational drug design by increasing our understanding of the mechanism of catalysis, establishing the constraints on structure-function relationships between proteins and cofactors and ligands, and assisting in the interpretation of the effects of targeted mutations. Excellent examples include the design of HIV protease inhibitors and influenza neuraminidase inhibitors on the basis of structural data. The pharmaceutical sector is particularly interested in the better definition and engineering of features that increase structural stability; the potential of proteins to form multicomplex interactions that may enhance or inhibit function; and in optimizing protein performance. Hegyi and Gerstein (1999) estimated that the average protein fold in SWISS-PROT has 1.2 distinct functions (1.8 for enzymes), while the average function can be performed by 3.6 different folds (2.5 for enzymes). Their findings suggest that structure will generally be quite useful for predicting function, and that novel structures might be expected to define novel functional families.

For proteins of unknown function, structures may suggest the location of active sites and hence promote rational site-directed mutagenesis to investigate function; or structures might suggest the use of particular enzyme assays based solely on the arrangement of residues at the putative active site. Ten common "superfolds," including TIM barrels and the αβ hydrolase, Rossmann, P-loop NTP hydrolase, and ferredoxin folds, account for several functions each. Further structural data enhances our ability to draw inferences using comparative methods. And, as the universe of known protein domain structures increases, possibilities for *in silico* docking and modeling also increase.

Protein structures are deposited upon publication in the worldwide Protein Data Bank (PDB; http://www.pdb.org), which since 1998 has been maintained by the Research Collaboratory for Structural Biology consisting of groups from Rutgers University, the San Diego Supercomputing Center at UCSD, and the University of Wisconsin. The PDB coordinates activities of the American, European, and Japanese protein structure data banks and ensures adoption of common data standards and archiving procedures. Database management is facilitated by an ADIT (AutoDeposit Input Tool) Web-based interface that uses an internationally agreed upon macromolecular crystallographic information file (mmCIF) dictionary of 1,700 terms to avoid ambiguity. ADIT helps the PDB meet the challenges associated with archiving an exponentially growing set of data, complete with functional annotation and control and verification of structure quality. In addition to atomic coordinates, structures are deposited with journal references, functional information, and attributes of the experimental procedures used to determine the structure—all of which must be checked and formatted in such a way that the data is compatible with sharing over the Internet.

Protein Structure Determination

Determining the structure of even a single protein is a labor-intensive effort that has traditionally required several years. Consequently, automation of each of the steps outlined in Figure 5.14 is essential to structural genomics (Manjasetty et al. 2008). Recent advances include the utilization of genome sequence information to define targets and design primers for PCR-based cloning of ORFs into expression vectors; improved affinity chromatography methods for purifying fusion proteins in sufficient quantities for crystal growth; development of high-throughput robotic methods for crystal growth; cryogenic storage and robotic retrieval and orientation of crystals at synchrotron facilities to minimize delays and human error during the gathering of structural data; and greater automation of the computational methods used to solve structures.

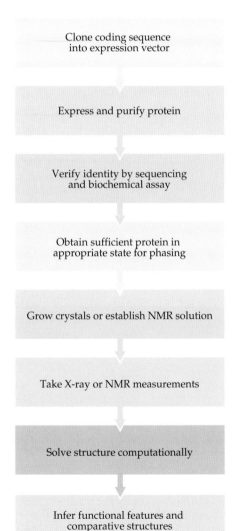

Figure 5.14 Flow diagram for solving a protein structure. Current efforts are focused on streamlining the process by automation, but failures and bottlenecks require human input at each step.

The two experimental methods used to solve protein structures are **X-ray crystallography** and **nuclear magnetic resonance (NMR) spectroscopy**. The solution of protein structures by crystallography proceeds in a series of steps, starting with data collection and moving through image processing, phasing, model building, model refinement, validation, and publication. Two types of data must be obtained in order to construct the electron density maps from which the structure is derived: the *amplitudes* and the *phases* of each diffracted X-ray. Amplitudes can be determined directly from the diffraction data, but phasing is a difficult problem. It can be solved computationally if extremely high-resolution data are available, or if data from very similar structures are used, but it generally requires additional diffraction data from crystals into which heavy atoms such as the transition metals mercury or gold have been incorporated. The PDB provides a range of online tools for interactive study of three-dimensional structures such as those shown in Figure 5.15.

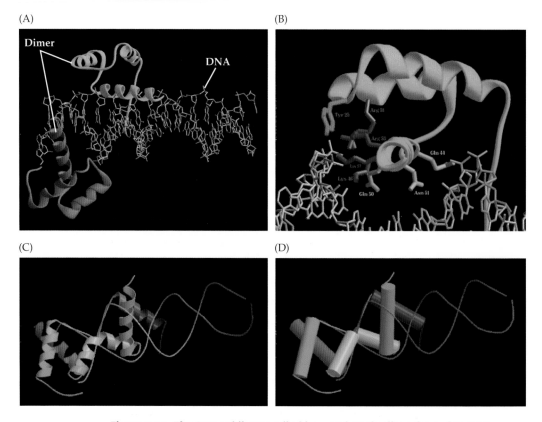

Figure 5.15 The *Drosophila* Engrailed homeodomain dimer bound to DNA.
Four different three-dimensional views. (A) Swiss-3D image from **http://ca.expasy.org/cgi-bin/sw3d-search-ac?P02836**. (B) Close-up of binding site for the third helix in the major groove of DNA. (C) Still ribbon and (D) cylinder views of the same interaction from the PDB Structure Explorer, accessed by entering the accession name 3HDD at **http://www.rcsb.org/pdb**.

The two major advantages of NMR spectroscopy over X-ray crystallography are that NMR is performed in solution, so there is no requirement that the protein crystallize (which considerably expands the number of structures that can be studied); and that the structure can be determined under conditions that resemble physiology and can readily be manipulated to mimic changes in pH or salt concentration. Typically, about 0.5 ml of 1-mM protein solution enriched in ^{13}C or ^{15}N is used to determine the structures of small proteins (10–30 kDa), but much larger structures can now also be determined. Further, protein-lipid micelle analyses allow structural determination of protein domains within membranes. Technical advances are bringing the resolution of NMR structures toward the 2.0–2.5 Å limit typical of X-ray crystal structures, and applications in the study of protein-small molecule interactions have been developed (Montelione et al. 2000; Yee et al. 2006).

Resolution of complex macromolecular structures such as the ribosome, the 12- protein subunit, 127-cofactor photosystem I of cyanobacteria, and channel or neurotransmitter receptor complexes demonstrates the power of crystallography to illuminate complex protein-protein and protein-drug interactions. High-resolution cryoelectron microscopy (Henderson 2004) also plays an important role in elucidation of the organization of macromolecular and membrane-bound complexes, which constitute over a quarter of all proteins.

Protein Structure Prediction and Threading

Protein structure prediction plays an important role in attempts to infer function from sequence data alone (Baker and Sali 2001). The lower limit for detection of potential structural homologs by sequence comparison is 30% sequence identity. Above this level, comparative modeling can be used to estimate the likely structure of a protein by overlaying the unknown structure onto the known. Below this level, at least three classes of strategy are used to try to fit an initial model of the most likely protein structure: *ab initio* prediction, fold recognition, and threading. The success of these methods is judged at biannual CASP (Critical Assessment in Structure Prediction) conferences at which theoretical and experimental solutions of previously unsolved structures are compared (Jones 2000).

Ab initio **protein prediction** starts with an attempt to derive secondary structure from the amino acid sequence, predicting the likelihood that a subsequence will fold into an α helix, β sheet, or coiled coil by using physicochemical parameters or HMM and neural net algorithms trained on existing probabilities that similar short peptides adopt the particular structures. Such methods have been claimed to accurately predict three-quarters of all local structures, and are being extended to predict the structure of membrane proteins. Subsequently, secondary structures are folded into tertiary structures, again using algorithms based on physical principles (Dunbrack 2006). Model quality is tested by fitting predictions against known structures and calculating the root mean square distance between the predicted and actual locations of α-carbon atoms on the peptide backbone. Attempts to fit the location and orientation of side chains are not yet within the realm of *ab initio* modeling, which is thus more concerned with generating hypotheses of the general shape of a polypeptide. Hypotheses can then be tested using site-directed mutagenesis and other biochemical approaches.

Fold recognition, or **structural profile**, methods attempt to find the best fit of a raw polypeptide sequence onto a library of known protein folds such as CATH or SCOP, or a "periodic table" of the possible forms that helices and sheets can fold into (Taylor 2002). A prediction of the secondary structure of the unknown is made and compared with the secondary structure of each member of the library of folds. The locally aligned sequences are also compared for sequence and/or profile similarity, and the two measures of structural and sequence similarity are condensed into a fold assignment confidence z-score that represents the probability of a match relative to random comparisons. A threshold chosen by application of the same algorithm to known structures is then used to identify likely matches in the manner in which an unknown domain folds relative to known domains (Figure 5.16).

Application of this approach to the predicted whole proteome of *Mycoplasma genitalium* resulted in putative functional assignment to an extra 6% of the ORFs that were not possible to assign based on sequence alignment alone (which improved the annotation of 16% of the 468 ORFs in the *M. genitalium* genome; Fischer and Eisenberg 1997). As the number of unique folds in the Protein Data Bank increases, the performance of this approach is expected to improve greatly.

Threading takes the fold recognition process a step further, in that empirical energy functions for residue pair interactions are used to mount the unknown onto the putative backbone in the best possible manner. Gaps are accommodated and the best interactions are maximized in an effort to derive the most likely conformation of the unknown fold, and to discriminate among different possibilities. Threading of the complete predicted proteomes of 33 prokaryotes and 4 animals involving over 165,000 genes with the THREADER algorithm assigned folds to 70% of the bacterial proteins and 60% of the animal ones (Cherkasov and Jones 2004). Similarly, Zhang and Skolnick (2004) added an extra side-chain fitting optimization step to their PROSPECTOR algorithm and accurately predicted the structure of two-thirds of 1,360 medium-sized proteins in *E. coli* relative to PDB structures. An example of a practical application of threading is the demonstration that the *fw2.2* gene, which has a quantitative effect on fruit size in tomatoes, is likely to encode a member of the heterotrimeric guanosine triphosphate-binding RAX family of proteins, and hence to play a role in controlling cell division during fruit growth (Frary et al. 2000).

The eventual aims of structural biology include not just the determination of protein structures, but also modeling of protein function at the atomic level. This will entail advances in modeling a number of aspects of protein

(A)

(B)

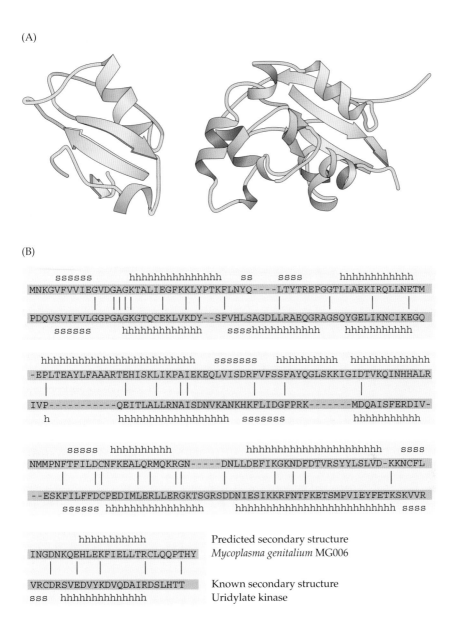

```
       ssssss       hhhhhhhhhhhhhhh   ss    ssss      hhhhhhhhhhh
MNKGVFVVIEGVDGAGKTALIEGFKKLYPTKFLNYQ----LTYTREPGGTLLAEKIRQLLNETM
          ||||       |     |   |             |          |       |
PDQVSVIFVLGGPGAGKGTQCEKLVKDY--SFVHLSAGDLLRAEQGRAGSQYGELIKNCIKEGQ
       ssssss      hhhhhhhhhhhh    sssshhhhhhhhhh     hhhhhhhhhh

  hhhhhhhhhhhhhhhhhhhhhhhhhhhhhh     sssssss   hhhhhhhhh   hhhhhhhhhhhh
-EPLTEAYLFAAARTEHISKLIKPAIEKEQLVISDRFVFSSFAYQGLSKKIGIDTVKQINHHALR
    |             |     |   ||            |                |
IVP----------QEITLALLRNAISDNVKANKHKFLIDGFPRK-------MDQAISFERDIV-
  h              hhhhhhhhhhhhhhhhhhh   sssssss        hhhhhhhhhh

    sssss  hhhhhhhhh            hhhhhhhhhhhhhhhhhhhhhh    ssss
NMMPNFTFILDCNFKEALQRMQKRGN-----DNLLDEFIKGKNDFDTVRSYYLSLVD-KKNCFL
       |    ||    |   |   ||            |   |  |              |
--ESKFILFFDCPEDIMLERLLERGKTSGRSDDNIESIKKRFNTFKETSMPVIEYFETKSKVVR
   sssss hhhhhhhhhhhhhhhh       hhhhhhhhhhhhhhhhhhhhhhhhhhhh ssss

       hhhhhhhhhh          Predicted secondary structure
INGDNKQEHLEKFIELLTRCLQQPTHY     Mycoplasma genitalium MG006
   |    |    |     |    |  |
VRCDRSVEDVYKDVQDAIRDSLHTT  Known secondary structure
sss  hhhhhhhhhhhhhhh       Uridylate kinase
```

Figure 5.16 Fold recognition. The objective of fold recognition and threading is to determine whether a domain is similar to a fold found in a known protein. Despite sequence identity of only 20% over the full length of the protein (A), this unidentified *M. genitalium* ORF MG006 (sequence shown in purple) was assigned as a probable member of the uridylate kinase family (sequence in green) on the basis of similarity of predicted secondary structure (B), which predicts a homologous folding pattern. (A after Fischer and Eisenberg 1997. B based on R. Altman, online lecture http://cmagm.stanford.edu/biochem218/16threading.html, no longer available.)

function, including: protein–small molecule interactions; the effects of site-directed mutagenesis on structural conformation; the movements in active sites during the milliseconds over which a catalytic event takes place; the effect of side-chain bulk and orientation on the specificity of molecular interactions; the formation of active sites at the interface between protein domains; and the docking of two or more proteins to form multisubunit complexes. Undoubtedly such endeavors will require integration not just of theory and experimental structural proteomics, but also constant feedback with functional genomics and molecular biology.

Functional Genomics

Functional genomics encompasses all research aimed at defining the function of each and every gene in a genome. Function can be defined at several ontological levels, from biochemistry to cell biology and on up to organismal phenotype. Two homologous genes that retain the same molecular structure and biochemical function may nevertheless have very different physiological roles in different organisms. They will often interact with a similar suite of proteins in different organisms, and at different phases of development, though the precise nature of the interactions can be highly labile. Nevertheless, the characterization of the function of a protein domain in one organism will generally provide a hint as to its function in another organism. Consequently, one of the first goals of functional genomics is to identify mutations that affect the activity of as many genes as possible in the major model organisms.

In this section we will consider approaches to functional genetics based on the generation and analysis of mutations. There are three basic approaches, each with its own goals:

1. The goal of **forward genetics** is to identify a set of genes that affect a trait of interest. This is pursued initially by random mutagenesis of the whole genome, screening for new strains of the organism that have an aberrant phenotype that is transmitted stably to subsequent generations in Mendelian proportions. Traits of interest include morphology, physiology, and behavior.

2. The goal of **reverse genetics** is to identify phenotypes that might be caused by disruption of a particular gene or set of genes. In this case, the starting point is the DNA sequence of interest—perhaps previously uncharacterized open reading frames, perhaps a cluster of co-expressed genes, but ultimately each and every predicted gene. A variety of strategies for the systematic mutagenesis of specific genes are now available.

3. The goal of **fine-structure genetics** is to manipulate the structure and regulation of specific genes in such a way that novel functions and interactions can be characterized, or so that hypotheses arising from in vitro analysis or structural comparison can be tested in vivo. This is a rapidly evolving field of research at the interface of molecular genetics and genomics.

Saturation Forward Genetics

Forward genetic strategies start with a phenotype and work toward identifying the gene or genes that are responsible for that phenotype. **Saturation mutagenesis** refers to genetic screens of such large scale that a point is reached where most new mutations represent second or multiple hits of previously identified loci (Figure 5.17). Classical genetic screens were based on identifying a phenotype so, to a broad approximation, the **saturation point** defines the subset of genes that are required individually for the development of a trait. For example, saturation mutagenesis screens for embryonic recessive lethal mutations in *Drosophila* defined a core set of fewer than 50 genes that are necessary for segmentation (Nüsslein-Volhard and Wieschaus 1980). It was later realized that such screens missed genes whose products are expressed maternally, as well as genes whose function is redundantly specified by other loci. Yet the subsequent molecular characterization of these loci provided the raw material for research that rapidly led to understanding of the process and the identification of other interacting loci by different strategies.

Many phenotypes, such as those related to behavior, physiology, or the morphogenesis of internal organs, cannot be studied by direct screening for visible aberrations. Consequently, screens in the genomic era increasingly rely on the isolation of random mutant chromosomes, followed by characterization of a battery of phenotypes. Because of the expense involved in mutagenesis of vertebrates, it is cost-effective to pool resources by screening hundreds of phenotypes simultaneously, in different laboratories that have expertise in different areas.

Mutant chromosomes can be isolated either by **insertional mutagenesis**—screening for the expression of a marker gene carried by a transposable element—or by administering a high enough dose of **mutagen** to guarantee the generation of at least one visible mutation in each gamete. Table 5.1 summarizes the range of phenotypes scored and frequencies of mutations recovered in a pair of mouse mutagenesis screens.

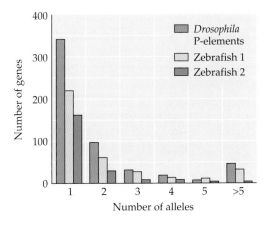

Figure 5.17 Frequency of mutations recovered in saturation screens. Even large screens (several thousand mutated genomes) typically fail to approach saturation. This figure shows the number of genes with 1, 2, 3, 4, 5, or more than 5 alleles recovered in two large zebrafish embryonic phenotypic screens (Driever et al.1996; Haffter et al.1996), as well as a series of P-element insertional mutagenesis screens in *Drosophila* that are summarized in Spradling et al. (1999).

TABLE 5.1 *Mouse Mutants from F_3 Recessive Screens*

Defect (phenotype)	Frequency of mutation in F_3 (% of offspring)	
	Germany[a]	England[b]
Craniofacial and skeletal defects	40	15
Coat color abnormalities	18	19
Eye defects, including cataracts	9	10
Behavioral abnormalities (circling, head tossing, etc.)	21	34
Skin or hair abnormalities	13	8
Growth, weight, size defects	4	17
Deafness	9	16
Clinical chemistry abnormalities	23	4
Immunological and allergic deficits	39	NA

[a]Data from Hrabe de Angelis et al. (2000).
[b]Data from Nolan et al. (2000).

Several types of mutagen are used to generate different classes of mutation, as shown in Figure 5.18. Point mutations (those affecting a single nucleotide) are best generated by chemicals such as ethylnitrosourea (ENU) or ethylmethanesulfonate (EMS). Such screens are easily performed, relatively unbiased, and result in multiple classes of effect, from non-sense and missense coding mutations (that is, introducing premature stop codons, frameshifts, or amino acid replacements) to splicing defects and disruption of regulatory sequences. Different point mutations affecting the same gene can result in very different phenotypes, potentially helping to define pleiotropic functions. The drawback of point mutations is that they are laborious to map, so cloning of the gene responsible for a mutant phenotype can take several years.

Larger insertions and chromosome rearrangements are commonly produced by irradiation with X-rays or gamma rays. These abnormalities either remove a gene (or genes) entirely, resulting in a null phenotype, or juxtapose novel regulatory sequences adjacent to a gene, resulting in loss of expression in the normal tissues and/or gain of expression in ectopic tissues. Cloning of the breakpoints can be performed by direct comparison of mutant and parental chromosomes, so does not necessarily require a large number of meioses for genetic mapping.

A popular class of mutagen comprises **transposable elements**. (Some features of these elements are discussed in more detail in the section below on fine-structure genetics.) **Transposon mutagenesis** offers the major advantages that (1) the phenotype can often be reverted to the wild-type state by inducing the inserted element to "jump" back out of the genome (providing formal proof that the insertion caused the phenotype); (2) cloning of the

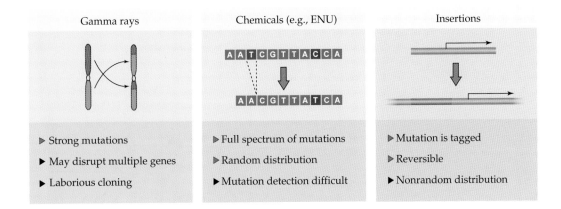

Gamma rays | Chemicals (e.g., ENU) | Insertions

AATCGTTACCA
⬇
AACGTTATCA

▶ Strong mutations

▶ May disrupt multiple genes

▶ Laborious cloning

▶ Full spectrum of mutations

▶ Random distribution

▶ Mutation detection difficult

▶ Mutation is tagged

▶ Reversible

▶ Nonrandom distribution

Figure 5.18 The three major types of mutagen.   Gamma irradiation induces chromosome inversions, translocations, and large deletions. Chemicals such as ENU induce point mutations and deletions of a few bases. Transposable element insertions can occur anywhere in a gene, but are often found in upstream regulatory regions. Advantages (red arrowheads) and disadvantages (black arrowheads) of each mutagen are indicated.

site of insertion can be completed in a few days by plasmid rescue (see Figure 5.24A); and (3) isolation of new insertions does not require a phenotypic screen. In this sense, transposon mutagenesis is unbiased, but on the other hand it is often biased by the biological fact that some transposons have insertion site preferences, or "hotspots," as well as "coldspots." These preferential sites exist at two levels: some genes are relatively refractory to certain transposons while others are unusually frequent targets; and some transposons preferentially insert close to the promoter of target genes, partially disrupting expression rather than knocking out the gene entirely.

A useful feature of transposons such as the *Drosophila* P element and maize Ds element is that their transposition can be controlled by introducing a source of the transposase enzyme through genetic crosses. This allows for collection of a large number of mutants in a single generation. Often the movement of the transposon is local—to a site within a few hundred kilobases—providing a means for "walking" along a chromosome and generating insertions in all the open reading frames in a small region of the genome.

Dominant mutations can be isolated in the progeny of the mutagenized parents (the F_1 generation), but recessive mutations require more laborious and complex designs in which the mutants are not recovered until at least the F_3 generation (Figure 5.19). Dominant mutations can be due either to gain of function (**hypermorphs**); generation of a novel function (**neomorphs**); production of a dominant negative function (**antimorphs**, in which the dominant protein interferes with wild-type function); or to haploinsufficient loss of function (**hypomorphs**).

(A)

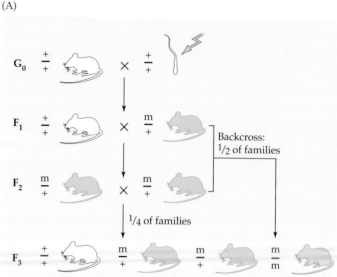

Figure 5.19 F₁ and F₃ genetic screens. After sperm mutagenesis, dominant mutations can be recovered directly in the F₁ progeny (A). For recessive mutations, crossing a heterozygous F₁ individual (orange) to a wild-type individual (white) results in transmission of each mutation to half of the F₂ grandchildren. Homozygotes (light red) are only recovered in one-quarter of the offspring produced by crossing two F₂ siblings, and only one-quarter of the individuals are mutant in such families. The probability of making any mutation homozygous is increased by setting up multiple families from each F₂. If a backcross between the heterozygous F₁ and an F₂ can be set up (B), the probability of homozygosity—and hence the efficiency of the screen—increases further.

(B)

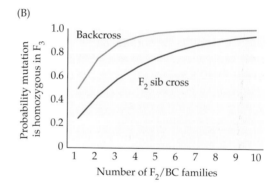

Recessive mutations are most often hypomorphs that either reduce protein activity level or downregulate transcription. There are, however, no simple rules relating dominance and recessivity to loss or gain of function. For example, two heritable recessive versions of human long-QT syndrome result from an array of mutations that disrupt either sodium or potassium channel functions, disrupting membrane polarization in a similar manner despite the fact that the channels pump ions in opposite directions (Splawski et al. 2000). Studies from model organisms indicate that multiple mutations are usually required to define the function or functions of a gene product at even a superficial level. Consequently, identification of a single mutation is often followed up by generation of a panel of new mutations that fail to complement the original allele.

The first step in mutant classification is the assignment of each mutation to a **complementation group**: a set of alleles that fail to complement (provide the function of) one another. Two independent mutations affecting the same locus will often (but not always) combine to produce a recessive phenotype that resembles that of at least one of the alleles when homozygous. Thus, complementation testing usually reduces a large number of mutations to a two- or threefold smaller set of loci. Intergenic noncomplementation can also arise when two mutations affect genes involved in the same biochemical pathway. Thus, physical or genetic map data is required to verify the inference that a complementation group defines independent mutations in the same gene.

Steps must also be taken to preserve the new mutations. Recessive mutations are often not viable as homozygotes, and even viable mutations can be lost in a few generations as a result of genetic drift. This problem is solved differently for different organisms. For plants, mutant seed can be stored; for mice, bacteria, and nematodes, germ cells (or whole animals) can be frozen; fly geneticists use "balancer" chromosomes (described below). Colony maintenance and stock tracking require great care and attention to detail and are not taken lightly. Most genome projects have stock centers with budgets in the tens of millions of dollars, yet even so must make difficult decisions about which stocks to keep in the face of the incredible volume of genetic resources being generated.

Once a set of loci affecting a trait have been defined and mapped to chromosomal regions on the basis of recombination frequencies, molecular genetic procedures must be employed to identify the gene that is disrupted. These can include searching for new single-nucleotide mutations in a chromosomal region that maps to the genetic lesion; looking for altered transcript expression profiles; and attempts to complement the mutation by making transgenic organisms with extra wild-type copies of a candidate genes. Subsequently, identifying the molecular nature of each of the alleles that define a locus can be very revealing as to the function of the gene product. For example, alleles that cause similar phenotypes often map to the same regulatory region or protein domain, hence serving to define those elements. These alleles may also provide clues as to the biochemical mechanisms by which distinct protein functions and interactions are mediated.

Somatic chromosomal rearrangements associated with many different cancers can be identified relatively quickly using comparative genome hybridization (CGH). Whereas large deletions and amplifications can be mapped to a resolution of around 5 Mb by hybridization to metaphase chromosome spreads, probing of DNA microarrays with labeled genomic DNA facilitates mapping of breakpoints to just tens of kilobases. The DNA spots on the microarray typically consist of fragmented and amplified BAC or P1 clones, but can also include cDNA clones or selected low-redundancy fragments. A difference in the ratio of fluorescence signal between normal and tumor DNA pinpoints a region of the genome that is reduced or duplicated in the tumor; comparison of hundreds of samples highlights loci that are repeatably associated with cancer progression (Kallioniemi 2008).

High-Throughput Reverse Genetics

There are many situations in which random forward mutagenesis is either ineffective or inefficient at identifying the genes involved in producing or inducing a trait. These include traits that are difficult to screen in live organisms, such as development of internal organs or complex aspects of physiology; quantitative continuous traits that are not disrupted in a Mendelian fashion, including life history and many behavioral traits; late-acting traits that cannot be screened directly because the genes that affect them disrupt an earlier aspect of development or kill the organism before the phenotype appears; and gene functions that are redundant either because there are multiple copies of the relevant genes or multiple pathways for achieving the observed phenotype. Some of these problems can be overcome with reverse genetic strategies that start with a gene of interest and work outward, toward a phenotype.

Systematic mutagenesis.   Systematic mutagenesis refers to the process of deliberately knocking out (mutating) a series of genes, one by one. It is typically employed in situations where the group of candidate genes to be targeted for mutagenesis is small enough that, despite the extra work involved per gene, mutations can be recovered more efficiently than by screening large volumes of random mutants. These include cases where a group of transcripts expressed in conjunction with the development or physiology of a trait have already been defined but for which no functional data is available; or where a region of the genome has been shown to be important for the ontology of a disease and disruption of each putative gene in the region is warranted. For some species, the resources now exist for creating a mutation in each and every gene in the genome, in which case the guarantee that each gene is disrupted provides a large advantage over random mutagenesis.

There are several ways to isolate a mutation in a known open reading frame. The classical genetic method is to screen for failure to complement a deficiency that removes a large portion of the genome that includes the gene of interest. A wild-type chromosome is mutagenized and crosses are performed to identify mutations that are sub-viable or that produce a mutant phenotype over the deficiency chromosome. The mutations are then placed in complementation groups and DNA sequencing is used to determine which set of mutations lie in the gene of interest.

Another method is to use PCR to screen a panel of mutant organisms for the insertion of a transposable element adjacent to the gene of interest. The forward primer is designed to hybridize to the terminus of the transposon, and a series of nested reverse primers are designed to hybridize to the gene. Genomic DNA from a number of different, potentially mutant stocks is pooled; if any of these carries an insertion near the gene, a unique amplification product will be observed. Sophisticated bar coding schemes have been designed for some model organisms to facilitate screening tens of thousands of insertion mutants such that just three or four reactions are required to precisely identify the relevant mutant stock. For *Drosophila* and *Caenorhabditis*, saturation transposon mutagenesis is performed in conjunction with direct sequencing of the insertion site by inverse PCR, in which case knowledge of the complete genome sequence leads directly to the assembly of a catalog of insertional mutants. Investigators need only query the database to obtain their mutation.

The most direct method for targeted mutagenesis is to specifically disrupt the ORF by **homologous recombination** (Thomas and Capecchi 1987). DNA that is carried into a cell by electroporation or chemical transformation tends to integrate at random locations, but if there is sufficient sequence homology with a piece of chromosomal DNA, it can align with the endogenous copy of the gene and replace it by two adjacent crossovers. The process of homologous recombination can be manipulated to allow precise targeting of mutations. The disruptive copy of the gene is engineered in bacteria, and typically carries a selectable marker that allows isolation of insertion-bearing chromosomes (Figure 5.20A). New insertional mutagenesis cassettes have been designed to direct *lacZ* expression in place of the endogenous gene, allowing visualization of expression of the gene at the same time as the knockout is performed. The technique works efficiently in yeast and in mouse embryonic stem cells because of a high ratio of homologous recombination to unequal recombination (that is, random insertion anywhere in the genome). Mouse embryonic stem (ES) cells can be grown in culture, screened for the correct insertion, and then injected into a mouse blastocyst in such a way that they will populate the germ line and give rise to transgenic mice carrying the targeted mutation. Knockout mutations are first generated in heterozygous condition, and mutant defects are examined in homozygous or haploid progeny. The techniques used for constructing transgenic animals and plants are summarized in Box 5.3.

Knock-ins. Homologous recombination can also be used to perform "knock-ins," in which the wild- type copy is replaced with a specific modification, as opposed to a generic insertion (Figure 5.20B). One application of knock-in technology is to place the expression of one gene under the control of the regulatory elements of another gene. For example, replacing the upstream regulatory region of the mouse *HoxD11* gene by the homologous region of the zebrafish *hoxd11* gene resulted in premature activation of expression of the gene and an anterior shift in the location of the sacrum (Gérard et al. 1997). Results from targeted replacement can be interpreted without concerns over whether sequences adjacent to the location of nontargeted insertions (that is, position effects) cause the premature activation or otherwise affect expression of the transgene.

Another application of knock-ins is to test the function of a particular point mutation *in vivo*, such as targeted mutagenesis of promoter sequences or testing of the requirements of specific amino acid residues in biological functions that cannot be measured in cell lines. **Gene therapy**, in which a mutant gene is replaced by a wild-type copy, also uses the knock-in approach to avoid potential deleterious effects of random mutagenesis or inappropriate regulation of transgenes that lie outside their normal chromosomal context.

(A)

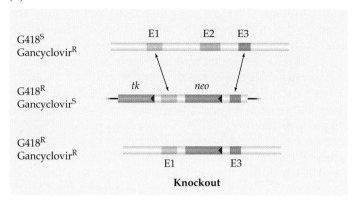

(B)

Figure 5.20 **Targeted mutagenesis.**   (A) Gene knockout mutations in yeast or mammalian cells are usually constructed using a targeting vector that carries both positive and negative selectable markers. Recombination between identical sequences in the vector and chromosomal DNA (somewhere in the vicinity of exons 1 and 3 in this example) results in the replacement of exon 2 with the *G418* gene that encodes resistance to the drug neomycin. Enrichment for homologous recombination events is carried out by selecting for absence of thymidine kinase (tk) activity that arises when the whole construct inserts elsewhere in the genome. (B) Knock-in mutagenesis is performed similarly, except that there are two selection steps. There is no enrichment for homologous recombination in the first step, so sufficient clones must be screened to ensure that the targeted insertion is present. Subsequent expression of Cre recombinase results in removal of the two selectable genes between the two *loxP* sites (yellow). This event can be selected for on gancyclovir, which is toxic in the presence of the TK enzyme. In this example, exon 2 has been replaced by a modified form of the exon.

RNAi screens. A simple alternative to homologous recombination is to induce transient loss of gene function using **inhibitory RNA (RNAi)** expression (Fire et al. 1998). The presence of just a few molecules of double-stranded RNA in a cell dramatically reduces the level of transcript of the associated gene, resulting in a loss-of-function phenotype. RNAi is a stretch of several hundred bases of RNA that is complementary to the sense strand of the gene of interest, and which hybridizes to that strand in a cell. Constructs that lead to expression of RNAi can be introduced into the embryo by injection, electroporation, or even (in the case of *Caenorhabditis*) by feeding the nematodes bacteria that transcribe both strands of the RNAi. Alternatively, transgenic animals can be generated that produce RNAi by transcription of a construct that contains the two strands separated by a short loop, so that the molecule will self-hybridize once synthesized (Figure 5.21).

The mechanism of RNAi function is related to that of small regulatory microRNAs that have normal functions in development and physiology (Hannon 2002). Long dsRNA is first cleaved into 21–25-nt small interfering siRNA molecules by the enzyme Dicer, and these are denatured and fed into an RNA-induced silencing complex (RISC) that aligns the antisense strand

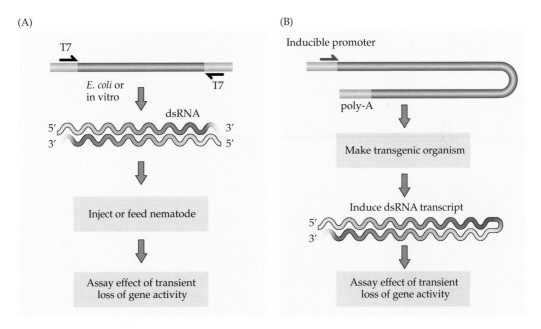

Figure 5.21 Inhibitory RNA expression (RNAi). (A) For transient infection assays, double-stranded RNA (dsRNA) is synthesized either in *E. coli* or in vitro using T7 promoters oriented in opposite directions on either side of the cloned gene. The two strands self-anneal in the bacteria, and can be fed directly to nematodes or isolated and injected. (B) For most other eukaryotes, RNAi is pursued by cloning two copies of the gene head-to-tail under the control of a tissue-specific or inducible promoter. Transcription in transgenic organisms results in assembly of a dsRNA that is often inhibitory.

BOX 5.3   Transgenic Animals and Plants

Transgenic animals and plants are organisms that have been transformed by a foreign piece of DNA that is stably integrated into the genome and transmitted from generation to generation. The foreign DNA may be from the same or a different species; it may be simply a reporter construct, or it may encode a novel product or a gene that will be expressed in a novel pattern; and, it may be present in multiple copies, as a single copy, or may even replace the endogenous copy of the gene. In all of these cases, four major steps must be taken in order to construct a transgenic organism:

1. The foreign DNA construct must be synthesized in a form that can be shuttled into the host.

2. The transgene must be delivered into the host animal or plant cells.

3. The transgene must integrate stably into a germ line chromosome of the host organism.

4. Appropriate lines that are viable and express the transgene at desired levels must be identified and established as a permanent stock.

Bacteria and yeast can be transformed with vectors that replicate autonomously. Transformation of higher eukaryotes, however, generally requires that a host chromosome take up the DNA. Any piece of DNA that gets past the host defense systems can integrate at random anywhere in a genome, through the process of nonhomologous (illegitimate) recombination. In some eukaryote transformation procedures, there is little control over the number of copies that insert and, particularly in plants, it is not uncommon for tens of copies to integrate as a tandem array. This problem can be circumvented by carrying the transgene into the genome using a disabled transposable element. Genetic engineering is used to replace the TE's transposase gene (that normally catalyzes transposition) with the foreign DNA, and the transposase enzyme is supplied transiently on a helper plasmid only during the transformation step. Consequently, once inserted, there is no way for the transgene to hop back out. For example, *Drosophila* transgenesis is performed using modified P-elements (Rubin and Spradling 1982) or mariner elements; this system is now being applied to other insects of agricultural importance and even to some vertebrates.

Transposable elements usually insert at a single location, but there is no control over

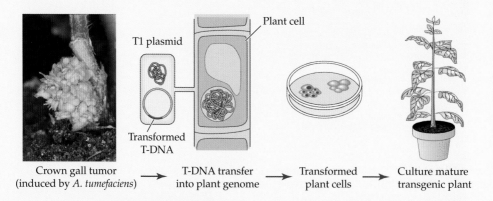

Crown gall tumor
(induced by *A. tumefaciens*) → T-DNA transfer into plant genome → Transformed plant cells → Culture mature transgenic plant

Figure A ***Agrobacterium*-mediated transformation.** *A. tumefaciens* induces gall cells on a plant, and injects T-DNA into the host cells. Transformed cells are then grown into whole plants by manipulating the hormones in the growth medium.

where this location is. Many insertion events knock out the function of an essential gene (and thus are deleterious), and most are affected by the transcriptional activity of the surrounding chromatin. The insertion site can be controlled by using artificial recombination systems such as the Cre-Lox system adapted from bacteriophage P1 for use in higher eukaryotes. As described in the text, homologous recombination can also be used to actually replace the endogenous gene.

A variety of different delivery systems are available for any given species. **Chemical transformation** or **electroporation** are convenient for delivering DNA into *cell cultures* such as protoplasts and stem cell lines, where there is no cell wall or chorion creating a barrier. *Large cells* such as oocytes can be directly **microinjected** with a DNA solution in fine glass needles, and gold particle guns have been developed to carry DNA into *plant cells and organelles* in a process known as **biolistics.**

Plant transformation is most simply performed using *Agrobacterium tumefaciens* bacteria as a vector (Birch 1997). The bacterium induces gall cells on the plant's tissue and injects part of Ti plasmid that has been engi-

neered to carry the transgene. This T-DNA then incorporates into the plant's somatic cells, from which whole plants can be regenerated by manipulating the growth medium with appropriate hormones. Originally, this technology was only applicable to dicots, but increased understanding of the process now allows it to be used to transform monocots, including agronomically essential grasses (Shen et al.1999).

Mammals can be transformed by injection of transforming DNA into their oocytes, as was exemplified dramatically by the production of the first transgenic rhesus monkey using a replication-defective retrovirus (Chan et al. 2001). Mice are more commonly transformed by first producing embryonic stem (ES) cells in culture (Capecchi 1989), then injecting the chosen cell line into a blastocyst (early embryo) from a host with a different visible genetic constitution (for example, coat color). Mature mice are chimeric, since the ES cells populate different tissue primordia. If they contribute to the germ line, transgenic progeny will be recovered and subsequent crosses are performed to establish a homozygous strain.

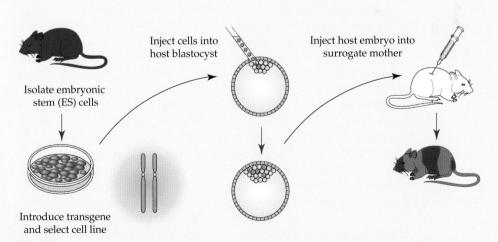

Figure B Transgenic mice via ES cells.
A transgene is introduced into donor ES cells, which are then injected into a host blastocyst. The blastocyst is implanted in a surrogate mother and produces chimeric offspring that can in turn be propagated, producing homozygous transgenic offspring.

with the complementary mRNA. This results in post-transcriptional down-regulation through a variety of mechanisms including mRNA degradation, and inhibition of translation, that vary among plants and animals.

An RNAi screen was applied to whole chromosome mutagenesis in *C. elegans* by designing 16,757 constructs to specifically knock out expression of 86% of the identified ORFs (Kamath et al. 2003). Mutant phenotypes were identified for 1,722 genes, two-thirds of which were previously undescribed. Annotation of cellular as well as whole-organism functions can be performed, depending on the choice of phenotypic assay. *Drosophila* and mammalian cell cultures have also been screened for various phenotypes with RNAi, and Kumar et al. (2003) reported a novel microarray platform for effective RNAi selection, in which spotted siRNAs are tranfected into cells that report loss of translation of a green fluorescent protein reporter gene. RNAi is not guaranteed to identify all mutant phenotypes (as evidenced by a small but significant failure to detect known phenotypes in the whole-chromosome screens), but these screens are at least as efficient as classical mutagenesis for identification of novel functions.

Gain-of-function mutagenesis. Systematic gain-of-function genetics can be employed in numerous situations where the disruption of gene expression or protein structure is not the best way to study gene function (Figure 5.22). These include cases in which redundancy masks the loss-of-function phenotype, or where the earliest effects of gene disruption preclude ascertainment of a later function—such as when an embryonic lethal mutation prevents study of the gene's function during adult organ development.

Generic methods for inducing ectopic gene expression employ inducible promoters, such as a heat-shock promoter or drug-inducible promoters. Some of these systems disrupt expression of a large fraction of the genome, which is a drawback; however, appropriate controls can be performed to demonstrate that a phenotype is dependent on gain of expression of the transgene. Follow-up screens for genes that suppress the gain-of-function phenotype will typically identify proteins with similar biochemical, developmental, or physiological functions. Gain-of-function can also be an efficient method for characterizing the effect of modification of particular parts of a protein, for example in defining the residues that are responsible for functional specificity of transcription factors or receptors.

Dominant loss-of-function is difficult to engineer, but is an important tool for dissection of pleiotropic functions that are masked by early defects of mutations. One approach to these mutations is to design proteins that are expected to perform some but not all of the functions of the wild-type protein. For example, receptors lacking the intracellular domain will often absorb the ligand that normally activates the receptor, thereby interfering with the transmission of the signal by the endogenous protein as well. Inducible expression of the dominant negative form allows the function of the gene to be studied at any phase of development. Another approach is to screen a panel of constructs in yeast or tissue culture cells to identify dom-

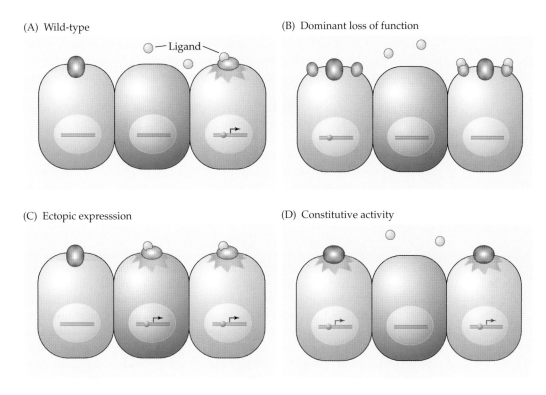

Figure 5.22 Targeted loss- and gain-of-function genetics.   Wild-type gene activity can be modified through the construction of transgenic organisms in a number of ways. (A) Suppose a protein is expressed only in the light blue cells, and that it requires the yellow ligand in order to become active. (B) Dominant negative proteins (red) can be expressed in these cells to inhibit the activity of the endogenous protein—for example, by soaking up the ligand. (C) Gain-of-function can be engineered by inducing ectopic expression of the gene in other cell types or (D) by introducing a modified protein that is constitutively active even in the absence of ligand.

inant negative proteins, and then to introduce these into the germ line of the model organism by way of a transgene.

Viral-mediated transfection. For many purposes, viral-mediated transfection can be more efficient than germ line transgenesis for introducing modified genes into a developing organism. Several broad host-range viruses have been developed for this purpose. Some viruses can be produced in both replication-competent and replication-incompetent forms. The former will infect a large field of cells in developing organisms, and thus can be used to study the effects of ectopic gene expression during embryogenesis of organisms for which traditional genetic analysis is not available. Notably, transfection with modified retroviruses has been used extensively to study appendage development and neurogenesis in chick embryos. Replication-incompetent virus is much safer to work with, and is more useful for studying the effects of ectopic expression on the differentiation of a small clone of cells. Viral transfection is better suited to specific hypothesis testing than general screening. Another application lies in comparative genomics, particularly for testing the evolution of gene function across a wide range of invertebrates. This tool is also being developed for gene delivery associated with human gene therapy and DNA vaccines.

Phenocopies. It is worth emphasizing that gene function can also be "phenocopied," which means that mutant phenotypes can be mimicked by environmental disturbance. In fact, in pharmacology, phenocopies may be a much more direct indicator of gene function than mutagenesis or overexpression studies, owing to the plasticity of neuronal systems. Monoamine receptor knockout mice show fascinating and dramatic behavioral defects, but these do not always agree with the phenotypes observed after administration of receptor antagonists (e.g., Holmes 2001). In some cases, this may be because the drug recognizes and interferes with more than one protein. It may also be a function of sweeping changes in regulation of other receptor family members that occur in the absence of a particular gene function in the synapse. So, while gene disruption is the most obvious and direct way to study gene function, there are situations where it can be misleading, and functional genomic analysis generally calls for a variety of strategies for biological annotation.

Fine-Structure Genetics

Whole-genome mutagenesis is a very efficient approach to identification of a core set of genes that function in a particular process, but it is just a first step. Characterizing the function of every gene in a genome also requires utilization of a suite of fine-structure genetic analysis techniques. Regional and pathway-specific screens have been devised to increase the efficiency of forward genetics. A battery of new procedures allow collections of genes to be expressed in situations where they are not normally active; facilitate expression of in vitro modified versions of genes; and are used to specifically disrupt gene function in particular tissues and at a certain times. Additionally, at least for microbes, the technique of genetic fingerprinting can be used to define subtle effects on growth that are not apparent in screens for heritable phenotypes.

Regional mutagenesis. Regional mutagenesis refers to genetic screens designed to saturate a small portion of the genome with insertions or point mutations. The aims are to generate lesions that disrupt each of the multiple functions of known genes; to identify genetic functions independent of evidence from expression data or prediction of coding potential; and to obtain a more thorough understanding of the genetic structure of chromosome regions than that offered by the preceding methods. For example, a genome center with the capacity to process 1,000 mutant mouse strains in a year, by focusing on a 10-cM region of a single chromosome, could be expected to generate several mutations in every one of up to several hundred genes in the region. This strategy was used to saturate the albino region of chromosome 7 with 31 lethal mutations that fell into 10 complementation groups (Rinchik and Carpenter 1999).

Whereas point mutations and transposable element (TE) insertions are generated at random throughout the genome, they can be targeted to a particular region by only choosing to retain mutations that map to that region. This is most simply achieved through the use of **balancer chromosomes**.

EXERCISE 5.4 *Designing a genetic screen*

Create a genetic screen designed to detect recessive mutations on chromosome 4 that are required in the female germline for viability of mice. Assume that you have available a strain of mice that carry a dominantly marked chromosome 4 with a large inversion (namely, a balancer chromosome).

ANSWER: *Progeny of mutagenized male mice are crossed to females that carry the chromosome 4 balancer, resulting in heterozygous mutant F_1 mice. These are crossed to another chromosome 4 balancer stock that carries another visible dominant marker on the other fourth chromosome (shown in white). All progeny of this cross that do not have the dominant marker (i.e., the white chromosome) must be heterozygous for the mutagenized chromosome and the balancer. A cross between two of these F_2 mice results in homozygous mutant F_3 progeny. You would cross these female mice to a wild-type strain, and screen for the absence of viable progeny.*

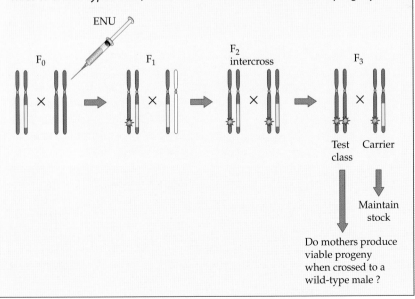

Do mothers produce viable progeny when crossed to a wild-type male ?

Balancers contain a recessive lethal that prevents the appearance of homozygous balancer individuals, one or more inversions that inhibit recombination with mutant chromosomes, and a dominant visible marker. *Drosophila* balancers contain multiple inversions that suppress recombination throughout the chromosome, but more restricted balancers can be constructed for other organisms in order to isolate a small region of the genome. Kile et al. (2003) used an inverson on mouse chromosome 11 to isolate 55 new mutations that are lethal at different stages of pre- and postnatal development, and 33 visible mutations with neurological, haematopoietic, morphological, and fertility defects.

When two heterozygotes for a mutant chromosome over the dominant marked inversion are crossed (mut/Bal × mut/Bal), failure to obtain progeny without the dominant phenotype indicates that a recessive lethal (mut/mut) is located opposite the inversion. Alternatively, complete co-segregation of the dominant marker and a dominant phenotype (for example, coat color) carried by a TE used to generate insertional mutants, indicates that the TE is inserted opposite the inversion and allows recovery of regional insertions in the absence of an overt phenotype. All other mutations or insertions can either be discarded, or sent to another group for analysis.

Mapping of regional mutations proceeds by deficiency complementation (Figure 5.23). Model organism stock centers are accumulating overlapping deficiency collections, otherwise known as **segmental aneuploids**. The deficiencies can be induced by radiation, or can be generated in a site-directed manner using the Cre-Lox recombinase system to delete all of the DNA between two flanking TE insertions. Since up to a quarter of the genes in higher eukaryotic genomes are thought to be essential for viability, deficiencies removing tens of genes are generally homozygous-lethal, and can even be sub-viable in heterozygotes as a result of the summation of haploinsufficient effects. With a set of overlapping deficiencies, it is possible to map

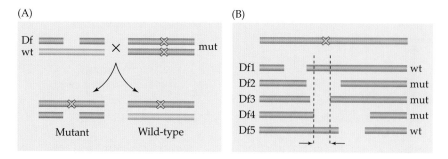

Figure 5.23 Deficiency mapping.   Recessive point mutations can be mapped by complementation testing. (A) Two classes of progeny—mutant and wild-type—are produced when a homozygous loss-of-function mutant is crossed to a line carrying a deficiency (Df) that removes the locus. If the deficiency does not remove the locus, all progeny will be wild-type. (B) By testing an ordered series of overlapping deficiencies, the extent of the region in which the mutation lies can be narrowed to tens of kilobases, at which point direct sequencing will often identify the mutation.

point mutations and TE insertions with high precision relative to both genetic and physical maps. This procedure also facilitates clustering of mutations into complementation groups.

Modifier screens. Where a phenotype is affected by a single known mutation, pathway-directed screens can be used to identify more genes that interact with the first mutation, in what are technically known as second-site modifier screens. Modifiers may either enhance or suppress a phenotype, and can do so in heterozygous or homozygous condition, depending on the sensitivity of the screen. Synthetic lethal screens identify pairs of loci that are singly viable but lethal in combination. Such screens have proven particularly useful in yeast: in one systematic series of crosses with 4,700 viable deletion strains, 291 interactions involving 204 genes were implicated in such processes as cytoskeletal organization, and DNA synthesis and repair (Tong et al. 2001; see Box 5.2).

Synthetic visible defects are likely to be particularly important in uncovering redundant functions, as well as more generally for identifying genes that interact in the same pathway. It is important to recognize that a genetic interaction (that is, a phenotype produced by an interaction between two or more genes) can have a large number of mechanistic causes. Sometimes this is because two proteins encoded by the genes physically interact; sometimes it is because they lie in the same pathway—for example, one gene regulates another, perhaps with several intermediaries in between the two products. Sometimes it is because both proteins utilize the same substrate. But the interactions can also be indirect, for example between a gene involved in neuropeptide secretion in the hypothalamus and one involved in liver or pancreatic homeostasis. Genetic backgrounds can also play a major role in modifying the observed interaction, and the more subtle a phenotype is the more care must be taken to ensure that the genetic background is uniform.

An extremely attractive feature of pathway-based screening for functional genomics is that it can be performed in **heterologous systems**. That is to say, a gene identified in one organism that cannot be studied using classical genetics (for example, human, dog, or forest tree) can be screened for interactions in another organism (mouse, fly, or *Arabidopsis*). The finding that over 50% of known human disease genes have homologs in invertebrate model organisms makes the fly and nematode popular systems for attempts to identify genes that interact with the primary disease locus. The strategy is either to knock out the homologous gene in the model organism, or to express the gene ectopically in the model system, and then screen for modifiers of any aberrant phenotype. As an example, mutant presenilin protein is a causative agent in the onset of Alzheimers' disease, and when introduced into *Drosophila* produces a neurodegenerative phenotype that has been subjected to modifier screens that suggest an interaction with the well-characterized Notch signaling pathway (Anderton 1999).

For the identification of biochemical interactions, it is not even necessary that the phenotype in the heterologous organism show any relation to the disease phenotype. The great conservation of protein structures implies that interactions should be conserved at the level of gene families across the entire range of multicellular organisms, from plants to animals. By choosing different heterologous systems, it may be possible to screen a much wider array of potential phenotypes and interactions, and to adjust the sensitivity of the interactions to enhance the probability of detecting them.

Enhancer trapping and *GAL4*-mediated overexpression. Genetic pathways that are not amenable to phenotypic characterization can be accessed by searching for genes that are specifically expressed at the time and in the place of interest. While this can be done with microarrays, there are advantages to being able to observe expression in the whole organism. An important tool that allows such observation is **enhancer trapping**. This technique has the added feature that the gene is tagged with a TE insertion that either mutates the gene or can be used to induce a mutation. Enhancer trapping was initially developed in *Drosophila* (O'Kane and Gehring 1987) and derivative procedures are being adapted to most other model organisms.

An enhancer trap is a transposable element vector that carries a weak minimal promoter adjacent to the end of the element, which is hooked up to a reporter gene such as *lacZ* (β-galactosidase) or GFP (green fluorescent protein). When the TE inserts into a gene, the nearby enhancers now drive expression of the reporter gene through the minimal promoter that is inactive on its own, but faithfully reports aspects of the expression of one of the genes adjacent to the site of insertion (Figure 5.24B). This method can be used to identify genes that are expressed in just a small subset of cells or at a precise time in development or after administration of a drug or behavioral regimen.

Various derivative methods, outlined in Figure 5.24C and D, have been developed for controlled gain-of-function genetics. Enhancers that are identified with enhancer traps can be harnessed to drive expression of other genes besides simply that of the innocuous reporter. Gain-of-function analysis is used to identify gene products that produce a novel phenotype when present at abnormal concentrations or in abnormal places. It can also be used to test the effects of in vitro modification of the gene, for example modifying the active site of an enzyme or ligand-binding domain of a receptor. Systematic gain of function is achieved by substituting the reporter gene of an enhancer detector construct with the *GAL4* gene, which encodes a potent yeast transcriptional activator that turns on any target gene with an upstream activator sequence (UAS) in its promoter. *GAL4* expressed in this way can be used to drive tissue-specific expression of any gene the investigator wishes to introduce under control of a UAS sequence. It can also be used to induce tissue-specific loss of function by driving expression of an RNAi construct; in *Drosophila*, almost 90% of the genes can be conditionally knocked out in this way now that a library of 22,270 transgenic RNAi lines has been developed (Dietzl et al. 2007).

For screening purposes, the first gene studied is a UAS-*lacZ* construct (since this shows where the enhancer is active), but once an enhancer that is active in some tissue of interest has been found, that line can be used for

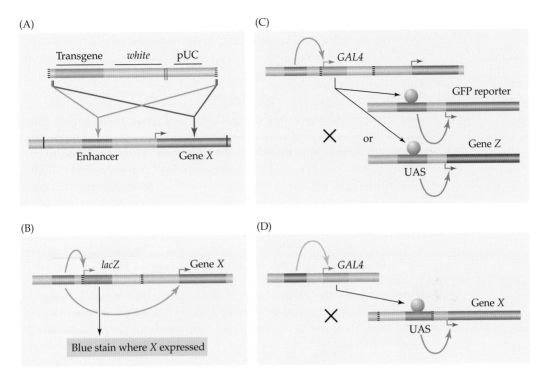

Figure 5.24 Transposon mutagenesis and enhancer trap screens.
(A) Transposon mutagenesis occurs when a mobile element, such as the *Drosophila* P-element, inserts into the regulatory or coding regions of a gene. The former position will generally disrupt transcription in one or more tissues; the latter will usually produce a null allele. Cleavage of the inserted genomic DNA by a restriction enzyme that cuts in the P-element *and* somewhere in the flanking genomic DNA (vertical red lines) facilitates "plasmid rescue" of the DNA adjacent to the site of insertion. Modified P-elements carry a selectable marker (such as the mini-*white* eye color gene); plasmid sequences (labeled pUC), including an origin of replication and selectable drug resistance gene; and the transgenic sequences of interest. (B) Enhancer traps are transposons that insert immediately upstream of a gene in such a way that a reporter gene on the transposon (often *lacZ* or GFP) comes under the control of endogenous enhancers that normally drive expression of the genomic locus. (C) Instead of using a reporter, the enhancer trap can be used to drive expression of the transcriptional activator *GAL4*, which, when crossed to another line carrying a *GAL4*-responsive UAS promoter hooked up to a gene of interest, results in activation of expression of that gene where the enhancer is active. In this way, any gene can be turned on specifically in any tissue at a given time. (D) Systematic gain-of-function genetics is performed by establishing a library of transposon insertions that bring a UAS promoter adjacent to a genomic locus, expression of which is driven with *GAL4* under control of any desired promoter.

controlled gain-of-function analysis (Brand and Perrimon 1993). For example, a *GAL4* driver active in a particular nucleus of the hypothalamus might be used to turn on expression of a particular modified neurotransmitter receptor in that portion of the brain and hence to test the function of the modified receptor. The only requirement is that a different strain with the desired transgene under control of a UAS promoter be available for crossing to the driver. Crossing this line to a panel of different *GAL4* drivers can also be used to screen for effects of ectopic expression of the gene of interest in different tissues, and even to screen directly for novel phenotypes.

An alternative method of screening for the effects of ectopic expression is the **enhancer-promoter (EP) method**. This technique is used to find genes that result in a novel phenotype when expressed in a tissue of interest, as opposed to screening for effects of a particular gene in a range of tissues. The strategy is to replace the minimal promoter of the enhancer detector construct with UAS sequences, so that when the TE inserts upstream of a gene, the gene can be turned on under control of a *GAL4* driver of choice introduced on a different chromosome (Rorth et al. 1998). By mixing and matching different constructs maintained in different lines, it is possible to study phenotypes that are lethal to the organism but can be regenerated at will, since the individual UAS and *GAL4* constructs have no effect on their own.

Floxing.   Loss-of-function genetics can be supplemented by techniques that allow gene function to be disrupted specifically at a certain time of development, and/or in specific tissues. This is crucial for separating the different pleiotropic functions of a gene, as well as for examining the function of those genes required for embryonic viability. The powerful technique used for this purpose in mouse genetics is known as floxing and is diagrammed in Figure 5.25.

Homologous recombination is used to introduce a *lox* site on either side of an essential exon of the gene. These sites are recognition-sequence targets for the yeast Cre recombinase enzyme, which is used to induce excision of a stretch of DNA between two adjacent *lox* sites. When mice homozygous for an exon flanked by *lox* sites are crossed to a line that expresses Cre under the control of a desired promoter, excision of the exon results in a temporally and spatially restricted mutation (Wagner et al. 1997). Viral infection has also been used to induce expression of the Cre recombinase and consequent excision of sequences flanked by *lox* sites (Akagi et al. 1997).

Floxing has enormous potential for the dissection of the roles of genes in the development and pharmacology of the brain, as it will be possible to remove gene expression precisely from individual regions of the cortex. Temporal activity of the recombinase has also been modulated by fusing it to a mutated ligand-binding domain of the human estrogen receptor, with the result that it is responsive to tamoxifen supplied in the diet (Metzger and Chambon 2001).

Genetic Fingerprinting

One of the remarkable findings of genome sequencing has been that at least a quarter of all genes are species-specific, in the sense that homologous genes are not readily detected in other organisms. Many of these genes are fast-evolving, and it has been hypothesized that they usually will not be asso-

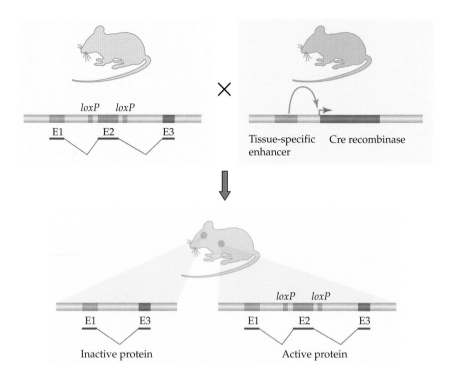

Figure 5.25 Floxing mice. One strain of mice is engineered by homologous recombination to carry *loxP* sites on either side of an essential exon of the gene that is being targeted for disruption. The other strain carries a transgene that expresses the Cre-recombinase gene specifically in the tissue of interest, and/or at a particular time. In F$_1$ progeny, excision of the exon between the *loxP* sites from genomic DNA in cells that express Cre recombinase results in a mutated gene that does not produce functional protein.

ciated with critical functions and so may be refractory to both forward and reverse genetic dissection. One possibility is that they encode phenotypic modifiers that are required for subtle adjustment of traits, or only for responses to extreme environmental circumstances (where extreme means anything outside of the laboratory). No approaches have yet been devised for testing this proposition in higher eukaryotes. However, the **genetic fingerprinting** approach diagrammed in Figure 5.26 has already been used to demonstrate subtle and environment-specific functions of a large fraction of yeast and microbial genomes.

The idea behind genetic fingerprinting is to generate a population of thousands of independent mutations on an isogenic background, and then to perform an artificial selection experiment that will result in specific loss of mutations in genes that affect fitness (Smith et al. 1996). The mutations are introduced by mobilization or transfection of a transposable element, and are detected by PCR-mediated multiplex amplification of genomic fragments, using one primer in the TE and another in the genome. The microbes

(A)

(B)

(C)

(D)

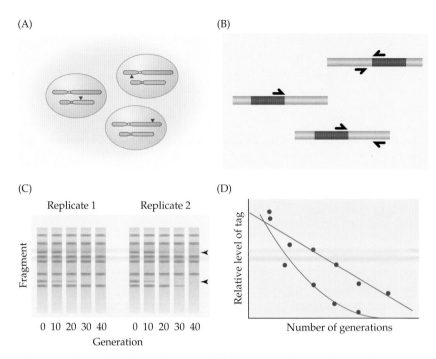

Replicate 1   Replicate 2

Fragment

0 10 20 30 40   0 10 20 30 40
Generation

Relative level of tag

Number of generations

Figure 5.26 Genetic fingerprinting. (A) A complex mixture of tens of thousands of clones is constructed in which a single transposon is inserted in or adjacent to a different locus. (B) Each insertion is tagged, since the combination of a primer within the transposon and one in the genome will amplify a fragment of a diagnostic length. (C) The population is grown for a hundred generations. Samples are extracted every 10 generations for genotyping. Multiplex PCR is performed, followed by separation of fragments by electrophoresis. If the transposon is within a gene that affects viability under the growth conditions, the corresponding bands will disappear with time in replicate experiments (arrowheads). Some tags will also disappear by random sampling in a single replicate. (D) Averaging over experiments, a plot of fragment intensity against time is indicative of the selection pressure against the insertion. In this hypothetical example, selection against the red gene is more intense than against the blue gene.

are grown in a chemostat, an apparatus that allows continuous feeding of exponentially growing cells with sterile medium for hundreds of generations. A selection differential as small as 1% will quickly amplify and will eliminate from the population a lineage carrying a mutated essential gene after just 20 generations of experimental evolution. The experiment is replicated multiple times to control for the fact that most mutations will actually be lost by genetic drift in one or a few of the replicates. Conducting the experiment under a variety of culture conditions allows detection of growth differentials on different nutrient sources, increasing the likelihood of identifying environment-specific functions. Mutations detected with the fingerprinting approach can subsequently be tested by more direct competition of wild-type and targeted mutant strains.

Summary

1. Prediction of protein function by comparison with the sequence and structure of other proteins often provides the first hint of gene function and is facilitated by the UniProt resource.

2. The simplest tool for characterizing the proteome of a cell is 2D electrophoresis, in which proteins are separated by charge in one dimension and mass in another. Methods for identifying each protein are available, but for comparisons of the same tissue type it is possible to compare an experimental gel with a canonical gel for the tissue that has been annotated previously.

3. High-throughput identification of proteins is performed by parallel peptide sequencing using time-of-flight mass spectrometry (MS). Each proteolytic fragment has a characteristic mass-to-charge (m/z) ratio, and comparison of a complex mixture of m/z peaks with the distribution predicted from the genome sequence allows inference of protein identities.

4. Tandem mass spectrometry (MS/MS) is a method for partial sequencing of complex mixtures of peptides, based on random degradation of individual peptides deflected into an ionization chamber. Combined with liquid chromatography, it has emerged as the preferred method of proteome characterization.

5. Quantification of protein levels can be performed by heavy isotope-labeling of peptide mixtures followed by MS, so that each peak appears side-by-side with a peak derived from an unlabeled sample, in proportion to the amount of protein. ICAT or iTRAQ reagents facilitate reproducible labeling.

6. Protein microarrays consist of in vitro purified proteins or antibodies spotted onto a glass slide. Specific protein-protein, protein-drug, protein-enzyme, and protein-antibody interactions can be detected by fluorescent labeling.

7. Protein interaction maps are networks of protein-protein links indicating which proteins interact physically and/or functionally. Yeast two-hybrid (Y2H) screens are a direct way to determine which pairs of proteins assemble complexes under physiological conditions.

8. Protein structures are deposited in the Protein Data Bank (PDB). Structural proteomics aims to determine a representative structure of each of the hypothesized 5,000 distinct protein folds.

9. X-ray crystallography and NMR spectroscopy are the two major methods used to determine the three dimensional structures of proteins and protein complexes.

10. Threading is a method for predicting protein tertiary structure by comparing predicted secondary structure and folding energies with those of solved structures. It allows structures to be predicted where there is less than 30% sequence identity, the lower cut-off for simple alignment.

11. Functional genomics is the process of assigning a biochemical and physiological/cell biological/ developmental function to each predicted gene.

12. Forward genetics refers to the random generation of mutations followed by mapping and localization of the mutated gene. Saturation genetic screens can be based either on detection of visible phenotypes or on generation of large collections of transposable element insertions that may or may not disrupt a gene.

13. Reverse genetics refers to the process of generating a mutation in a specific open reading frame and then searching for a phenotype. Homologous recombination is a general method for knocking out gene function, but is difficult to perform on a high-volume scale and only works efficiently for mice and yeast.

14. Whereas a mutation that disrupts the function of a gene is called a "knockout," transgenic organisms can also be constructed in which the wild-type gene is replaced either by a different gene or by a modified form of the gene. These are called "knock-in" genes.

15. RNAi is a method for abolishing gene function based on the ability of siRNAs—double-stranded RNA diced into "short interfering" sequences in the cell—to interfere with the synthesis of the gene product. Both transient and transgenic approaches can be taken.

16. An ever-increasing arsenal of methods for systematic manipulation of gene expression in vivo allow genes to be turned on or off in specific tissues, at specific times, and in particular genetic backgrounds. This manipulative ability will allow dissection of the pleiotropic functions of genes in development, physiology, and behavior.

Discussion Questions

1. Why aren't transcript and protein expression profiles always in agreement?

2. What is a protein domain? To what extent is it possible to infer the function of a protein by sequence and phylogenetic comparison with domains present in other proteins?

3. Protein interaction networks can be assembled from yeast two-hybrid experiments, protein microarrays, and using bioinformatic methods. Discuss the limitations and biases inherent in each of these approaches.

4. What types of trait are amenable to dissection using random mutagenesis to identify genes that affect the phenotype? Under what circumstances is reverse genetic analysis more likely to lead investigators to "the genes that matter"?

5. How can functional genomics be applied to characterize the roles of human genes that do not have a known mutation?

Web Site Exercises

The Web site linked to this book at http://www.sinauer.com/genomics provides exercises in various techniques described in this chapter.

1. Build a zinc-finger protein profile and execute a PSI-BLAST exercise.

2. Characterize the secondary structure and modification sites of several related proteins.

3. Perform a simple threading exercise.

4. Identify the complementation groups in a hypothetical genetic data set.

Literature Cited

Akagi, K. et al. 1997. Cre-mediated somatic site-specific recombination in mice. *Nucl. Acids Res.* 25: 1766–1773.

Anderton, B. H. 1999. Alzheimer's disease: Clues from flies and worms. *Curr. Biol.* 9: R106–R109.

Apweiler, R. et al. 2004. UniProt: The Universal Protein Knowledgebase *Nucl. Acids Res.* 32: D115–D119.

Baker, D. and A. Sali. 2001. Protein structure prediction and structural genomics. *Science* 294: 93–96.

Bass, B. L. 2002. RNA editing by adenosine deaminases that act on RNA. *Annu. Rev. Biochem.* 71: 817–846.

Berman, H. M. et al. 2000. The Protein Data Bank. *Nucl. Acids Res.* 28: 235–242.

Birch, R. G. 1997. Plant transformation: Problems and strategies for practical application. *Annu. Rev. Plant Physiol. Plant Mol. Biol. 48:* 297–326.

Brand, A. and N. Perrimon. 1993. Targeted gene expression as a means of altering cell fates and generating dominant phenotypes. *Development* 118: 401–415.

Burley, S. K. 2000. An overview of structural genomics. *Nat. Struct. Biol.* 7 (Suppl.): 932–934.

Capecchi, M. R. 1989. Altering the genome by homologous recombination. *Science* 244: 1288–1292.

Celis, J. E. et al. 2000. Gene expression profiling: Monitoring transcription and translation products using DNA microarrays and proteomics. *FEBS Lett.* 480: 2–16.

Chalmers, M. and S. Gaskell. 2000. Advances in mass spectrometry for proteome analysis. *Curr. Opin. Biotechnol.* 11: 384–390.

Chan, A. W., K. Chong, C. Martinovich, C. Simerly, and G. Schatten. 2001. Transgenic monkeys produced by retroviral gene transfer into mature oocytes. *Science* 291: 309–312.

Cherkasov, A. and S. J. Jones. 2004. Structural characterization of genomes by large scale sequence-structure threading. *BMC Bioinformatics* 5: 37.

Dietzl, G. et al. 2007. A genome-wide transgenic RNAi library for conditional gene inactivation in *Drosophila*. *Nature* 448: 151–156.

Driever, W. et al. 1996. A genetic screen for mutations affecting embryogenesis in zebrafish. *Development* 123: 37–46.

Dunbrack, R. L. Jr. 2006. Sequence comparison and protein structure prediction. *Curr. Opin. Struct. Biol.* 16: 374–384.

Fenyö, D. 2000. Identifying the proteome: Software tools. *Curr. Opin. Biotechnol.* 11: 391–395.

Fields, S. and O. Song. 1989. A novel genetic system to detect protein-protein interactions. *Nature* 340: 245–246.

Fire, A., S. Xu, M. Montgomery, S. Kostas, S. Driver and C. Mello. 1998. Potent and specific genetic interference by double-stranded RNA in *Caenorhabditis elegans*. *Nature* 391: 806–811.

Fischer, D. and D. Eisenberg. 1997. Assigning folds to the proteins encoded by the genome of *Mycoplasma genitalium*. *Proc. Natl Acad. Sci. (USA)* 94: 11929–11934.

Frary, A. et al. 2000. *fw2. 2*: A quantitative trait locus key to the evolution of tomato fruit size. *Science* 289: 85–88.

Gérard, M., J. Zákány and D. Duboule. 1997. Interspecies exchange of a *Hoxd* enhancer in vivo induces premature transcription and anterior shift of the sacrum. *Dev. Biol.* 190: 32–40.

Gygi, S., B. Rist, S. Gerber, F. Turecek, M. Gelb and R. Aebersold. 1999. Quantitative analysis of complex protein mixtures using isotope-coded affinity tags. *Nat. Biotechnol.* 17: 994–999.

Gygi, S., B. Rist and R. Aebersold. 2000. Measuring gene expression by quantitative proteome analysis. *Curr. Opin. Biotechnol.* 11: 396–401.

Haffter, P. et al. 1996. The identification of genes with unique and essential functions in the development of the zebrafish, *Danio rerio*. *Development* 123: 1– 36.

Hannon, G. J. 2002. RNA interference. *Nature* 418: 244–251.

Hegyi, H. and M. Gerstein. 1999. The relationship between protein structure and function: a comprehensive survey with application to the yeast genome. *J. Mol. Biol.* 288: 147–164.

Henderson, R. 2004. Realizing the potential of electron cryo-microscopy. *Q. Rev. Biophys.* 37: 3–13.

Holmes, A. 2001. Targeted gene mutation approaches to the study of anxiety-like behavior in mice. *Neurosci. Biobehav. Rev.* 25: 261–273.

Hrabe de Angelis, M. H. et al. 2000. Genome-wide, large-scale production of mutant mice by ENU mutagenesis. *Nat. Genet.* 25: 444–447.

Ito, T. et al. 2000. Toward a protein-protein interaction map of the budding yeast: A comprehensive system to examine two-hybrid interactions in all possible combinations between the yeast proteins. *Proc. Natl. Acad. Sci. (USA)* 97: 1143–1147.

Jones, D. T. 2000. Protein structure prediction in the postgenomic era. *Curr. Opin. Struct. Biol.* 10: 371–379.

Kallioniemi, A. 2008. CHG microarrays and cancer. *Curr. Opin. Biotechnol.* 19: 36–40.

Kamath, R. S. et al. 2003. Systematic functional analysis of the *Caenorhabditis elegans* genome using RNAi. *Nature* 421: 231–237.

Kannicht, C. (ed.) 2002. *Posttranslational Modification of Proteins: Tools for Functional Proteomics*. Volume 194 of *Methods in Molecular Biology*. Humana Press, New York.

Kelso, R. J., M. Buszczak, A. T. Quinones, C. Castiblanco, S. Mazzalupo and L. Cooley. 2004. Flytrap, a database documenting a GFP protein-trap insertion screen in *Drosophila melanogaster*. *Nucl. Acids Res.* 32: D418–420.

Kile, N. T. et al. 2003. Functional genetic analysis of mouse chromosome 11. *Nature* 425: 81–86.

Koegl, M. and P. Uetz. 2007. Improving yeast two-hybrid screening systems. *Brief. Funct. Genom. Prot. Adv.* 6: 302–312.

Kumar, R., D. S. Conklin and V. Mittal. 2003. High-throughput selection of effective RNAi probes for gene silencing. *Genome Res.* 13: 2333–2340.

Lehner, B. and A. G. Fraser. 2004. A first-draft human protein-interaction map. *Genome Biol.* 5: R63.

Liu , H., D. Lin and J. R. Yates III. 2002. Multidimensional separations for protein/peptide analysis in the post-genomic era. *Biotechniques* 32: 898–911.

MacBeath, G. and S. Schreiber. 2000. Printing proteins as microarrays for high-throughput function determination. *Science* 289: 1760–1763.

Manjasetty, B. A., A. Turnbull, S. Panjikar, K. Büssow and M. R. Chance. Automated technologies and novel techniques to accelerate protein crystallography for structural genomics. *Proteomics* 8: 612–625.

Marcotte, E., M. Pellegrini, H. Ng, D. Rice, T. Yeates and D. Eisenberg. 1999. Detecting protein function and protein-protein interactions from genome sequences. *Science* 285: 751–753.

Metzger, D. and P. Chambon. 2001. Site- and time-specific gene targeting in the mouse. *Methods* 24: 71–80.

Minden, J. 2007. Comparative proteomics and difference gel electrophoresis. *Biotechniques.* 43: 739–743.

Montelione, G., D. Zheng, Y. Huang, K. Gunsalus and T. Szyperski. 2000. Protein NMR spectroscopy in structural genomics. *Nat. Struct. Biol.* 7 Suppl.: 982–985.

Nolan, P. M. et al. 2000. A systematic, genome-wide, phenotype-driven mutagenesis programme for gene function studies in the mouse. *Nat. Genet.* 25: 440–443.

Nüsslein-Volhard, C. and E. Wieschaus. 1980. Mutations affecting segment number and polarity in *Drosophila*. *Nature* 287: 795–801.

O'Kane, C. and W. J. Gehring. 1987. Detection *in situ* of genomic regulatory elements in *Drosophila*. *Proc. Natl Acad. Sci. (USA)* 84: 9123–9127.

Panchaud, A., M. Affolter, P. Moreillon and M. Kussmann. 2008. Experimental and computational approaches to quantitative proteomics: Status quo and outlook. *J. Proteomics* 71: 19–33.

Pellegrini, M., E. Marcotte, M. Thompson, D. Eisenberg and T. Yeates. 1999. Assigning protein functions by comparative genome analysis: protein phylogenetic profiles. *Proc. Natl Acad. Sci. (USA)* 96: 4285–4288.

Ramachandran, N., E. Hainsworth, B. Bhullar, S. Eisenstein, B. Rosen, A. Lau, J. C. Walter and J. LaBaer. 2004. Self-assembling protein microarrays. *Science* 305: 86–90.

Rinchik, E. M. and D. A. Carpenter. 1999. *N*-ethyl-*N*-nitrosourea mutagenesis of a 6- to 11-cM subregion of the *Fah-Hbb* interval of mouse chromosome 7: Completed testing of 4557 gametes and deletion mapping and complementation analysis of 31 mutations. *Genetics* 152: 373–383.

Rorth, P. et al. 1998. Systematic gain-of-function genetics in *Drosophila. Development* 125: 1049–1057.

Ross, P. L. et al. 2004. Multiplexed protein quantitation in *Saccharomyces cerevisiae* using amine-reactive isobaric tagging reagents. *Mol. Cell. Proteomics* 3, 1154–1169.

Rubin, G. M. and A. C. Spradling. 1982. Genetic transformation of *Drosophila* with transposable element vectors. *Science* 218: 348–353.

Shen, W., J. Escudero and B. Hohn. 1999. T-DNA transfer to maize plants. *Mol. Biotechnol.* 13: 165–170.

Smith, V., K. Chou, D. Lashkari, D. Botstein and P. O. Brown. 1996. Functional analysis of the genes of yeast chromosome V by genetic footprinting. *Science* 274: 2069–2074.

Sonnhammer, E. L., S. R. Eddy, E. Birney, A. Bateman and R. Durbin. 1998. Pfam: Multiple sequence alignments and HMM-profiles of protein domains. *Nucl. Acids Res.* 26: 320–322.

Splawski, I. et al. 2000. Spectrum of mutations in long-QT syndrome genes *KVLQT1, HERG, SCN5A, KCNE1,* and *KCNE2. Circulation* 102: 1178–1185.

Spradling, A. C. et al. 1999. The Berkeley *Drosophila* Genome Project gene disruption project: Single P-element insertions mutating 25% of vital *Drosophila* genes. *Genetics* 153: 135–177.

Stevens, R. C., S. Yokoyama and I. A. Wilson. 2001. Global efforts in structural genomics. Science 294: 89–92.

Taylor, W. R. 2002. A "periodic table" for protein structures. *Nature* 416: 657–660.

Thomas, K. and M. Capecchi. 1987. Site-directed mutagenesis by gene targeting in mouse embryo-derived stem cells. *Cell* 51: 503–512.

Tong, A. H. et al. 2001. Systematic genetic analysis with ordered arrays of yeast deletion mutants. *Science* 294: 2364–2368.

Tong, A. H. et al. 2004. Global mapping of the yeast genetic interaction network. *Science* 303: 808–813.

Uetz, P. et al. 2000. A comprehensive analysis of protein-protein interactions in *Saccharomyces cerevisiae. Nature* 403: 623–627.

Unlü, M., M. E. Morgan and J. S. Minden. 1997. Difference gel electrophoresis: a single gel method for detecting changes in protein extracts. *Electrophoresis* 18: 2071–2077.

Unwin, R. D. and A. D Whetton. 2006. Systematic proteome and transcriptome analysis of stem cell populations. *Cell Cycle* 5: 1587–1591.

Uttamchandani, M., J. Wang and S. Q. Yao. 2006. Protein and small molecule microarrays: powerful tools for high throughput proteomics. *Mol. BioSyst.* 2: 58–68.

Wagner, K. et al. 1997. Cre-mediated gene deletion in the mammary gland. *Nucl Acids Res.* 25: 4323–4330.

Washburn, M. P., D. Wolters and J. R. Yates III. 2001. Large-scale analysis of the yeast proteome by multidimensional protein identification technology. *Nat. Biotech.* 19: 242–247.

Wingren, C. and C. A. Borrebaeck. 2004. High-throughput proteomics using antibody microarrays. *Expert Rev. Proteomics* 1: 355–364.

Yee A., A. Gutmanas and C. H. Arrowsmith. 2006. Solution NMR in structural genomics. *Curr. Opin. Struct. Biol.* 16: 611–617.

Zhang, Y. and J. Skolnick. 2004. Automated structure prediction of weakly homologous proteins on a genomic scale. *Proc. Natl Acad. Sci. (USA)* 101: 7594–7599.

Zhu, H. et al. 2001. Global analysis of protein activities using proteome chips. *Science* 293: 2101–2105.

6 Integrative Genomics

An uncharitable view of genome research is that it is little more than a collection of technologies that can be used to pursue classical molecular genetics at an accelerated pace. Throughout this book, we have adopted the contrary viewpoint that in fact the conjunction of the new technologies with statistical and bioinformatic reasoning fosters an entirely different way of looking at biological systems. By extension, genome science has emerged as an independent discipline with a novel framework for studying and understanding biological complexity. Specifically, as comprehensive catalogs of genomic and cellular constituents are combined with computational strategies for studying how the parts assemble into working wholes, genome science is beginning to address problems such as:

- Predicting microbial and cellular phenotypic properties.
- Ordering of genes and gene products into biochemical and developmental networks.
- Formulating *in silico* hypotheses for gene function that can be tested in vivo.
- Manipulating whole genomes to perform novel applications of biotechnological and biomedical importance.
- Integrating genome science with ecology and evolutionary biology.

An appreciation for the magnitude of this task can be seen from a glance at Figure 6.1, which plots the growth of knowledge in the sphere of genomics since the early 1990s. While the whole-genome sequencing phase has been completed for most model organisms, we can expect an explosion of new sequence data. Likewise, the annotation of proteomes and transcriptomes is just ramping up, and the construction of protein-protein and protein-DNA interaction maps is still in its infancy. There is a vast literature on biochem-

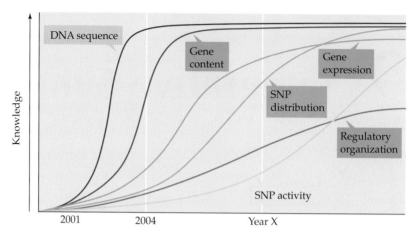

Figure 6.1 The future of genomics.   By 2004, genome scientists had assembled a reasonably thorough picture of the sequence and gene content of representative genomes and were making inroads into description of gene expression profiles and protein function. Knowledge of SNP function and the organization of metabolic and other regulatory pathways is in its infancy. "Year X" is a debatable number of years into the future. (After Schilling et al. 1999.)

ical and metabolic pathways in bacteria and yeast, much of which is likely to apply to higher eukaryotes, but knowledge of the structure and design of developmental networks is arguably sporadic.

Yet some of the core questions in the biology of complexity—such as the molecular basis for homeostasis, the roles of constraint and pleiotropy in morphological evolution, and the structure of genetic variation in natural populations—all relate to the interactions among genes, their products, and the environment. To some observers, genome science offers the only real hope of tackling these deep biological problems.

This chapter begins by adding to the genomic triumvirate of genome, transcriptome, and proteome a description of methods for studying the metabolome. The **metabolome** may be defined as the sum total of substrates, metabolites, and other small molecules that are present in a population of cells. The second section of the chapter presents an overview of some of the *in silico* approaches being taken to turn parts lists into systems of dynamic equations and networks of interactions that allow us to predict and describe the properties of cells.

The title "integrative genomics" is misleading in the sense that very little work synthesizing the different aspects of genome science has yet been performed. This is largely because the discipline is so young; microarrays and tandem mass spectrometry are still emerging and expensive technologies, and there are only a few research groups in a position to integrate genomic platforms. The name also reflects the predominantly reductionist framework of contemporary genetics, in which genomics is viewed as a means for identifying genes of interest which can then be studied using classical methods, rather than being a discipline in its own right. Bioinformatics has emerged as an independent discipline, but often uses datasets that were generated by collaborators prior to consideration of the analytical issues. Bioinformatics is most powerful when incorporated at the experimental design stage in a manner that leads naturally to integrative science. The methodologies discussed in the second half of this chapter are simply not part of standard genetics curricula, so are not yet sufficiently well known to support systems analysis.

A further reason for the disjoint nature of genomics is that a major part of the puzzle is still missing: namely, descriptions of the metabolites and regulatory interactions found in any organism. These topics are dealt with in the first part of this chapter, but as a separate set of research methods, because this research is also not yet integrated with other areas of genome science. It is possible only to speculate on how the integration will occur, but one area almost certainly will be in the prediction of gene function based on correlation between changes in expression profile and metabolite profiles, either in cohorts of individuals, a panel of mutant lines, or after environmental perturbation.

A major goal of evolutionary and ecological genomics is describing and interpreting genomic responses to genetic variation, and to interactions between organisms. Host-parasite and host-symbiont interactions are generally mediated by small molecules, many of which have been characterized using painstaking biochemical approaches. But **metabolomics**—profiling the structure and distribution of metabolites (small molecules), particularly organic compounds—promises dramatic advances in this area. On the biochemical front, toxicology and dietary imbalance are just two areas where integration of transcript, protein, and small-molecule data will not just increase our understanding of metabolic mechanisms, but may quickly translate into clinical and nutritional recommendations. **Pharmacogenomics**, the study of interactions between drugs and genomes, is already represented by several new journals, and the subdiscipline of **psychogenomics**, dealing with the effects of mood-modulating chemicals, will see rapid advances once in vivo methods for parallel analysis of chemicals and gene products become established.

Metabolomics

Analysis of Cellular Constituents

Several spectrometric techniques (Table 6.1) are being applied to profile the complement of both organic and inorganic compounds found inside cells. The first three techniques listed (pyrolysis, infrared, and Raman spectrometry) are used to characterize differences in the total spectrum of cellular constituents, and have applications in sorting panels of mutants, natural isolates, or lines grown under different environmental conditions. The gas and liquid chromatographic techniques, as well as classical biochemical purification, when combined with mass spectrometry provide a more focused view of which particular compounds vary among treatments. These

TABLE 6.1 *Metabolomic Methods*

Acronym	Method
PyMS	Pyrolysis mass spectrometry
FT-IR	Fourier transform-infrared spectrometry
DRS	Dispersive Raman spectrometry
NMR	Nuclear magnetic resonance spectroscopy
GC-MS	Gas chromatography mass spectrometry
LC-ESI-MS	Liquid chromatography electron spin ionization mass spectrometry
MDA	Multivariate data analysis: Chemometrics, artificial neural networks, genetic algorithms

techniques can be used to identify enzyme substrates and metabolites or to characterize quantitative variation among networks of compounds.

Pyrolysis refers to the thermal degradation of materials into volatile fragments (the pyrolysate) that can be separated according to mass-charge ratio (m/z) and analyzed by mass spectrometry, providing a chemical profile of the constituents in the complex sample (Irwin 1982). A few microliters of sample are dried onto a ferromagnetic foil, which is then rapidly heated either by a laser or by a current passing through a coil surrounding the sample. A temperature between 350° and 770°C is chosen to achieve a balance between lysis of carbohydrates and proteins. Products with $m/z < 50$ (e.g., methane, ammonia, water, and methanol) and large products ($m/z > 200$) are usually ignored, so that the biologically informative material lies between these ranges. Peak heights can be quantified and used to distinguish samples, for example to identify microbial species in mixed isolates; to characterize the effects of growth conditions; to identify extreme metabolic outliers in pharmaceutical screens; to detect contaminants in food products such as orange juice, Scotch whiskey, and virgin olive oil; and in support of forensic analysis. The identities of the peaks, or of the products that give rise to peaks, are not always known, but most analyses are more concerned with detecting differences in the distributions.

Infrared spectrometry measures the absorbance of energy at specific wavelengths by highly polar bonds (such as C—N, N—H, C=O, etc.), which provides a fingerprint of the presence of the chemical groups present in particular molecules in a sample. Samples are typically dried onto the surface of a metal plate, or are mixed with KBr as a fine, dried powder that reflects infrared radiation passed through the sample. Automation of the procedure allows up to 100 samples to be analyzed in parallel, with hundreds of data points collected over the range of wavelengths from 4,000 to 600 cm^{-1}. Applications include monitoring effects of drugs on microbial and fungal metabolism, pharmaceutical screening, and quantitative typing of plant cell wall differences (often due to variation in lignins).

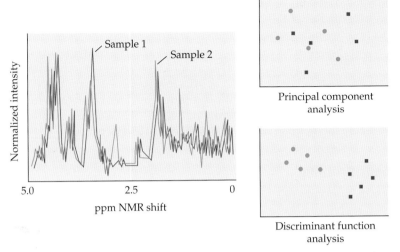

Figure 6.2 Nuclear magnetic resonance spectroscopy with chemometrics. NMR spectra capture the intensity of free-induction decays due to specific atoms, typically protons. After Fourier transformation and normalization, principal component analysis is applied to identify the major components of variation in the overall spectra in a set of samples. Further statistical manipulation such as discriminant function analysis may separate the samples by removing within-group factors, in this case clearly showing that the orange and blue samples tend to cluster. (After Raamsdonk et al. 2001.)

A promising approach to detecting a full range of metabolites is direct **nuclear magnetic resonance spectroscopy** of cellular extracts, or even of cellular suspensions. Different resonance spectra can be used to focus on carbon or phosphate metabolism, or to observe changes in ion gradients and energy metabolism (Stoyanova and Brown 2001). Proton NMR has been used to characterize overall changes in the metabolome without prior specification of which compounds are being observed (Figure 6.2; Raamsdonk et al. 2001). The idea is that many mutations may not produce a visible growth phenotype, since metabolic networks have the capacity to re-route metabolites to restore biomass production; yet such mutations might be detected through their effect on intracellular substrate concentrations. Statistical analysis of NMR spectra obtained from several replicates of each of six strains resulted in clustering of two strains known to disrupt 6-phospho-fructo-2-kinase activity. Although neither mutant had a visible phenotype, they apparently shared a distinct metabolic profile. In principle, this method might be used to sort through the entire genome, clustering mutants that disrupt similar metabolic processes.

Chemometric procedures, namely the application of statistical and mathematical methods to chemical data, are used both to extract information from metabolomic spectra (a process known as spectral analysis) and to search for patterns in the data (Wishart 2007; Xu et al. 2008). Analyses can focus on individual peaks, or on bins of peaks, generally with the objective of identifying the major components of variance in the entire spectrum. They can lead to estimates of the number of chemical compounds in a mixture, identification of outlier samples, and separation of individuals or clusters of individuals into distinct categories. Assfalg et al. (2008) provide an example of how binning of peaks in NMR spectra followed by a combination of advanced statistical procedures such as principal component analysis (PCA) and canonical analysis (CA) identifies unique profiles of metabolites in human urine samples. Artificial neural networks (ANNs) are an alternate strategy for chemometric analysis, as described in Goodacre and Kell (1996).

Metabolic Profiling

No single method currently allows each of the hundreds of different metabolite classes within cells to be assayed together, but the combination of chromatographic and mass spectrometry profiling under different conditions gives a remarkable and ever improving display of the metabolome (Dunn 2008). The terms *metabolomics* and *metabonomics* are both used to describe efforts to quantify the profile of low-molecular weight metabolites up to 1,500 Da in a sample. These include carbohydrates, lipids, amino acids, drugs, vitamins, and intermediates of central metabolism.

Identification of individual molecules within a sample—typically blood serum, urine, cerebrospinal fluid, or plant tissues—requires techniques in which the various compounds are first separated using standard methodologies such as gas, liquid, or HPLC chromatography, and then visualized either by stains or absorbance spectra, or more comprehensively using mass spectrometry. Different systems are most appropriate for different classes of molecule. For example, amino acids do not require MS for characterization; lipids and lignins can be separated by gas chromatography (GC), complex carbohydrates by capillary electrophoresis (CE), and complex steroids and derivatives by HPLC. Where well characterized standards exist—for example in relation to lipid or monoamine profiles, which are crucial for research into cardiovascular disease and pharmacology, respectively—there is often no need for an MS step. In addition, certain metabolites and small hormones, such as albumin, cortisol, insulin, melatonin, and prolactin, can be detected immunologically, while a variety of enzymatic assays exist for detection of individual compounds such as glucose, ketones, lactate, and alcohols.

A typical sample run linking liquid or gas chromatography to mass spectrometry takes around 20 minutes per sample and leads to the identification of between 200 and 300 peaks. These can be identified by reference to the standard retention times of known metabolites, or inferred by tandem MS fragmentation, analogous to that described for protein sequencing in Chapter 5. Detection limits are of the order of just 10^{-12} M, and sensitivity is such that groups as small as a sulfate or carboxyl group can be distinguished. Deconvolution of complex spectra is not a straightforward process, however, and despite the availability of many open-source and commercial software options, considerable manual processing is currently required to move from peak heights to a robust description of the concentrations of hundreds of metabolites. Technical variability is typically low relative to biological variability, so replication of sample collection is necessary.

The integration of transcriptomic, proteomic, and metabolomic data (Figure 6.3) provides fascinating challenges. One of the first studies to attempt such integration examined the effects of carbon, nitrogen, phosphate, and sulfate limitation on growth of the yeast *Saccharomyces cerevisiae* (Castrillo et al. 2007). Patterns of change were observed at each of the three "omic" levels, but methods for correlating specific alterations in transcript or protein abundance with metabolite flux are still under development. Similar studies contrasting two of the "omic" levels have appeared in organisms as diverse as corals, grapes, and enteric bacteria, suggesting that the era of systems biology has truly arrived.

Equally important, associations between genomic variation and metabolite abundance are being applied in the context of plant breeding and human disease susceptibility, among other applications. Schauer et al. (2008) identified 332 QTL for metabolite abundance in tomato fruit, three-quarters of which show dominant or recessive modes of transmission, with a tendency for adjacent metabolites to share modes of inheritance. Similar observations were made by Keurentjes et al. (2006) in *Arabidopsis*; both papers discuss the implications for engineering of plant nutrient profiles based on joint genetic and metabolic profiling. Passador-Gurgel et al. (2007) defined transcripts whose abundance is significantly correlated with a phenotype in a population of organisms as **quantitative trait transcripts** (**QTT**) and demonstrated that nicotine resistance in *Drosophila* is associated with ornithine amino transferase expression, which in turn correlates with ornithine levels measured by LC-MS. Taking the systems approach a step further, Zhu et al. (2008) combined genotypes, gene expression profiles, and knowledge of transcription factor binding sites as well as the protein-protein interaction network in yeast to reconstruct functional gene networks that could be verified by targeted mutagenesis.

One of the first studies of this type in humans has established a significant relationship between expression of the *SLC2A9* gene and serum uric acid concentrations (Döring et al. 2008). A genome-wide association study of 1,644 German individuals demonstrated that polymorphism in this gene, which encodes a likely fructose transporter that regulates uric acid levels in the kidney, explains 6% of variance for the metabolite in women, although substantially less in men. These SNPs were subsequently associated with gout in both sexes, though their effect on transcription of the gene seems limited.

Studies that report genome-wide association with thousands of metabolites measured by GC-MS or LC-MS profiling will be particularly interesting in light of the fact that metabolic syndrome (a state of predisposition to diabetes and cardiovascular disease) has been ascribed to about one-third of all adults in developed countries. Steps in this direction have been reported by Kathiresan et al. (2008), who associated 18 genes with blood LDL cholesterol, HDL cholesterol, and triglyceride concentrations in Europeans, collectively accounting for at least 5% of the variance in the metabolites; and by Schaeffer et al. (2006), who reported association of two *FADS*

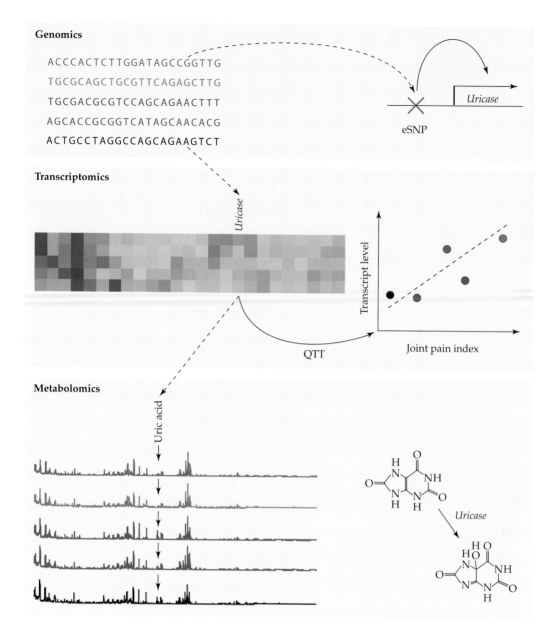

Genomics

```
ACCCACTCTTGGATAGCCGGTTG
TGCGCAGCTGCGTTCAGAGCTTG
TGCGACGCGTCCAGCAGAACTTT
AGCACCGCGGTCATAGCAACACG
ACTGCCTAGGCCAGCAGAAGTCT
```

Uricase

eSNP

Transcriptomics

Uricase

Transcript level

QTT

Joint pain index

Metabolomics

Uric acid

Uricase

Figure 6.3 Integration of genomic, transcriptomic, and metabolomic data. The primary objective of metabolomic analysis is to identify peaks in a mass spectrum that vary among individuals (bottom panel; small arrows above uric acid peak in each profile). Enzymes involved in the synthesis or catabolism of variable metabolites may vary in protein sequence or expression, possibly resulting in an association between transcript abundance and metabolite abundance, or of either with a phenotype of interest (middle panel). Further addition of genetic polymorphism data may identify eSNPs or linkage peaks (*cis*-eQTL) that explain the variation in transcript abundance, and possibly metabolite levels and/or phenotype. Ideally, the three domains of analysis reinforce one another, as in this hypothetical example suggesting that a regulatory polymorphism in the *uricase* gene affects levels of gene transcript, and hence the concentration of the protein (uric acid), resulting in a specific phenotype (joint pain) associated with gout.

desaturase genes with levels of various polyunsaturated fatty acids. Joint genotype and metabolite profiling is expected to play an important role in studies of the molecular basis of genotype-environment interactions in promoting common diseases.

Metabolic and Biochemical Databases

The Human Metabolome Database (HMDB; http://www.hmdb.ca) provides a comprehensive record of over 2,000 metabolites derived by curation of the literature. It includes sample spectra from MS or NMR analyses, tabulation of estimated metabolite abundances in various tissues, and an average of 90 separate data fields for each entry providing the chemical structure, physicochemical data, disease associations, and links to other genomic databases (Wishart et al. 2007). Extensive search and query tools interface with many of the other Web-based compilations of metabolic and biochemical pathways, including those listed in Table 6.2. Very useful overviews of metabolism and cellular biochemistry are provided by ExPASy in digital wall charts, along the lines shown in Figure 6.4. The BRENDA, MetaCyc, LIGAND, and UMBBD databases present aspects of metabolism in search-

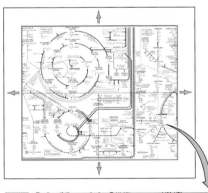

Figure 6.4 Visualization of metabolic pathways.   Two percent of the ExPASy representation of microbial metabolism, assembled from the adjacent tiles E7 and F7, showing aspects of fatty acid and arginine biosynthesis. Compounds are in black, cofactors in red, and enzymes in blue. (From http://biochem.boehringer-mannheim.com/ prodinfo_fst.htm?/techserv/metmap.htm.)

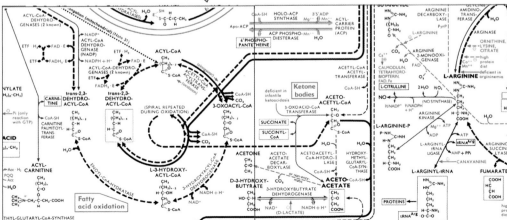

TABLE 6.2 *Databases of Biochemical Pathways and Networks*

Database	Description and URL
Allen	Allen Brain Atlas of anatomical gene expression in the brain http://www.brain-map.org
BRENDA	Comprehensive enzyme information system http://www.brenda-enzymes.info
DBD	Transcription factor prediction database http://www.transcriptionfactor.org
EcoCyc	Encyclopedia of E. coli genes and metabolism http://ecocyc.org
EPD	Eukaryotic promoter database http://www.epd.isb-sib.ch
ExPASy	Thumbnail sketches of metabolism and cellular biochemistry http://www.expasy.ch/cgi-bin/search-biochem-index
GNRF	Gene Expression Atlas of the Genomics Institute of Novartis Research Foundation http://symatlas.gnf.org/SymAtlas
GOLD	Genomes Online Database http://www.genomesonline.org
JASPAR	Transcription factor binding profile database http://jaspar.genereg.net
KEGG	Kyoto encyclopedia of genes and genomes http://www.genome.ad.jp/kegg/kegg2.html
LIGAND	Database for enzymes, compounds, and reactions http://www.genome.ad.jp/dbget/ligand.html
MetaCyc	Encyclopedia of metabolic pathways (HumanCyc is specialized for humans) http://metacyc.org and http://humancyc.org
MIRAGE	Molecular informatics resource for the analysis of gene expression http://www.ifti.org
Protein Lounge	Databases of protein-protein interactions, pathways, transcription factors, etc http://www.proteinlounge.com
MIPS	Computational analysis of the Munich Information Center for Protein Sequences http://mips.gsf.de
REACTOME	Curated pathway database with advanced query features http://www.reactome.org
TRANSFAC	Transcription factor and gene regulatory database http://www.gene-regulation.com
TRRD	Transcriptional regulatory regions database http://www.bionet.nsc.ru/trrd
UMBBD	University of Minnesota Biocatalysis/Biodegradation Database http://umbbd.ahc.umn.edu/index.html

able formats that allow cross-genome comparison. The KEGG and GOLD databases cover whole-genome annotations more generally.

A second important class of bioinformatic databases in the context of integrative genomics are those that document protein-protein and protein-DNA interactions. The Protein Lounge presents tools and databases that allow browsing and assembly of protein-protein interactions established by a variety of methods, particularly physical interactions detected in a whole genome two-hybrid screens. The MIPS protein-protein interaction database (http://mips.gsf.de/proj/ppi) links to many other resources that tend to have short-lived links on the internet. Protein-DNA interactions are supported by the TRANSFAC, MIRAGE, and TRRD sites, which in addition to providing searchable databases of known transcription factors and transcription-factor binding sites, link to an ever expanding suite of tools for detection of novel DNA binding sites and conserved putative regulatory motifs in the regulatory regions of genes. These incorporate information from expression profiling, comparative genomics, and molecular biology assays, and are too numerous to list here. As explained in Box 4.4, the major difficulty in compiling databases of regulatory sites is that DNA binding sites are often short (less than 10 bp) and binding specificity can be heavily influenced by cofactors and flanking DNA sequences.

In Silico Genomics

There are two major approaches to modeling metabolic and genomic networks. The first is to attempt to derive the properties of enzymatic pathways from a knowledge of kinetic parameters associated with the activity of each component and the order in which the components act. The second is to search for global properties of networks based on static properties of the system such as stoichiometry and Boolean logic. The closing section of this book presents a brief introduction to some of the first applications to genome data of both classes of modeling approach.

Metabolic Control Analysis

Metabolic control analysis (**MCA**) refers to the study of the sensitivity of global properties of metabolic pathways (most simply flux and metabolite concentrations) to perturbation of individual enzyme activity. *Flux* refers to the rate at which metabolites pass through a pathway and is a function of the velocity of the reaction at each enzymatic step, which in turn can be computed using Michaelis-Menten kinetics. Kacser and Burns (1973) first developed the concept of a control coefficient, which is a relative measure of the effect of perturbation of one component on the system. At steady state, the summation of all the control coefficients in a system is unity, since an increase in activity of one enzyme must be compensated for by a decrease in another. As the number of enzymes increases, the contribution that each makes to regulation of the total flux diminishes, as does the effect of a mutational perturbation in the activity of an enzyme. For an online discussion and mathematical derivation of MCA, see http://dbkgroup.org/mca_home.htm.

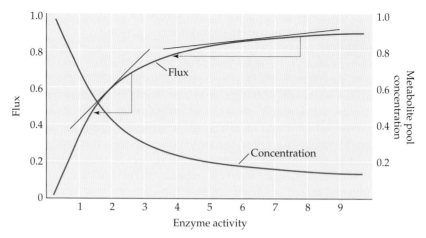

Figure 6.5 Metabolic control analysis of linear pathways. Parameters such as metabolic flux (blue) and metabolite pool concentrations (red) are not expected to be linear functions of enzyme activity. Metabolic systems are expected to evolve to a point where a large change in activity of the enzyme has only a very small effect on flux. The two black tangents to the flux curve show that the effect of halving enzyme activity has a much smaller effect on flux at higher enzyme activities. See Clark (1991) for a discussion of how fitness may affect the distribution of enzyme activities in natural populations.

Two of the first applications of MCA were in providing explanations for the dominance and near-neutrality of gene activity. The predicted relationships between enzyme activity and flux or metabolite concentrations are nonlinear, as shown in Figure 6.5. If a microbe has evolved to optimize its growth rate, then metabolic flux should be as high as possible, and the activity of each enzyme (which is a function of its level of expression as well as a catalytic rate constant set by the protein structure) will take on a value that places it close to the point at which the maximal flux reaches a plateau. Under these conditions, halving the dosage of the enzyme (as in a heterozygote) will have little effect on the total flux, hence the gene tends to be dominant (Kacser and Burns 1981). In addition, slight changes in activity due to new mutations and polymorphisms will not greatly affect flux, so the genetic variants should have little effect on fitness, and their dynamics will be in accordance with the expectations of neutral theory.

A considerable body of experimental data from *E. coli* in particular corroborates this conclusion (Dykhuizen et al. 1987). Theoretical modeling becomes more complicated when it is realized that most biochemical pathways are branching, and that permease activity and external factors such as drugs and hormones can have a major impact on metabolic rates. It is also clear that certain loci, notably phosphoglucose isomerase in diverse organisms, exert unexpectedly high levels of control over metabolism, perhaps because they sit at branch points where selection is more concerned with the regulation of power and efficiency than raw flux (Watt 1986). A particularly interesting application of MCA involves investigating the reasons for covariation in enzyme activities and steady-state substrate concentrations, both in natural populations and as a result of mutation accumulation.

A number of tools for such modeling are available, notably GEPASI (http://www.gepasi.org; Mendes 1993), a highly developed program for solving series of ordinary differential equations in the context of metabolic control analysis. Users build their own models of biochemical pathways, including the kinetic parameters of interest, and simulate systems with stable states, limit cycles, or a tendency toward chaos. A dozen other modeling tools are reviewed in Alves et al. (2006) and offer options for compartmentalizing reactions into modules or even parts of a virtual cell.

The extension of metabolic control analysis to other regulatory networks such as receptor-ligand interactions, signal transduction, transcription, cell cycle regulation, and synaptic function is much less advanced. Two major difficulties encountered are that measurement of kinetic parameters for protein-protein interactions in vivo is extremely difficult, and mathematical models of reactions that do not involve metabolite conversion (for example, phosphorylation or initiation of transcription) are still being developed. Nevertheless, some notable successes have been obtained:

- Barkai and Leibler (1997) identified a mechanism for the robustness of the bacterial chemotaxis network to perturbation of the concentrations and activities of components of the chemosensory apparatus.

- Ferrell (1997) demonstrated how a graded concentration of ligand can be converted into a switchlike intracellular response and has modeled the effects of variation on the transmission of the response down an intracellular kinase cascade.

- Threshold-dependent transcriptional regulation has been modeled as a consequence of the cooperativity of protein-protein-DNA interactions and applied to a thorough description of the lambda genetic switch (Ptashne 1992).

- Similarly, conversion of a morphogen gradient into a robust, stable, and accurate transcriptional response has been modeled using detailed knowledge of the molecular biology of the retinoic acid receptor (Kerszberg 1996).

More comprehensively, Bhalla and Iyengar (1999) have promoted attempts to combine models of multiple biochemical pathways in series (Figure 6.6). They argue that networks of interaction will often give rise to emergent properties, such as the ability to integrate signals at different time scales and to generate distinct outputs dependent on the strength and duration of the input signal.

Numerous authors, starting with Jacob and Monod as early as 1962 and reviewed by McAdams and Arkin (1998), have remarked on similarities between biological regulatory pathways and electronic circuit construction. Principles such as serial and parallel processing, feedback, cooperative combinatorial control, and fan-out control of multiple subsystems are thought to be prevalent in the design of biological networks. Genome science promises to contribute to our understanding of biological circuits by cataloging

(A)

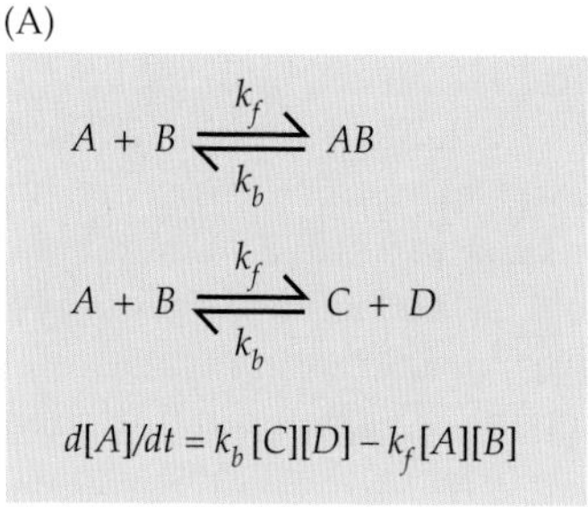

$$A + B \underset{k_b}{\overset{k_f}{\rightleftharpoons}} AB$$

$$A + B \underset{k_b}{\overset{k_f}{\rightleftharpoons}} C + D$$

$$d[A]/dt = k_b [C][D] - k_f [A][B]$$

Figure 6.6 Regulatory network analysis.
(A) Like metabolism, regulatory networks can be modeled as systems of differential equations representing simple two-component interactions and reactions. (B) Two examples from Bhalla and Iyengar (1999) are shown, involving the phospholipase C and Ras signaling pathways, both of which are integrated through the epidermal growth factor receptor (EGFR).

(B)

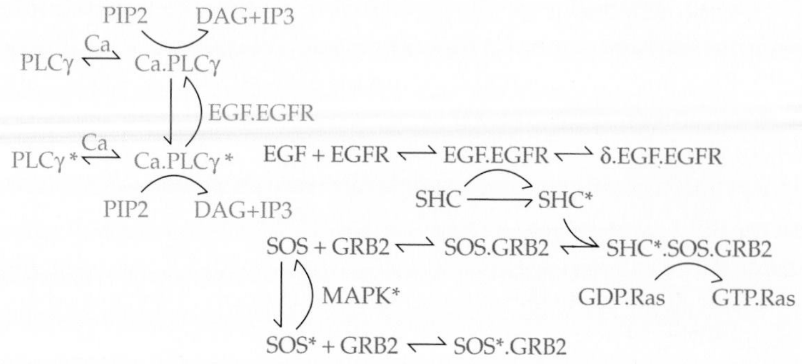

temporal programs of gene activation, identifying the design principles of transcriptional enhancers and repressors, and establishing networks of protein-protein and protein-DNA interactions.

A practical extension of this work is in biological engineering. Examples include the introduction of bistable genetic switches and oscillatory networks in microbes; the design of efficient multistep degradative pathways into which checkpoints and feedback loops are incorporated; and the introduction of advanced biosensors. In most cases, the control of noise, sensitivity, and robustness associated with circuits designed *de novo* is unlikely to meet industrial specifications, but may be enhanced and optimized by artificial selection (Nielsen 1998). Subsequent research into the genetic changes accompanying selection promise to yield new insight into biological circuit construction.

An even more ambitious task for modeling based on the structure of known gene products is to simulate the cellular interactions that are required for pattern formation during development. A step in this direction has been taken by von Dassow et al. (2000) in their model of the *Drosophila* segment polarity network. They established a system of almost 50 nonlinear ordinary differential equations that describe parameters of the function of five genes expressed in a repetitive series of stripes in the cells of each embryonic segment (Figure 6.7A). These parameters are designed to allow for variation in factors such as the half-lives of the mRNA and proteins, their binding rates, and cooperativity coefficients over at least three orders of

(A)

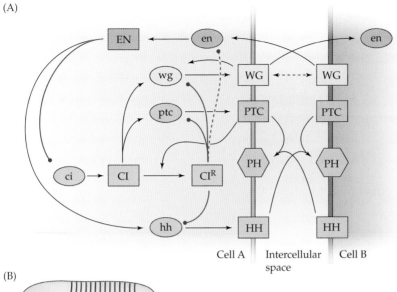

(B)

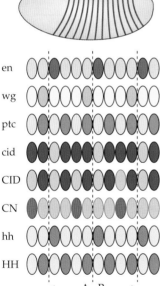

Figure 6.7 Modeling of pattern formation using Ingeneue.   (A) Von Dassow and colleagues (2000) established a model for patterning of segment polarity in *Drosophila* embryos that used five genes (engrailed, *en*; wingless, *wg*; patched, *ptc*; cubitus interruptus, *ci*; and hedgehog, *hh*; ovals) and seven protein products (including the repressor fragment of CI, and the PTC-HH complex; boxes). Black arrows indicate activating interactions, and red lines indicate repressing interactions, each of which were modeled as a differential equation. Only when the two interactions shown as dashed lines (involving the CI repression of engrailed, and WG autoactivation) were included did Ingeneue simulations reproduce the target pattern of expression (dark shading indicates expression) of each of the genes and proteins shown for a real embryo (B).

magnitude. Since the true values of the parameters are unknown, the objective was to determine whether random permutations of the parameters could support the emergence of the observed stripe patterns as the system runs to equilibrium, starting from a reasonable set of initial conditions. Astonishingly, 1,192 solutions were found among a set of 240,000 random parameter assignments, corresponding to 1 in 200 trials, and over 90% of individual possible parameter values were represented in the set of solutions. Just as remarkably, perturbation of individual parameters once a solution had been found showed that in the majority of cases a single variable could be perturbed over 10% of its allowed range—an indication of great robustness.

Such analyses suggest that the notion of a single optimal configuration of kinetic parameters for genetic components of complex regulatory networks is biologically unnecessary, and imply that developmental modules must be quite free to absorb genetic variation. The Ingeneue software developed by von Dassow et al. (2000) can be adapted to describe and model all manner of gene networks. The software is available online at http://www. ingeneue.com.

Systems-Level Modeling of Gene Networks

The second broad class of genomic modeling eschews any explicit reference to the kinetic properties of defined pathways, aiming instead to find generic systems properties that emerge from the logic and connectivity of interacting networks. The most general of these approaches are not even concerned with the structure of actual networks, dealing rather with universal properties such as the propensity to evolve. The results are consequently less obviously amenable to experimental verification than are models that take data in the form of the stoichiometry and order of reactions as their starting point. Truly integrative genomics aims to merge observation with modeling along the lines of classical hypothetico-deductive science.

The generic approach is perhaps most starkly illustrated by Kauffman's innovative ideas concerning the centrality of spontaneous order in biological homeostasis and evolution. Reducing genetic interactions to a network of Boolean switches, in which each gene interacts with k other genes, Kauffman (1993) argues first that stable attractor states are an expected property of complex systems and hence that there is a pervasive tendency toward order independent of natural selection. Second, he shows that as k increases, a phase transition occurs from a tendency for the generation of order to a tendency for chaos, such that there must be a limit to the amount of interaction that genomes can support. Third, between order and chaos is a phase in which biological systems may be poised on the edge of chaos, yet are most capable of adaptation and coevolution; and the average number of 3–5 interactions per gene that support this state is at least consistent with empirical data. Critics have argued that it is not clear that biological switches obey Boolean logic, and that the theory does not help explain quantitative phenomena, but the metaphors of rugged fitness landscapes and order out of chaos are persuasive and may attract more attention as mathematical and biological sciences merge in the coming years.

Another example of the exploration of the large-scale organization of genomic systems based largely on their degree of connectivity lies in the demonstration that metabolic systems are scale-free rather than exponential (Jeong et al. 2000; see Barabási 2003 for a general overview). As shown in Figure 6.8, this means that rather than each node (substrate) in a metabolic network being connected randomly with k other nodes with Poisson-

(A)

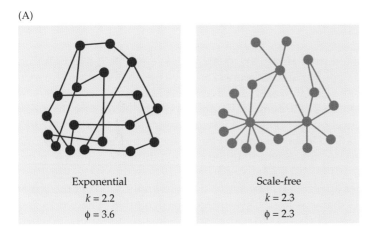

Exponential
$k = 2.2$
$\phi = 3.6$

Scale-free
$k = 2.3$
$\phi = 2.3$

(B)

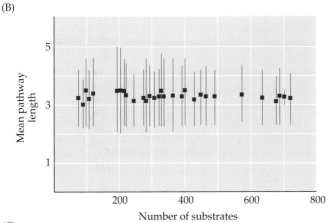

Mean pathway length

Number of substrates

(C)

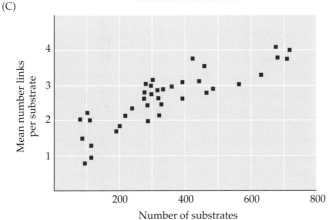

Mean number links per substrate

Number of substrates

Figure 6.8 Large-scale organization of metabolic networks. (A) Whereas random linkage of nodes into a network with just over 2 links per node results in most nodes having $k = 1, 2,$ or 3 connections (left), metabolism seems to be organized as a scale-free system, in which a few nodes with many more connections are found (right). In this hypothetical example of 18 substrates, the average pathway diameter ϕ (number of links between any two substrates/nodes) is significantly reduced in the scale-free network relative to the random/exponential one. (B) Forty-three organisms with varying number of known metabolic substrates were nevertheless found to have a constant mean pathway diameter, which comes about because there are more reactions predicted per substrate in more complex metabolomes (C). (After Jeong et al. 2000.)

distributed probability $P(k) \sim e^{-k}$, the fraction of nodes with many links is actually much higher and follows a power law, $P(k) \sim ke^{-\gamma}$. These highly connected nodes are called "hubs," and their presence was demonstrated for a sample of 43 organisms that included archaea, bacteria, and a eukaryote, despite the fact that the number of known metabolic substrates among these organisms ranged from about 200 to 800. The average path length (number of catalytic steps required to connect any two metabolites) was also seen to be constant at a "small world" linkage of just over three—which means that increases in substrate complexity are bought by increasing the mean number of connections per node. A consequence is that metabolic networks are quite resistant to metabolic disturbance, since removal of one enzyme can be compensated for by shunting the reaction through a different pathway.

Considerable attention has been given to the implications of the scale-free topology of many biological systems for the evolution of networks and for the robustness of networks to perturbation. Some authors argue that network topology evolves under natural selection, while others view the emergence of nonrandom distributions of network linkages and the presence of metabolic hubs as a natural consequence of gene duplication followed by specialization (Pfeiffer et al. 2005). Biological robustness is reviewed generally by Wagner (2005). Since disruption of strongly connected hubs should perturb many linkages, hubs are often thought more likely to be "weak" points of genetic networks. Empirically, most studies find only a weak correlation between network connectivity and lethality. Siegal et al. (2007) argue on theoretical grounds that specific rather than general properties of hub genes are more likely to influence their importance to an organism. The practical utility of network thinking for real-world problems in evolutionary biology, agriculture, and biomedicine thus remains to be demonstrated.

Constraint-based modeling provides another approach to computational simulation of biological systems. Edwards and Palsson (2000) demonstrated that the assembly of *in silico* genotypes (Figure 6.9) facilitates optimization of metabolic pathways and understanding of how the rerouting of substrate utilization occurs in response to mutation, or alteration of nutrient sources. Combining information from genome sequence and metabolic databases, it is possible to assemble a universal stoichiometric matrix of m metabolites by n reactions, as well as a vector of all known fluxes through the network, from which an organism-specific matrix can be extracted. For example, the minimal *E. coli* matrix consists of 436 metabolites and 720 reactions. Various constraints can be imposed based on established features of metabolism and desired inputs. A desired metabolic objective such as maximal growth rate (flux) is also imposed, and linear programming is then used to find a feasible and/or optimal set of parameters describing the flux through each reaction. Setting chosen fluxes to zero allows prediction of how the organism will respond to a mutation by redistributing fluxes, in the majority of cases with no major impact on the metabolic objective. Of 79 predictions made by this metabolic reconstruction approach, 68 were confirmed in vivo, with just 7 enzymes involved in central metabolism being indispensable, and only 9 others reducing biomass yield by more than 10%. The same approach can be used to predict the growth requirements and metabolic properties of any microbial genome in a variety of environments (Becker and Palsson 2008), with obvious applications in bioengineering.

(A)

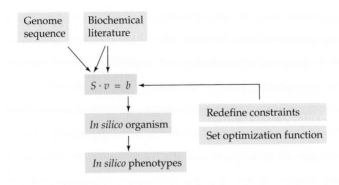

(B)

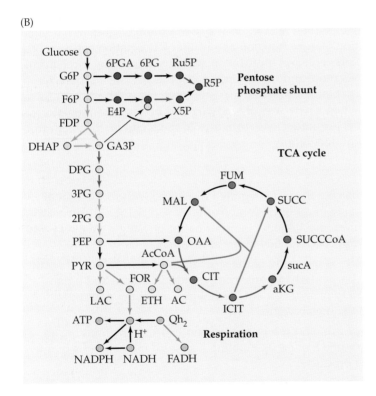

Figure 6.9 Metabolic reconstruction *in silico*. (A) *In silico* metabolomes are constructed by extracting the enzymes predicted to occur in a complete genome, synthesizing a stoichiometry matrix (S) and vector of possible metabolic fluxes (*v*). Solving for steady state defines the possible and optimal set of parameters for flux through each enzyme. (B) The diagram of *E. coli* central metabolism shows which enzymes are predicted to be essential for life (red arrows), which retard growth when deleted from the genome (orange), which are dispensable (black), and which are not utilized under aerobic conditions (gray). The green arrows indicate how flux is shunted across the TCA cycle when the *sucA* gene is deleted. (After Schilling et al. 1999.)

Finally, a natural progression in genome science is the integration of genome sequencing, gene expression profiling, proteomics, functional genomics, and theory. Ideker et al. (2001) define a four-step procedure, which they used to explore galactose utilization in yeast:

1. Characterize the genome, proteome, and metabolome of the organism and assemble a preliminary model of a cellular process of interest.

2. Perturb the system through mutation or by changing the growth conditions, and monitor the changes in mRNA and protein expression responses.

3. Integrate the expression profiles with the model, including data on genetic and physical interactions.

4. Modify the original hypothesis to accommodate differences between prediction and observation, and use standard molecular biological procedures to test the new hypothesis.

This approach, sometimes referred to as **systems biology**, goes well beyond simple estimation of gene function as a result of association between profiles of gene expression and physical interactions, to probing mechanisms of regulation. As more and more "omics" approaches are adopted across the disciplines of biology, the need for computational synthesis has grown, and it has spawned a new way of studying and understanding biological systems. In time, genome science promises to shed light on the deep structure of biological integration.

Summary

1. Genomics has become an integrated discipline with its own set of research questions, and the combination of technological advances with computational biology/bioinformatics is changing the way that geneticists approach a wide range of research.

2. Metabolomics is the study of the structure and distribution of small molecules, particularly organic compounds, in tissues and organisms. A variety of spectrometric techniques are being developed for analysis of the metabolome.

3. Databases of protein-protein and protein-DNA interactions are an important online tool for assembling and studying networks of interacting gene products.

4. *In silico* genomics refers to computational research aimed at predicting the metabolic and physiological performance of cells, using genomic data to define the initial capabilities of the biological system that is being modeled.

5. Computational approaches can either explicitly attempt to model individual reactions and known networks of interactions (for example, metabolic control analysis), or be concerned with global properties of generic systems, such as the organization and robustness of networks.

6. Integrative genomic approaches will coordinate all areas of genome research with classical molecular biology, in a framework of hypothesis testing and formulation.

Discussion Questions

1. Describe how you would go about profiling the complete set of novel organic compounds found in a unique tropical plant that is found to have anticancer activity.

2. Experimental and theoretical physics have always been tightly linked disciplines, but that has not been the case in cell and developmental biology. How do you think the relationship between theoretical modeling and empirical research will change in the post-genomic era?

3. What is the relationship between robustness, redundancy, and homeostasis in genetic systems? How can it be studied?

4. How far into the future do you think Year X in Figure 6.1 will be? Is this a reasonable representation of the growth of knowledge in genome science?

Literature Cited

Alves, R., F. Antunes and A. Salvador. 2006. Tools for kinetic modeling of biochemical networks. *Nat. Biotech.* 24: 667–672.

Assfalg, M., I. Bertini, D. Colangiuli, C. Luchinat, H. Schäfer, B. Schütz and M. Spraul. 2008. *Proc. Natl. Acad. Sci. (USA)* 105: 1420–1424.

Barabási, A. L. 2003. *Linked: How Everything Is Connected to Everything Else and What It Means.* Plume, New York.

Barkai, N. and S. Leibler. 1997. Robustness in simple biochemical networks. *Nature* 387: 913–917.

Becker, S. A. and B. O. Palsson. 2008. Context-specific metabolic networks are consistent with experiments. *PLoS Comput. Biol.* 4: e1000082.

Bhalla, U. S. and R. Iyengar. 1999. Emergent properties of networks of biological signaling pathways. *Science* 283: 381–387.

Castrillo. L. I. et al. 2007. Growth control of the eukaryote cell: A systems biology study in yeast. *J. Biology* 6: 4.

Clark, A. G. 1991. Mutation-selection balance and metabolic control theory. *Genetics* 129: 909923.

Döring, A. et al. 2008. SLC2A9 influences uric acid concentrations with pronounced sex-specific effects. *Nat. Genet.* 40: 430–436.

Dunn, W. B. 2008. Current trends and future requirements for the mass spectrometric investigation of microbial, mammalian, and plant metabolomes. *Phys. Biol.* 5: 11001.

Dykhuizen, D., A. Dean and D. Hartl. 1987. Metabolic flux and fitness. *Genetics* 115: 25–31.

Edwards, J. and B. Palsson. 2001. The *Escherichia coli* MG1655 *in silico* metabolic genotype: Its definition, characteristics, and capabilities. *Proc. Natl. Acad. Sci. (USA)* 97: 5528–5533.

Ferrell, J. E. 1997. How responses get more switch-like as you move down a protein kinase cascade. *Trends Biochem. Sci.* 22: 288–289.

Goodacre, R. and D. Kell. 1996. Pyrolysis mass spectrometry and its applications in biotechnology. *Curr. Opin. Biotechnol.* 7: 20–28.

Ideker, T. et al. 2001. Integrated genomic and proteomic analyses of a systematically perturbed metabolic network. *Science* 292: 929–933.

Irwin, W. J. 1982. *Analytical Pyrolysis: A Comprehensive Guide.* Marcel Dekker, New York.

Jacob, F. and J. Monod. 1962. On the regulation of gene activity. *Cold Spring Harbor Symp. Cell Reg. Mech.* 193–209.

Jeong, H., B. Tombor, R. Albert, Z. Oltvai and A. Barabási. 2000. The large-scale organization of metabolic networks. *Nature* 407: 651–654.

Kacser, H. and J. Burns. 1973. The control of flux. *Symp. Soc. Exp. Biol.* 27: 65–104.

Kacser, H. and J. Burns. 1981. Thermolecular basis of dominance. *Genetics* 97: 639–666.

Kathiresan, S. et al. 2008. Six new loci associated with blood low-density lipoprotein cholesterol, high-density lipoprotein cholesterol, or triglycerides in humans. *Nat. Genetics* 40: 189–197.

Kauffman, S. A. 1993. *The Origins of Order.* Oxford University Press, Oxford.

Kerszberg, M. 1996. Accurate reading of morphogen concentrations by nuclear receptors: A formal model of complex transduction pathways. *J. Theor. Biol.* 183: 95–104.

Keurentjes, J. J. B. et al. 2006. The genetics of plant metabolism. *Nat. Genet.* 38: 842–849.

McAdams, H. H. and A. Arkin. 1998. Simulation of prokaryotic genetic circuits. *Annu. Rev. Biophys. Biomol. Struct.* 27: 199–224.

Mendes, P. 1993. GEPASI: A software package for modelling the dynamics, steady states, and control of biochemical and other systems. *Comput. Appl. Biosci.* 9: 563–571.

Nielsen, J. 1998. Metabolic engineering: Techniques for analysis of targets for genetic manipulations. *Biotechnol. Bioeng.* 58: 125–132.

Passador-Gurgel, G., W-P. Hsieh, P. K. Hunt, N. Deighton and G. Gibson. 2007. Quantitative trait transcripts for nicotine resistance in *Drosophila melanogaster. Nat. Genet.* 39: 264–268.

Pfeiffer, T., O. S. Soyer and S. Bonhoeffer. 2005. The evolution of connectivity in metabolic networks. *PLoS Biol.* 3: e228.

Ptashne, M. 1992. *A Genetic Switch*, 2nd Ed. Cell Press, Cambridge, MA.

Raamsdonk, L. M. et al. 2001. A functional genomics strategy that uses metabolome data to reveal the phenotype of silent mutations. *Nat. Biotech.* 19: 45–50.

Schaeffer, L. et al. 2006. Common genetic variants of the *FADS1/FADS2* gene cluster and their reconstructed haplotypes are associated with the fatty acid composition in phospholipids. *Hum. Mol. Genet.* 15: 1745–1756.

Schauer, N. et al. 2008. Mode of inheritance of primary metabolic traits in tomato. *Plant Cell* 20: 509–523.

Schilling, C., J. Edwards and B. Palsson. 1999. Toward metabolic phenomics: Analysis of genomic data using flux balances. *Biotechnol. Prog.* 15: 288–295.

Siegal, M. ., D. E. L. Promislow and A. Bergman. 2007. Functional and evolutionary inference in gene networks: Does topology matter? *Genetica* 129: 83–103.

Stoyanova, R. and T. R. Brown. 2001. NMR spectral quantitation by principal component analysis. *NMR Biomed.* 14: 271–277.

von Dassow, G., E. Meir, E. Munrow and G. Odell. 2000. The segment polarity network is a robust developmental module. *Nature* 406: 188–192.

Wagner, A. 2005. *Robustness and evolvability in living systems.* Princeton University Press, Princeton, NJ.

Watt, W. B. 1986. Power and efficiency as indexes of fitness in metabolic organization. *Am. Natur.* 127: 629–653.

Wishart, D. S. 2007. Current progress in computational metabolomics. *Brief. Bioinf.* 8: 279–293.

Wishart, D. S. et al. 2007. HMDB: The human metabolome database. *Nucl. Acids Res.* 35: D521–D526.

Xu, L. et al. 2008. Chemometric methods for evaluation of chromatographic separation quality from two-way data—a review. *Analytica Chimica Acta* 613: 121–134.

Zhu, J. et al. 2008. Integrating large-scale functional genomic data to dissect the complexity of yeast regulatory networks. *Nat. Genet.* 40: 854–861.

Glossary

ab initio **gene discovery**   A method for identifying likely genes in a stretch of genomic DNA sequence that does not depend on prior information such as similarity to a gene in another species or identity to a transcript. Most *ab initio* approaches use a **hidden Markov model** to search for sequence motifs that are commonly found in genes, such as long open reading frames, intron-exon boundary signatures, and conserved upstream regulatory motifs.

ab initio **protein structure prediction**   Any attempt to determine the tertiary structure of a protein (the manner in which helices, sheets and coils fold together) based on the principles of physical biochemistry. Alternate methods include X-ray crystallography, NMR spectroscopy, and fitting modeling by homology.

acrylamide   A polymeric compound used to make gels for electrophoretic separation of proteins or nucleic acids; hence PAGE (polyacrylamide gel electrophoresis).

affinity chromatography   A method for purifying proteins and complexes of proteins based on their affinity for some compound that is crosslinked to a matrix in a column. Proteins wash through the column when a particular buffer disrupts the interactions between proteins on the column.

alignment   The process of lining up two or more DNA or protein sequences so as to maximize the number of identical nucleotides or residues while minimizing the number of mismatches and gaps.

alternative splicing   The combination of different sets of exons to produce two or more primary mature messenger RNAs from the same primary transcript. Commonly observed in higher eukaryotes, with the result that a single gene can generate multiple protein isoforms.

annotation   In the most general sense, linking information from the literature to database entries for genes or proteins. In the context of genome sequencing, annotation refers to the identification of putative genes using a combination of *ab initio* methods, homology searches, and physical evidence.

antibody   A secreted immunoglobulin molecule that specifically recognizes a stretch of up to ten amino acids (or other similarly sized molecule) known as an **epitope**. Polyclonal antibodies are a group of different antibodies all of which recognize different parts of the same protein; monoclonal antibodies recognize a single epitope and are produced by permanent hybridoma cell lines.

association mapping   The search for genes that affect disease susceptibility by testing whether each of the alleles at a series of DNA polymorphisms tend to be present in affected individuals more or less commonly than expected by chance. Linkage disequilibrium between the sites both assists and complicates the mapping process.

balancer chromosomes   Chromosomes that have been engineered to carry multiple inversions that suppress crossing over, and so can be used to maintain recessive mutations in genetic stock. Balancers usually also carry a recessive lethal and a dominant visible genetic marker.

base-calling   The process of calling the series of nucleotides from a sequence trace. Usually automated, but with the facility for manual curation to resolve ambiguities.

Bioconductor project   A project that brings together open-source software written in the *R* programming language for purposes of statistical analysis of genomic data.

case-control association mapping   An approach to screening for associations between genetic markers and disease status, based on a comparison of allele frequencies in a group of affected individuals, and in a similar control group of unaffecteds.

cDNA   A DNA molecule that is complementary to an mRNA (messenger RNA) molecule. The first strand of cDNA is synthesized by **reverse transcriptase**, but the term is also used to describe a double stranded DNA clone that is derived from a transcript.

cDNA clone   A complementary DNA copy of a full-length transcript. Synthesized by reverse transcription of mRNA.

cDNA library   A collection of cDNA clones, typically isolated from a single tissue.

cDNA microarray   An array of cDNAs on a glass microscope slide of nitrocellulose filter, hybridized to labeled mRNA for the purposes of profiling gene expression.

centiMorgan (cM)   The standard unit of genetic map distance, corresponding to a 1% probability of a crossover occurring between two sites in any meiosis.

chain termination sequencing   The most commonly used method for sequencing clones of DNA up to one kilobase in length, based on a method first devised by Fred Sanger, in which molecules of all possible lengths are produced by random termination of DNA polymerization when a dideoxynucleotide (ddNTP) is incorporated.

chemometrics   A series of analytical methods for quantifying chemical profiles, including principal component analysis (PCA) and artificial neural networks.

chemostat   An apparatus used for long-term exponential growth of microbial cultures, into which fresh medium is introduced at the same rate as liquid culture is removed.

chromatin immunoprecipitation microarrays (ChIP chips)   Microarrays consisting of DNA corresponding to potential regulatory regions of genes that are used to detect sequences that bind to transcription factors.

chromosome painting   A cytological procedure for aligning the chromosomes of two different eukaryotic species, based on fluorescence in situ hybridization (FISH). A set of chromosome-specific probes from one species are prepared using unique combinations of fluorescent dyes, and these "paint" the chromosomes in a mitotic chromosome spread from cells of the second species.

chromosome walking   A procedure for cloning a large contiguous portion of a chromosome, in which a probe prepared from the end of one clone is used to identify overlapping genomic clones in a library. The process is iterated for as many steps as it takes to cover the region of interest.

clusters of orthologous genes (COGs)   Sets of genes from a collection of species that are proposed to encode the same gene product, based on pairwise best-match sequence similarity.

complementation group   A set of alleles that fail to complement (substitute for the function of) one another, often indicating that they are mutations in the same locus.

consensus sequence   A hypothetical sequence consisting of the most common amino acid at each position in a multiple alignment of DNA or protein sequences.

copy number variation (CNV)   Polymorphism in the number of copies of a stretch of DNA, including deletions and duplications of whole genes.

contig   A contiguous stretch of cloned DNA. May refer simply to a scaffold of overlapping clones that have been mapped physically, or to a long stretch of DNA sequence assembled by merging two or more sequences.

cosmid   Large insert plasmids that generally exist as a single copy within host bacterial cells, but which contain cos sites that allow in vitro packaging of inserts as phage molecules if desired.

CpG islands   Stretches of vertebrate DNA typically between 1 and 2 kb long that contain a tenfold higher frequency of the doublet nucleotide CG than occurs elsewhere in the genome. CpG islands are commonly found at the 5′ end of genes.

Cre-Lox recombination system A combination of site specific recombinase (Cre) and its recognition site (*lox*) from the bacteriophage P1 that has been engineered into yeast, mouse and other eukaryotic genomes to facilitate targeted recombination.

C-value paradox The absence of any correlation between the number of genes and the amount of DNA in a genome. A corollary is that there is high variation in the DNA content of even closely related organisms, and no obvious relationship between complexity and DNA content.

cytological map A map of the location of genes or other DNA features relative to the banding patterns of the chromosomes of a species.

data normalization The process of removing systematic biases from microarray data that otherwise contributes to misinterpretation of apparent differences in transcript abundance.

deficiency complementation mapping A method proposed for fine-scale mapping of QTL based on the variable ability of wild-type alleles to complement the effect of hemizygotes for a deletion of a gene or genes.

dideoxynucleotide A nucleotide that is missing the hydroxyl groups on both the 2′ and 3′ carbon atoms of the sugar backbone, and which is incapable of covalently linking to the next nucleotide in a growing molecule of DNA. Used in chain termination sequencing.

DNA-binding motif A short stretch of DNA, usually between 8 and 12 nucleotides, that is thought to be recognized by a DNA-binding protein. Such motifs can be represented by a profile of the frequency of occurrence of each of the four nucleotides at each position in the sequence. They typically identify sequences important for the regulation of gene expression.

DNA library A collection of hundreds of thousands of clones, each of which contains a different piece of genomic or cDNA. If the clone is oriented in such a way that the clone can be transcribed and translated, the library is called an expression library.

ectopic expression Activation of expression of a gene in cells outside the normal domain of ex-

pression, that is, in an abnormal cellular context.

embryonic stem (ES) cell A cell line that can be transformed and manipulated in culture, and then injected into the blastula (early embryo) where it integrates with and contributes to the development of the adult animal. Injected embryos are chimeric, but if the ES cells populate the germ line, a transgenic organisms is produced in the next generation.

enhancer An orientation- and distance-independent regulatory sequence that increases levels of transcription in a spatial and temporal manner. Enhancers may occur anywhere in a gene, can act over hundreds of kilobases, and can sometimes affect the transcription of several genes.

enhancer trap A transposable element that has been modified so that when it inserts into the genome adjacent to a gene, the enhancers that drive expression of that gene also drive expression of a reporter gene carried on the enhancer trap transposable element.

Ensembl genome browser The major repository and resource for genomic data in Europe, run jointly by the Sanger Center in Cambridge, England and the European Bioinformatics Institute within the European Molecular Biology Laboratory.

epistasis In quantitative genetics, an interaction between two or more loci that results in nonadditive effects of one allele as a function of the genotype at the other locus. Note that in developmental or physiological genetics, by contrast, the term is often used to describe a mutation whose phenotype is unaffected by (is epistatic to) another mutation.

epitope The portion of a protein, carbohydrate, or other molecule that is specifically recognized by an antibody.

E-value The expected number of sequences in a database that would by chance produce an equivalent or better alignment score than the one under consideration.

expressed sequence tag (EST) A sequenced piece of cDNA. Whereas a full-length cDNA sequence defines the structure of a transcript, an EST is merely a tag that indicates that the particular sequence is a part of a transcribed gene.

expression library A library of cDNA clones in a vector that allows the gene products to be expressed (transcribed and translated) in a controlled manner.

expression vector   A cloning vector that facilitates transcription and translation of a cDNA fragment that is inserted into the multiple cloning site.

expressivity   The severity of a disease, or degree to which a trait is observed in affected individuals. Often affected by the environment.

F_3 design   A genetic screen designed to isolate recessive mutations. Requires that the phenotype be measured in third-generation progeny of the mutagenized individual.

floxing   A method for inducing a mutation at a precise time and place in an organism. When a mouse containing *loxP* binding sites on either side of an exon of the gene to be mutated (placed there by homologous recombination) is crossed to a strain expressing the Cre recombinase in the tissue of interest, the exon is excised solely in that tissue.

fold recognition   In structural genomics, a method for predicting the tertiary structure of a protein. Secondary structure prediction is combined with limited sequence similarity to find the previously described domain fold that most closely fits the unknown protein structure.

forward genetics   Genetic analysis that starts with a phenotype and moves toward isolation of the gene that causes the phenotype.

functional genomics   The study of the function of each and every gene encoded in a genome. "Function" may refer to biochemical activity, cell biological function, or organismal function. Functional genomics encompasses genetic analysis, microarrays, proteomics, and computational biology.

fusion protein   A hybrid protein produced by fusing parts of two genes together in an expression vector. The N-terminal part is often a tag such as polyhistidine or the small glutathione S-transferase (GST) protein domain, while the C-terminal portion is usually the protein of interest.

GAL4   A potent transcription factor from yeast that enhances gene expression only through a UAS sequence adjacent to the promoter. *GAL4* in the absence of UAS sequences has no effect on transcription in heterologous genomes, so it can be used specifically to drive expression of transgenes introduced into that genome.

gene knock-in   Replacement of the endogenous gene with a different functional piece of DNA such that the inserted gene is expressed in place of the original gene. Germline gene therapy uses knock-in technology to replace a defective gene with an active copy. The replacements are performed using a positive-negative double selection strategy in embryonic stem cells.

gene knockout   A mutation that targets a specific gene, produced by using homologous recombination to replace an exon of the target gene with a piece of foreign DNA (which is sometimes the reporter gene *lacZ*). Insertional mutations can also cause gene knockouts.

genetic fingerprinting   A strategy for testing subtle effects of mutations on the fitness of microbial strains in competition with other strains during long-term culture.

genetic heterogeneity   The observation that the same disease or phenotype can have multiple different genetic causes. If the different variants are within a single locus, the effect is known as allelic heterogeneity.

genetic map   A map of the order of and distance between genes in a genome based on the frequency of recombination between markers. The markers may be physical (molecular variants) or visible (Mendelian loci). Mapping populations may be pedigrees, crosses between lines, or radiation hybrid cell panels. Genetic distances are given in Morgans (or centiMorgans, cM).

genome-wide association study (GWAS)   A study designed to scan the entire genome for SNP and CNV that are associated with a disease or trait. Typically, at least 500,000 different genetic variants are measured in several thousand disease cases and a similar number of healthy controls.

germ line   The population of cells in a multicellular eukaryote that are destined to undergo meiosis and become oocytes or sperm. The germ line is set aside very early in animal development, but can be specified at the time of flowering in plants.

haplotype   A multi-site genotype consisting of two or more polymorphisms on the same chromosome. For example, individuals that are homozygous at one site for a G allele and heterozygous at a nearby site for A and T would have GA and GT haplotypes.

Hardy-Weinberg equilibrium   The mathematically based expectation that genotype frequencies in a population will tend to be stable and predictable as a simple function of individual allele frequencies *unless* there is some evolutionary force (such as migration, inbreeding, mutation or selection) leading to increase or decrease in the number of heterozygotes.

heavy isotope labeling A method for quantification of protein expression differences between two samples. One protein sample is labeled with a heavy isotope such as deuterium, so that peptide fragments move slightly more slowly through a TOF spectrometer than do corresponding fragments from the unlabeled sample. ICAT reagents are used for uniform labeling of protein mixtures after extraction from cells.

heteroduplex DNA A double-stranded DNA molecule containing a polymorphism, formed by renaturation of PCR products from two different alleles.

heuristic search Algorithms that use time-saving methods to search for the most likely solution, usually reducing the search space by excluding unlikely solutions from the analysis. Heuristic methods are not guaranteed to find the optimal solution, but are often the only feasible way to perform a phylogenetic analysis or sequence alignment involving a large number of sequences.

hierarchical sequencing An approach to whole-genome sequencing based on the principle that the genome is first divided into an ordered set of clones. As initially envisaged, the process consisted of cloning a genome into artificial chromosomes, then cosmid or BAC clones several hundred kilobases long, then plasmids up to 10 kb long. In practice, BAC clones are now more commonly directly sequenced using the shotgun strategy.

hidden Markov model (HMM) A class of bioinformatics procedures for identifying sequence features. HMMs exploit the fact that many features have local amino acid or nucleotide usage patterns that are distinct from random sequence.

HKA test Statistical test of neutrality named after its developers (Hudson, Kreitman, and Aguadé) and based on the expectation that divergence between species and polymorphism with species are highly correlated in the absence of selection.

homogenous assay A SNP genotyping assay in which all of the steps are performed in a single sample tube or well of a microtitre plate well.

Since there is no need to transfer products from one reaction to another, there are savings in labor and materials.

homolog Biological features (ranging from molecules to traits) that show a similar structure due to the fact that they derive from a common ancestor. Homology thus refers to identity by descent. Similarity of structure may or may not reflect homology. Unfortunately, the term is often used merely to imply that two DNA or amino acid sequences are similar.

horizontal gene transfer Transfer of a gene from the genome of one individual to that of another by means other than sexual (i.e., vertical) transmission. Usually used in the context of microbial genomes that can be shown to have incorporated genes from a different species.

hybridization The annealing of one strand of DNA to its complementary strand to form a double helix. Most often used in a context where one strand is labeled by the investigator as part of a procedure for detecting arrayed transcripts or clones.

immunohistochemistry Detection of proteins in a tissue sample based on chemical detection by antibodies that recognize the protein.

inbreeding The process of mating siblings or close relatives repeatedly, leading to loss of genetic variability in the line. Near-isogenic lines (NIL) are almost homozygous throughout the genome as a result of 10 or more generations of inbreeding, and are commonly used in quantitative genetic analysis.

indel An insertion-deletion polymorphism. Indels range in size from one or a few bases to several kilobases. Large indels often involve transposable elements (TEs).

in situ hybridization Detection of a specific mRNA in a tissue sample by hybridization of a section of whole-mount of a tissue to a DNA or RNA probe that is complementary to the mRNA, and which is labeled with a fluorescent or radioactive group, or with a small compound such as biotin or digoxygenin that can be recognized by an antibody.

insertional mutagenesis The process of creating mutations by controlled insertion of a transposable element in the vicinity of a gene of interest.

interference The observation that recombination is suppressed by the occurrence of nearby recombination events, with the result that the genetic map distances between more than two markers do not necessarily equal the sum of the distances between each pair of adjacent markers.

interval mapping   A method for QTL mapping that uses the genotypes of two adjacent genetic markers to estimate the likely genotype at each point in the interval between the markers.

introgression   Introduction of a small portion of one genome into the genetic background of another genome, by repeated backcrossing with selection for the region of interest.

isogenic   Homozygous for the entire portion of the genome under consideration.

laboratory information management system (LIMS)   An automated tracking system implemented with bar codes, robotics, and software checkpoints to ensure accurate tracking of samples and data in high-throughput genomics laboratories.

linkage disequilibrium (LD)   The nonrandom segregation of genetic markers in a population. LD decays over time as a result of recombination, so tends to diminish as physical distance between the markers increases; however, LD can also be caused by a variety of other forces, including founder effects, admixture, and epistatic selection.

linkage disequilibrium mapping   An approach to identifying the genes that correspond to a quantitative trait locus; based on detection of LD between a marker, the trait (or disease), and the causal polymorphism in an outbred population.

LOD score   Logarithm of the odds score. A measure of statistical significance in association studies and linkage mapping, essentially the logarithm of the ratio of the probability of observing the data given an association to the probability under the null hypothesis.

MALDI-TOF   Matrix-assisted laser-desorption ionization time-of-flight spectrometry. A method for ionizing and then separating small fragments of DNA or protein for the purpose of identifying the corresponding sequence.

mapping function   A mathematical function that converts recombination frequencies into genetic map distances by accounting for the incidence of double crossovers between markers. Two commonly used functions are due to Haldane and Kosambi; the latter adjusts the data for interference.

marker-assisted selection   An approach to improved animal and plant breeding based on selection in each generation DNA markers that are associated with some desired trait (as opposed to selection for the trait itself).

mass spectrometry (MS)   A technique for determining the identity of a molecule based on comparison of the spectrum of molecules separated by mass/charge ratio, with a theoretical standard. In genomics and proteomics, separation is based on time-of-flight of an ionized fragment through a vacuum, and is sensitive to mass differences as small as a few parts per million.

mate-pair sequences   A pair of sequences derived from the two ends of a single clone. An essential component of shotgun sequencing as the distance between the pairs assists in resolving repetitive DNA sequences and in verifying the sequence assembly.

McDonald-Kreitman statistic   A test for selection acting on protein sequences, based on comparison of levels of synonymous and replacement polymorphism and divergence.

metabolome   The set of metabolites present in a cell or tissue—including such molecules as lipids, carbohydrates, steroids, and amino acids—that mediate many of the physiological properties of organisms.

metabolomics   High-throughput methods for characterization of the metabolome.

metabolic control analysis (MCA)   An approach to modeling of metabolism, based on biophysical and biochemical principles, that seeks to understand and predict the effects of genetic or environmental perturbation.

microsatellite   A stretch of repetitive DNA made up of a variable number of several to one hundred or more tandem repeats of a small number of nucleotides, most commonly di- or tri-nucleotides. For example (AG)n or (CAG)n. Microsatellites tend to be highly polymorphic and heterozygous, and occur at high density (several per hundred kilobases) in the genomes of higher eukaryotes.

minimal genome   The smallest number of genes required to sustain life.

modifier   A mutation or polymorphism that slightly modifies (either enhancing or suppressing) a phenotype associated with a different mutation.

monoclonal antibodies (MAbs)   Antibodies derived from a single immunoglobulin gene that recognize a single epitope on a protein. Generated by creating an immortal hybrid cell line that secretes the MAb.

motif   A short conserved sequence of nucleotides or amino acids, often suggesting conservation of function.

multiple cloning site   Also known as a **polylinker**; the site in a plasmid into which foreign DNA is inserted at one of a number of unique restriction enzyme sites.

multiplex PCR   Simultaneous amplification of multiple different fragments of DNA by using several pairs of gene-specific primers in the PCR reaction.

National Center for Biotechnology Information (NCBI)   The major depository and resource for genomic data in the United States. NCBI is part of the National Library of Medicine (NLM) within the National Institutes of Health (NIH).

neutral theory   The null hypothesis explaining the distribution of molecular variation in natural populations in the absence of natural selection. Factors affecting rates of neutral evolution include mutation pressure, migration rate, population size, breeding structure, and recombination rate.

next-generation DNA sequencing   Currently emerging technologies (e.g., pyrosequencing; reversible terminator technology; ligation sequencing) that are replacing traditional dideoxy-based methods due to their speed and economy.

Northern blotting   A method for characterizing gene expression based on transfer of mRNA to a nylon or nitrocellulose filter, which is then probed with chemically or radioactively labeled DNA corresponding to the gene of interest.

nucleotide diversity   The average proportion of nucleotide differences between all pairs of sequences in a sample. A measure of polymorphism that is a function of the number and frequency of variable alleles.

Online Mendelian Inheritance in Man (OMIM)   A Web site maintained by the NCBI and Johns Hopkins University that documents genetic and genomic information relating to human disease.

open reading frame (ORF)   A reading frame is a stretch of genomic DNA that encodes at least 20 codons without a stop codon; there are six possible reading frames on any stretch of DNA (three in each orientation). A reading frame is *open* if it supports translation of a peptide sequence, and may be conceptually assembled from exons by splicing out introns.

orphan gene   A predicted gene that does not show sequence similarity to any other gene in the databases, and hence which cannot be assigned to a gene family.

orthologs   Two genes in separate species that derive from a common ancestor without duplication.

paralogs   Two genes that arose by duplication of an ancestral gene.

penetrance   The frequency of individuals with an allele who show the phenotypic trait.

PERL   A simple programming language used to perform basic bioinformatic procedures such as extracting DNA sequences from a database.

pharmacogenomics   The study of the effect of the genomic base on response to drugs, toxins, and other pharmacological agents.

phenocopy   An environmentally induced phenotype that mimics a known mutation.

phylogenetic analysis   In the context of comparative genomics, a methodological approach to annotation of gene function based on the supposition that evolutionary history is a more reliable indicator of likely function than sequence similarity alone.

phylogenetic footprinting   The process of aligning the sequences of a stretch of DNA from multiple divergent species, typically for the purpose of detecting evolutionarily conserved elements that may encode genes or other important DNA sequences.

phylogenetic shadowing   The process of aligning the sequences of a stretch of DNA from several closely related species, typically for the purpose of detecting unusually highly conserved DNA elements that may encode regulatory elements.

physical map   A map of a genome consisting of an ordered series of large insert clones in which the distance between molecular features such as restriction enzyme sites and sequence-tagged sites (STS) is expressed in kilobases.

polony PCR colonies made up of cell-free clones of a single molecule of DNA that accumulate as a concentrated spot of amplified DNA within an acrylamide matrix on a glass microscope slide.

population stratification Differences in allele or haplotype frequencies between populations. In many situations, the identity of the populations is not obvious as the population structure may be hidden, as for example historical populations that no longer correspond with geographic location or phenotypic attributes.

positional cloning Cloning of a gene that is responsible for a disease or trait on the basis of its position in the genome, generally using recombination mapping.

position effect A phenomenon often observed in transgenic animals and plant, in which the site of insertion has a large effect on the level of expression of the transgene.

profile A list of the frequencies of each amino acid in each position in a multiple alignment of protein sequences.

promoter The region immediately 5′ to the start site of transcription of a gene that serves as a binding site for the RNA polymerase initiation complex and generally also includes regulatory sequences.

protein domain A structurally distinct region of a protein, generally less than 150 amino acids in length, that often performs a particular subset of the functions of the whole protein. Examples include DNA-binding domains, kinase domains, and extracellular domains.

protein interaction map A description of the network of interactions among a group of proteins, including physical associations detected using two-hybrid screens and protein microarrays, as well as interactions inferred from biochemical and genetic analysis.

proteome The full complement of proteins that are found in a particular cell or tissue under a particular set of circumstances. May include information on their relative or absolute abundance.

pseudogene A DNA sequence that shares many of the structural features of true genes, but is not active. Many pseudogenes are produced by reverse transcription and so lack introns. A pseudogene may never have been active, or it may be in the process of decaying in the absence of any selection pressure to maintain function.

psychogenetics The study of behavior and psychosis using genomic approaches.

pyrolysis The thermal degradation of materials into volatile fragments, one of several spectrometric techniques used in profiling of the metabolome.

QTL mapping Determination of the location of quantitative trait loci using statistically sophisticated procedures for detecting nonrandom associations between genetic markers and trait values. Conceptually similar to recombination mapping of several loci simultaneously.

quantitative trait locus (QTL) A region of the genome that has a quantitative effect on a trait, meaning that it is only responsible for a portion of the genetic variance. QTL may affect continuous traits, or liability to discrete traits including diseases.

quantitative trait nucleotides (QTNs) The SNPs or CNV that actually contribute to the effect of a quantitative trait locus.

radiation hybrid mapping A method for assembly of genetic maps of vertebrates or plants in which fragments of the genome of one species are propagated in hybrid cell lines with another species. Co-segregation of sequences in multiple lines indicates that the two sequences are physically linked.

random mutagenesis The process of generating a large collection of new mutations, generally involving a screen for aberrant phenotype(s) or the insertion of transposable elements.

recombinant inbred line (RIL) A line derived from two genetically distinct parents that has been bred to be nearly homozygous throughout the genome (nearly isogenic) by several generations of inbreeding. Each member of a panel of recombinant inbred lines contains a different combination of fragments from the two parents, so RIL are very useful for mapping QTL.

recombination mapping Determination of the location of a gene that is responsible for a particular phenotype, disease, or quantitative trait, based on the co-segregation of linked genetic markers with the trait in a pedigree.

redundant gene A gene whose function can be supplied by another gene or genes if it is mutated. Redundancy can be due to multiple copies of the gene in a genome, or because the protein activity can be supplied by a different type of gene, or because the enzyme can perform the function with another genetic pathway.

reverse genetics   Genetic analysis that starts with a gene and moves to determine the phenotype it generates.

restriction fragment length polymorphism (RFLP)   A polymorphism that is detected as a difference in the length of the fragments formed when a piece of DNA is digested with a restriction enzyme.

reverse transcriptase   The enzyme that converts RNA to single-stranded DNA; usually encoded in the genome of RNA viruses.

Rosetta Stone approach   A bioinformatic approach to assembly of protein interaction maps based on the notion that two genes are likely to encode interacting proteins if they exist as a single fused gene in another species.

saturation random mutagenesis program   A forward genetic screen of a sufficiently large number of mutagenized chromosomes to guarantee that all genes affecting the trait of interest will be hit by at least one mutation.

sequence-contig scaffold   An alignment of sequenced contigs against a physical and eventually a cytological map of the genome. The step in genome sequencing immediately preceding the finishing stage (in which gaps between scaffolds are filled in).

sequence-tagged site (STS)   Any sequenced fragment of DNA derived from a library of clones that is placed on the physical map of the genome.

serial analysis of gene expression (SAGE)   A method for profiling gene expression based on the sequencing of very large numbers of unique tags corresponding to each gene in the genome.

shotgun sequencing   Determination of the sequence of a long stretch of DNA by randomly breaking it into a redundant set of small clones that are sequenced en masse so that each fragment is represented between 5 and 10 times. The contig is then assembled by computer alignment of the overlapping sequences. **Whole-genome shotgun sequencing** refers to the process of sequencing the entire genome in this way, without first dividing it into large clones as in hierarchical sequencing.

single-base extension (SBE)   A SNP detection method based on minisequencing reactions

that only detect the identity of the base adjacent to the sequencing primer.

single nucleotide polymorphism (SNP)   A site in the genome at which a single nucleotide is found to have two or more states in a collection of individuals of the same species. Most SNPs (pronounced "snips") are substitutions involving just two nucleotides (for example A and G), but the term also applies to single nucleotide indels.

single-sperm typing   The determination of the genotypes of multiple markers from a single spermatozoan, usually for the purpose of directly measuring recombination rates.

site-directed mutagenesis   Deliberate modification using recombinant DNA technology of a cloned sequence so that the protein is altered when expressed in a transgenic organism, allowing testing of hypotheses as to the role of particular residues in protein function.

Southern blotting   A method for detecting DNA sequences in a restriction digest of genomic DNA that has been separated by electrophoresis and transferred to a nylon or nitrocellulose membrane. Used to detect differences in the genomic DNA encompassing a gene, and to see whether a gene is present in the genome of another species.

structural proteomics   Also referred to as **structural genomics**, it refers to the study of the tertiary structures of the complete set of proteins encoded in genomes, generally using X-ray crystallography, NMR spectroscopy, and *ab initio* or homology-based prediction.

synteny   The conservation of gene order between divergent lineages.

synthetic lethal   Two mutations that are alone homozygous viable, but together are inviable. Synthetic lethality generally indicates the two genes function in a similar process.

systems biology   An integrative approach to genome biology that involves theoretical modeling of large genomic, transcriptomic, proteomic, and metabolomic datasets to guide the generation and testing of hypotheses.

tagging SNPs   Single nucleotide polymorphisms used to represent haplotypes. Chosen so as to capture the majority of DNA sequence variation in a population.

threading   An approach to predicting protein structure, based on a combination of secondary structure similarity and assessment of likely binding energies of potential folds.

tiling path   In an aligned set of large insert clones, the choice of a subset of clones that completely covers the contig with minimal redundancy.

transcriptome   The complete set of transcripts expressed in a particular cell or tissue under defined conditions. A full description of a transcriptome includes data on the abundance of each class of transcript.

transient expression   Activation of gene expression for a limited period of time, for example from plasmids injected into embryos, or after infection of a tissue with a replication-defective virus.

transmission disequilibrium testing   A test of association based on an unequal ratio of allele frequencies in the affected children of heterozygous parents.

transposable elements (TEs)   Mobile pieces of DNA that can "jump" from one location on the genome to another, usually utilizing an enzyme encoded on the natural transposable element.

transposon mutagenesis   Mobilization of modified transposable elements to create mutations that are tagged with the inserted DNA fragments.

UAS   The binding site for the *GAL4* transcription factor, only found upstream of yeast genes. The *UAS-GAL4* combination can thus used to drive transgene expression in plants and animals.

unigene set   A set of unique cDNA clones derived from a cDNA library by filtering out duplicate copies of the same transcript.

unitig   A unique contig, namely a set of overlapping DNA sequences that correspond to a single piece of genomic DNA that is represented multiple times in a shotgun sequence library.

Western blotting   A molecular biology technique for detecting protein expression. Based on transferring proteins from an acrylamide gel to a membrane that is then probed with a labeled antibody.

yeast two-hybrid screen (Y2H)   A method for detecting protein-protein interactions based on the reconstitution of transcription factor activity when one protein domain fused to a DNA-binding domain (the "bait") interacts with another protein domain fused to an activation domain (the "prey"). Initially developed in and most commonly performed in yeast cells.

List of Abbreviations

ADIT AutoDeposit Input Tool

BAC Bacterial Artificial Chromosome

BLAST Basic Local Alignment Search Tool

CASP Critical Assessment in Structure Prediction

CE Capillary Electrophoresis

CNP Copy Number Polymorphism

CNV Copy Number Variation

DASH Dynamic Allele-Specific Hybridization

ddNTP dideoxyNucleotide Triphosphate

DGGE Denaturing Gradient Gel Electrophoresis

DHPLC Denaturing High Performance Liquid Chromotography

EBI European Bioinformatics Institute

EC Enzyme Commission

ELISA Enzyme-Linked ImmunoSorbant Assay

ELSI Ethical, Legal, and Social Implications

EP Enhancer-Promoter

EST Expressed Sequence Tag

FISH Fluorescent In Situ Hybridization

FRET Fluorescence Resonance Energy Transfer

GAL4 Galactose-4 transcription activator protein

GC Gas Chromatography

GST Glutathione S-Transferase

GWAS Genome-Wide Association Study

HGP Human Genome Project

HKA test Hudson-Kreitman-Aguadé test of neutrality

HMM Hidden Markov Model

HPLC High-Performance Liquid Chromotography

ICAT Isotope Coded Affinity Tag

IDAT ImmunoDetection by Amplification with T7

IHGSC International Human Genome Sequencing Consortium

IPG Immobilized pH Gradient

LC-MS/MS Liquid Chromatography-Tandem Mass Spectrometry

LD Linkage Disequilibrium

LIMS Laboratory Information Management System

LINE Long Interspersed Nuclear Element

LOD Logarithm of the Odds

LTR Long Terminal Repeat

MAb Monoclonal Antibody

MAGEML MicroArray Gene Expression Markup Language

MALDI-TOF Matrix-Assisted Laser-Desorption Ionization Time-of-Flight Spectrometry

MCA Metabolic Control Analysis

MIAME Minimal Information for the Annotation of Microarray Experiments

mmCIF macromolecular Crystallographic Information File

MPSS Massively Parallel Serial sequencing

MS Mass Spectrometry

MudPIT Multidimensional Protein Identification Technology

NCBI National Center for Biotechnology Information

NextGen Next-Generation DNA sequencing

NMR Nuclear Magnetic Resonance

OMIA Online Mendelian Inheritance in Animals

OMIM Online Mendelian Inheritance in Man

ORF Open Reading Frame

OST ORF Sequence Tag

P1 P1 large-insert bacteriophage clone

PAGE PolyAcrylamide Gel Electrophoresis

PCR Polymerase Chain Reaction

PDB Protein Data Bank

Q-PCR Quantitative Polymerase Chain Reaction

QTL Quantitative Trait Locus

QTN Quantitative Trait Nucleotide

RefSeq Reference Sequence in GenBank

RFLP Restriction Fragment Length Polymorphism

RH Radiation Hybrid

RIL Recombinant Inbred Line

SAGE Serial Analysis of Gene Expression

SBE Single Base Extension

SDS Sodium Dodecyl Sulfate (detergent)

SFP Single Feature Polymorphism

SINE Short Interspersed Nuclear Element

SNP Single Nucleotide Polymorphism

SQRL Short Quantitative RNA Library

SSCP Single-Stranded Conformation Polymorphism

SSR Simple Sequence Repeat

STS Sequence Tagged Site

TE Transposable Element

TOF Time-of-Flight

UAS Upstream Activator Sequence (used by Gal4)

VDA Variant Detector Array

VNTR Variable Number Tandem Repeat

Y2H Yeast Two Hybrid

Index

ab initio gene discovery, 98–103
ab initio protein prediction, 291–292
ABI, see Applied Biosystems
.ace files, 72
Actinobacteria, *55*
ADIT (AutoDep Input Tool), 288
admixture, 161
affinity chromatography, 269–270
Affymetrix GeneChips, 175–176, 202–203, *204, 206*
Africa, 147
African sleeping sickness, *57*
Agilent Technologies, 201
Agricultural Research Service, 35
agriculture. *See* crop plants; farm animal genome
    projects
Agrobacterium tumefaciens, 304, 305
Ahab software, 227
alfalfa *(Medicago),* 44
AlignACE software, 229, 230
alignment algorithms, 73, 74–77
allelic heterogeneity, 161
Allen database, *332*
alternative splicing, 96, *97*
Alzheimer's disease, 311
American College of Medical Genetics, 19
American Society of Human Genetics, 19
AmiGO browser, 126
AMOS (A Modular Open Source), 89
ampholytes, 267
analysis of variance (ANOVA), 213, 215–216,
    217–218
ancestry informative markers (AIMs), 166
"anchor enzyme," 231
animal breeding projects, 35
animal genome projects
    breeding projects, 35
    International Sequencing Consortium, 28
    invertebrate model organisms, 36–39

primates, 28, 30
rodent, 30–34
vertebrate biomedical models, 34–35
Ankyrin repeat, *116*
annotation, 2
    See also functional annotation; gene annotation;
        genome annotation; protein annotation
Anopheles gambiae (mosquito), 29
ANOVA (analysis of variance), 213, 215–216,
    217–218
antibodies, 276, *277*
antimorphs, 297
Antirrhinum majus (snapdragon), 47
APC gene, 155
Apis mellifera (honeybee), *29,* 248
apo(A) promoter, 103
Apollo visualization tool, 36
Applied Biosystems (ABI)
    Prism DNA sequences, 68
    SNPlex, 184, *185*
    SOLiD system, 80–81
    TaqMan assay, 238–239
Aquificae, *55*
Arabidopsis thaliana
    comparative gene content, *17*
    genome studies, 40–44
    integrating metabolomic and genomic data, 329
    synteny with rice, *45*
Arachne, 89
Archaea, 48, 52, *55*
ArrayExpress, 253
artificial neural networks, 328
artificial selection
    crop plants and, 45–46
    QTL mapping and, 156–157
Aspergillus, 54
association mapping. *See* linkage disequilibrium
    mapping

association studies
case-control, 164–165
family-based, 167–168
theoretical basis, 159–162
asthma, 152
Atlas, 89
automated DNA sequencing
contig assembly, 71–79
high-throughput, 68
"next generation" methods, 79–83
reading sequence traces, 68–71
Sanger sequencing, 65–67
See also genome sequencing

Bacillus, 49
B. subtilis, 50
backcross design, 155, *156*
BACs. *See* bacterial artificial chromosomes
bacteria
genome projects, 48–51
organisms sequenced, *55*
Rosetta Stone analysis of proteomes, 282, 285
See also microbial genomics
bacterial artificial chromosomes (BACs)
aligning by hybridization and fingerprinting, *87*
in hierarchical sequencing, 84, *85, 86,* 87–88
Bacteroides, *55*
balancer chromosomes, 309–310
balancing selection, 136
base calling, 68–71
Baylor College of Medicine, 89
bead arrays, 201–202
BeadChips, 184
BeadXpress reader, 184
behavioral genomics, 248
Beijing Genomics Institute, 44
best point estimate, 214
biochemical databases, 331–333
Bioconductor, 219, 221
biolistics, 305
biological circuits, 335–336
biological engineering, 336
biological replicates, 194
biological switches, 338
BioMart, 23
biomedical models, 34–35
biotin-affinity chromatography, *270*
bipolar disorder, *172*
black cottonwood (*Populus trichocarpa*), 47
blackfly, *57*
BLAST (Basic Local Alignment Search Tool)
in database searches, 23
domain clustering and, 114, *115*
in functional annotation, 113
interpreting output, 90–91
introduction to, 90
scoring matrices, 91
BLASTn, 113
BLASTp, 113
BLASTx, 113
BLAT, 25
BLOSUM scoring matrices, 91
Bonferroni correction, 164
Boolean switches, 338

bootstrapping, 121
bottom-up clustering, 221
brain tissue
gene expression, 243
transcript diversity, *244*
branch length, 120
Brassica oleracea, 47
Brassicaceae, 47
BRCA2 gene, 155
breakpoints, 156
breast cancers, 244, 246
breast epithelium, *244*
breeding
animal, 35
tests of neutrality and, 151
BRENDA database, 331, *332, 333*
Broad Institute, 58, 89
BsmFI, 231
budding yeast. *See* Saccharomyces cerevisiae
bulked segregant analysis (BSA), 231
bushbaby (*Otolemur garnetti*), 30

C_2H_2 zinc finger, *116*
C/D box snoRNAs, 105
C-value paradox, 107
Caenorhabditis elegans (nematode)
comparative gene content, *17*
genome studies, 36–37, *38*
miRNAs, 107
orthologs to genes from human disease, 39
RNAi screen, 306
systematic mutagenesis, 300–301
CAMERA (Community Cyberinfrastructure for Advanced Marine Microbial Ecology Research and Analysis), 60
cancers
gene expression profiling, 243–246
human genes in model organisms, *39*
identifying chromosomal rearrangements, 299
See also individual cancers
Candida albicans, 54, 242
candidate complex gene, *163*
candidate genes, 192
Canis familiaris (domestic dog), *29, 34*
canonical analysis, 328
capillary electrophoresis, 68
cardiomyocytes, *244*
case-control association studies, 164–165
case-control population sampling, 159
cat genome, *12*
CATH, 292
cation exchange/reversed phase liquid chromatography, 274
CD-CV proposition, 159
cDNA
clones, 2
databases, 98
EST sequencing and, 95–98
labeling and hybridization, 205–206
cDNA libraries, 95, 96
cDNA microarray data
Bioconductor Project, 219
normalization, 208–210, 212

reference sample analysis, 207–208
significance testing, 213, 217–219
cDNA microarrays
  advantages and disadvantages, 204
  statistical analysis of data, 207–219
  technology of, 198–200
  uses, 191
CDS, 98
Celera Genomics, 20, *21*
cellular constituents, analysis of, 325–328
centiMorgan (cM), 5
centromeres, 111
Chagas disease, *57*
chain termination sequencing, 65–67
chemical mutagens, 296, *297*
chemical transformation, 305
chemometrics, 327–328
chicken *(Gallus gallus),* 34
chimeric mice, 305
chimpanzee *(Pan troglodytes),* 28, *29,* 30
Chinese macaques, 28
ChIP chips (chromatin immunoprecipitation microarrays), 225–227
Chlamydiae, *55*
chondrocytes, *244*
Chou-Fassman algorithm, *266*
chromatin immunoprecipitation microarrays (ChIP
  chips), 225–227
chromatogram files, 72
chromosome duplications, 42, *43*
chromosome painting, 11–12
chromosome walking, 8, *9*
chromosomes
  balancers, 309–310
  centiMorgan length, 5
  saturation mutagenesis, 295, 296, 299
  synteny, 4
Ciona savignyi (sea squirt), *29*
cistrons, 285
clades, 121
clone-by-clone sequencing. *See* hierarchical
  sequencing
clones, cDNA, 2
cloning vectors, 84, *85*
clustering, 220–224
clusters of orthologous genes (COGs), 116–119
co-immunoprecipitation, *270*
coalescent theory, 148–150
coding SNPs, 134
codon biases, 105, *106*
colon tissue, *244*
color-coding, 220–221
colorectal cancer, *233*
comparative genome hybridization (CGH), 227, 299
comparative genomics, 10–12
compendium of expression profiles, 57, 242, *243*
complementation group, 298–299
composite interval mapping, *157*
Comprehensive Microbial Resource (CMR), 51
conditional probability, 101
confidence intervals, 214–215
confidentiality, 18
connectivity, in networks, 283–284

consed graphic editor, 71–73, 78
consensus sequence, 78, 264
conserved non-genic sequences (CNGs), 111
constraint-based modeling, 340, *341*
contigs (contiguous DNA segments)
  assembly, 8, *9,* 71–79
  contig N450, 89
  defined, 2, 89
  measure of genome sequence quality, 89
  physical maps and, 8
  sequence scaffolds, 92–94
control coefficient, 333
copy number variation (CNV), 133, 134
"core" genome. *See* minimal genome
coronary artery disease, *172*
cosmids, 84
CpG dinucleotides, 109–110
Cre-Lox system, 305
Cre recombinase, 314, *315*
Crenarchaeota, *55*
Crohn's disease, *172*
crop plants
  genome projects, 44–46, 47, 48
  issues of proprietary rights, 48
cryoelectron microscopy, 291
crystallography. *See* X-ray crystallography
Cy3 dye, 205
Cy5 dye, 205
Cyanobacteria, *55*
cystic fibrosis, 28
cytological maps, 8–9, *10*

2D-difference gel electrophoresis (DIGE), 269
2D-PAGE (two-dimensional polyacrylamide gel
  electrophoresis), 267–269
Danio rerio (zebrafish), 34–35
databases
  biochemical, 331–333
  gene expression, 252–253
  metabolic, 331–333
  microarrays, 252–253
  microbial genomics, 51–53
  object-oriented, 40–41
  protein domains, 260–261
  queries to, 252–253
  See also individual databases
DBD database, 332
dbEST database, 96
dCAPS (derived cleavable amplified polymorphic
  sequences), 179–180
ddNTPs (dideoxynucleotides), 66
deCODE Company, 155
deficiency complementation mapping, 158
Deinococcus, 55
Dendrome Web, *47,* 48
Dengue hemorrhagic fever, 57
derived cleavable amplified polymorphic sequences
  (dCAPS), 179–180
diabetes, 166, 169, *172*
diauxic shift, 241
Dicer, 303
dideoxy sequencing, 65–67
dideoxynucleotides (ddNTPs), 66

differential transcription, analyzing, 225–227
diffuse large B-cell lymphomas (DLBCL), *245*
DIGE (2D-difference gel electrophoresis), 269
dioxin receptor, 124
disease control, public health and, 58
disease genes
    linkage disequilibrium mapping (*see* linkage dise-
        quilibrium mapping)
    pedigree mapping, 152–153
    recombination mapping, 152–155
distance methods, 120, 121
ditags, 231
4300 DNA Analysis system, 68
DNA libraries, in hierarchical sequencing, 84
DNA sequencing
    Human Genome Project and, *14*
    See also automated DNA sequencing; genome se-
        quencing
DNA transposons, 108–109
dog (*Canis familiaris*), *29,* 34
domain, 286
domain clustering, 114–116
dominant mutations, in saturation mutagenesis, 297,
    298
dot/slot blots, 237
Drosophila (fruit fly)
    balancer chromosomes, 310
    comparative gene content, *17*
    Engrailed homeodomain, *290*
    gene annotation, *38, 246–248*
    gene evolution, 124
    genome studies, 36–39
    microsatellites, 110
    modeling of pattern formation, 336–338
    OAMB protein structural profile, *266*
    orthologs to genes from human diseases, 39
    P-elements, 297, 304
    presenilin and, 311
    proportion of genome in scaffolds, 93
    saturation mutagenesis, 295
    systematic mutagenesis, 300–301
    transgenesis, 304
drug design, 288
Ds element, 297
dynamic programming, 75

E-value, 90–91
EASE (Expression Analysis Systematic Explorer)
    software, 219
EBI. *See* European Bioinformatics Institute
EcoCyc database, *332*
ectopic gene expression, 306, *307*
edges, in networks, 283
Eigenstrat software, 166, 174
electroporation, 305
electrospray ionization (ESI) devices, 271
elephantiasis, *57*
ELISA (enzyme-linked immunosorbant assay), 276
ELSI program, 18–19
emergent properties, 335
ENCODE (Encyclopedia of DNA Elements) Project,
    112–113
end-sequencing, 87–88
endocrine diseases, *39*

enhancer-promoter (EP) method, 314
enhancer trapping, 312–314
Ensembl EST database, 96
Ensembl Gene View report, 27
Ensembl Genome Browser, 22, 23
Entrez Gene tool, 126
Entrez Genome Project, 52, *53*
Entrez Life Science Search Engine, 22, 23
environmental sequencing, 59–60
Enzyme Commission (EC), 115
enzyme-linked immunosorbant assay (ELISA), 276
enzymes, hierarchical classification, 115
EPD database, *332*
epistasis, 158, 162
epitope, 276
eQTL (expression quantitative trait loci), 250, *251*
error probability, 71
error sum of squares, 216
Escherichia coli (E. coli)
    comparative gene content, *17*
    genome plasticity, 51
    genome studies, 48, 49
    minimal matrix, 340, *341*
    production of proteins and, 278
eShadow, 103
Estonia, 18
ESTs. *See* expressed sequence tags
ethics, 18–19
ethylmethanesulfonate (EMS), 296
ethylnitrosourea (ENU), 296
Eucalyptus, 47
European Bioinformatics Institute (EBI), 22, 23, 40,
    59, 259
European Molecular Biology Organization, 23
Euryarcheota, *55*
evolution, experimental studies, 240–242
evolutionary and ecological genomics, 248–251, 325
evolutionary distances, 121
evolutionary trees, 120
exons, alternative splicing, 96, *97*
ExPASy (Expert Protein Analysis System), 259, 260,
    263, 331, *332*
expressed sequence tags (ESTs)
    cDNA microarrays, 198–200
    defined, 2, 95
    sequencing, 95–97
Expression Connection, 253
Expression Console software, 202
expression library, 270
expression profile, 221
expression quantitative trait loci (eQTL), 250, *251*
expressivity, 160

F_2 design, 155, *156*
F_1 genetic screen, *298*
F_3 genetic screen, *298*
false discovery rate (FDR), 219
family-based association tests, 167–168
family-based linkage disequilibrium mapping,
    166–170
FANTOM cDNA database, 98
farm animal genome projects, 35
FASTA, 90
Federal Bureau of Investigation (FBI), 110

FGENES, 98
fine-structure genetics
  enhancer trapping and *GAL4*-mediated overexpression, 312–314
  floxing, 314, *315*
  goal of, 294
  modifier screens, 311–312
  regional mutagenesis, 308–311
fingerprinting, in hierarchical sequencing, 86–87
Firmicutes, 55
FISH (fluorescent in situ hybridization), *11*
Fisher exact test, 140–141
fission yeast (*Schizosaccharomyces pombe*), 54, 240
floxing, 314, *315*
fluconazole, 242
fluorescence resonance energy transfer (FRET), 238–239
fluorescent in situ hybridization (FISH), *11*
flux, 333, 334
FlyBase, 36, 126, 128
fold recognition, 292, *293*
forest trees, genome projects, 47–48
4300 DNA Analysis system, 68
forward genetics, 294, 295–299
fosmids, 84
454 Life Sciences platform, 79, *80*
Fragile X syndrome, *111*
FRET (fluorescence resonance energy transfer), 238–239
Friedreich ataxia, *111*
FTO gene, 176
Fugu rubripes (Japanese pufferfish), 10, 35
Fulani people, 246
functional annotation
  clusters of orthologous genes, 116–119
  domain clustering, 114–116
  first-pass classification, 113
  gene ontology, 124–128
  overview, 113
  phylogenetic analysis, 119–122
functional genomics
  defined, 4, 259
  overview, 294
functional proteomics
  immunochemistry, 276–277
  mass spectrometry, 270–276
  protein annotation, 259–267
  protein interaction maps, 280–286
  protein microarrays, 277–280
  protein separation, 267–270
Fundulus, 249
fw2.2 gene, 292

G-protein WD-40 repeat, *116*
gain-of-function genetics, 312–314
gain-of-function mutagenesis, 306–307
GAL4-mediated overexpression, 312–314
GAL4 transcription factor, 280
galago (*Otolemur garnetti*), 30
Gallus gallus (chicken), 34
gametic phase disequilibrium. *See* linkage disequilibrium
gamma rays, 296, *297*
gas chromatography, 325, 328, 329

GBrowse visualization tool, 36
GC content, 109–110
GC-MS profiling, 328, 329
GE Healthcare, 68
GenBank database, 23, 40
GenBank files, 26–27
gene annotation
  defined, 2
  Drosophila, 38
  with microarrays, 192
  prokaryotes, 48
  See also functional annotation; genome annotation
gene clustering
  orthologous genes, 116–119
  by sequence similarity, 114–116
gene expression
  compiling atlases of, 3
  databases, 252–253
Gene Expression Omnibus (GEO), 253
gene expression profiling
  development studies in *Drosophila*, 246–248
  microarrays (*see* microarrays)
  RNA sequencing, 231–236
  single-gene, 236–239
gene finding, 102
gene knockouts
  determining gene function and, 99
  overview, 301, *302*
  reverse, 99
  yeast transcriptome and, 242
gene networks, systems-leveling modeling, 338–341
gene ontology, 124–128
Gene Ontology (GO) consortium, 124–128
gene regulation, analyzing with ChIP chips, 225–227
Gene Sorter, 25
gene therapy, 301
GeneFinder, 98
Generic Model Organism Database (GMOD) database, 36, *37*
genes
  ab initio discovery, 98–103
  alternative splicing, 96, *97*
  annotation (*see* gene annotation)
  non-protein coding, 104–107
  "plant-specific," 42
  synteny, 4
genetic fingerprinting, 56, 314–316
genetic heterogeneity, 161
genetic maps
  alignment with other maps, *10*
  assembly, 4–7
  defined, 4
  Human Genome Project and, 13, *14*
  uses, 2
genetic pathways, defining with microarrays, 192–193
genetical genomic analysis, 250
Genie, 98
genome annotation
  ab initio gene discovery, 98–103
  EST sequencing, 95–98
  non-protein coding genes, 104–107
  regulatory sequences, 103

structural features of genome sequences and, 107–113
genome assembly programs, 89
genome mapping
 aligning of maps, *10*
 comparative genomics, 10–12
 cytological maps, 8–9
 genetic maps, 4–7
 physical maps, 8
genome projects
 aims of, 1–4
 managing and distributing data, 40–41
 microbial, 48–51
 See also animal genome projects; Human Genome Project; plant genomes
genome scans, linkage disequilibrium and, 159
genome science
 aims of, 1–4
 future of, *324*
 the growth of knowledge and, 323–324
 problems addressed by, 323
 unique status of, 13
genome sequencing
 general strategies in, 83
 hierarchical, 84–88
 sequence verification, 94–95
 shotgun, 88–89, 92–94
 See also automated DNA sequencing
genome-wide association studies (GWAS), 171–176
genomes
 minimal, 48–51
 plasticity in prokaryotes, 51
genotyping, by hybridization, 175–176
Genscan, 98
Gensiphere 3DNA Submicro system, 206
GEO BLAST, 25, 28
GEPASI program, 335
Gibbs sampling, 103, 227, 228–230
Giemsa, 109
glucose depletion experiment, 240–242
GNRF database, *332*
GO Consortium, 219
GOLD cDNA database, 98
GoSurfer, 127
gout, 329, *330*
Grail, 98
Gramene resource, 45
graph theory, 283
grasses, genome projects, 44–46
great apes, 11
GST-fusion chromatography, *270*

H/CA snoRNAs, 105
H-InvDB, 98
Haemophilus influenzae, 48, 49
hairpin RNAs, 107
Haldane mapping function, 6
haplotypes
 blocks, *145*
 defined, 3, 135
 maps, 20, 144–146
 phasing methods, 185–186
 polymorphism and, 135–136
Hardy-Weinberg equilibrium, 139

Haw River syndrome, *111*
Helianthus spp. (sunflowers), 47
hematological diseases, *39*
heterochromatin
 content, 111
 Human Genome Project and, 20
 sequencing and, 94
heterodimers, 286
heterologous systems, 311–312
heteromultimerics, 286
HGNC GeneID, 27
hidden Markov models (HMMs), 98, 100–103, 230, 264–265
hierarchical clustering, *220*, 221–222, 244, *245*
hierarchical sequencing
 end-sequencing, 87–88
 fingerprinting, 86–87
 hybridization, 85–86, *87*
 introduction to, 84
 versus shotgun sequencing, 83
high-throughput genotyping platforms, 183–185
high-throughput reverse genetics
 defined, 4
 gain-of-function mutagenesis, 306–307
 knock-ins, 301, *302*
 phenocopies, 308
 reasons for, 300
 RNAi screens, 303, 306
 systematic mutagenesis, 300–301, *302*
 viral-mediated transformation, 307–308
high-throughput sequencing, 68
high-volume sequencing, 2
hits, 90
HIV infection, 152
HKA test, 151
HMMgene, 98
homeodomain, *116*
homodimers, 286
homologous recombination, 301, *302*
homologous sequences, 74
homologs, 10
homology, 120
honeybee *(Apis mellifera)*, *29*, 248
horizontal gene exchange, 51
HT12 array, 202
hubs, 339, 340
human diseases
 microsatellite-associated, 110, *111*
 orthologs in invertebrates, 39
 See also cancers; disease genes; *individual diseases*
human genome
 alignment of chromosomes 5 and 22, *95*
 chromosome arms, 9
 chromosome banding, 9
 comparative gene content, *17*
 content from the Human Genome Project, 16–17, 20–21
 GC content of chromosome 1, *109*
 haplotype frequencies, 145, *146*
 human geographic distribution and, 145, *147*
 linkage disequilibrium distribution, *142*
 microsatellites, 110
 mouse synteny and sequence conservation, *32*

non-coding RNAs, *116*
polymorphisms, *135*
proportion in scaffolds, 93
repetitive sequences, *108*
segmental duplications, 110
synteny with cat genome, *12*
tRNA content, *106*
Human Genome Project (HGP)
    content of the human genome, 16–17, 20–21
    core libraries, 20, *21*
    GenBank files, 26–27
    Internet resources, 22–25, 28
    legal and ethical aspects (ELSI), 18–19
    objectives, 13–16
Human Genome Resources site, 23
Human Metabolome Database (HMDB), 331
HumanRef8 array, 202
humans
    divergence from primates, 152
    diversity, 145, *147*
    evolutionary and ecological functional genomics,
        249, 250–251
    "out-of-Africa" theory, 147, 150
    transcript diversity, *244*
Huntington disease, *111*
hybridization, in hierarchical sequencing, 85–86, *87*
hydrophobicity plots, 263, 266
hydrothermal vents, 59
hypermorphs, 297
hypomorphs, 297, 298
hypothesis test, 215

Iceland, 18, 155, 250
IDDM diabetes, 166, 169
Igf1 gene, 34
IL4 cytokine, 152
Illumina
    Genome Analyzer, 79–80, *80*
    GoldenGate genotyping assay, 184, *185*
    Infinium genotyping assays, 176, *177*
    oligonucleotide bead arrays, 201–202
image processing, of microarrays, 211–212
immobilized pH gradients (IPGs), 267
immune disorders, *39*
immunochemistry, 276–277
immunoglobin, *116*
immunogold labeling, 277
immunohistochemistry, 277
in silico genomics
    metabolic control analysis, 333–338
    systems-level modeling of gene networks, 338–341
indels. *See* insertion-deletion polymorphisms
Indian macaques, 28
Indiana University, 38
infrared spectrometry, 325, 326
Ingeneue software, *337*, 338
inhibitory RNA (RNAi), 303
INNER NO OUTER gene, 126
InParanoid, 119
insertion-deletion (indel) polymorphisms, 28, *70, 71,*
    74, 133
insertional mutagenesis, 295, 296
insulin-like growth factor gene, 34

Integrated Microbial Genomes resource, 53
integrative genomics, 324–325
interaction networks, 283–285
interactome maps, 282
interference, 6
International HapMap Project, 20, 144, 173
International Human Genome Sequencing
        Consortium (IHGSC), 20, *21*
International Sequencing Consortium (ISC), 28
Internet, managing genome data and, 40–41
interval mapping, 157–158
invertebrate model organisms, 36–39
ionization device, 271
isochores, 109
isoelectric point, 267
isotope-coded affinity tag (ICAT) reagents, 274, *275*
iTRAQ reagents, 274–275

J. Craig Venter Institute, 51, 58
Jackson Laboratory, 31
Japanese pufferfish *(Fugu rubripes)*, 10, 35
JASPAR database, *332*
JCVI Celera, 89
Joint Genome Institute, 52–53

k-means clustering, 222–223
KEGG database, *332*, 333
Kennedy disease, *111*
keratinocytes, *244*
kidney epithelium, *244*
knock-ins, 301, *302*
knockout mutations, 301, *302*
        See also gene knockouts
Kosambi mapping function, 6
Kyoto, Japan, 38

laboratory information management systems
        (LIMS), 40
lactate dehydrogenase, 124
large insert arrays, 227
Lawrence Berkeley Laboratory, 19
LC-MS profiling, 328, 329
LD. *See* linkage disequilibrium
LDL receptor gene, 155
legume genome projects, 44–46
leishmaniasis, *57*
lens crystallins, 124
leprosy, *57*, 58, 246
LET4 gene, 107
leukemias, 246
Li-Cor, 68
454 Life Sciences platform, 79
LIGAND database, 331, *332*, 333
ligation sequencing, 80–81
likelihood methods, 120, 121
LIN7 gene, 107
LINEs, *108*, 109
linkage disequilibrium (LD)
    defined, 3, 138
    distribution in the human genome, *142*
    overview, 138, 141
    quantifying, 139–140
    statistical tests for, 140–141

linkage disequilibrium mapping
  genome-wide association studies, 171–176
  linkage mapping compared to, 167
  overview, 158–159
  pedigree-based analysis, 166–170
  population-based case-control design, 162–166
  theoretical basis, 159–162
linkage, distinguished from association, 160
linkage mapping, 167
  See also recombination mapping
"linkage phase," 186
lipoprotein lipase *(LPL)* gene, *142, 146*
liquid chromatography, 274, 325, 328, 329
local alignment, 90, 91
Loess procedure, 210
logarithm of the odds (LOD), 158
logistic regression, 174–175
long-oligonucleotide microarrays, 200–202
long-QT syndrome, 153, 298
loop designs, 195
lox site, 314, *315*
LPL gene, *142, 146*
LS/MS/MS, 274
LTR elements, *108,* 109
lung epithelium, *244*
lymphoblastoid cell lines, 250

M13-derived phagemid, *85,* 88
Macaca mulatta (Rhesus macaques), 28, *29*
Macropus eugenii (Tammar wallaby), *29*
Magnaporthe grisea (rice blast), 54
maize *(Zea mays)*
  Ds element, 297
  genome projects, 44, 45–46
MaizeGDB, 45
malaria, *57,* 242, 246
malformation syndromes, *39*
Mammalia, conserved non-genic sequences, 111
manual sequencing, 66
map-based sequencing. *See* hierarchical sequencing
mapping function, 6
mariner elements, 304
marker-assisted selection, 156
MAS5 algorithm, 202
mass spectrometry, 270–276, 325, 328, 329
mass spectrophotometers, 271–272
mass spectrophotometry, 82
mass-to-charge ratio, 271
MassARRAY iPLEX Gold, 185
massively parallel genome sequencing, *80*
matrix-assisted laser desorption ionization
  (MALDI), 271
maximum likelihood estimate, 121
McDonald-Kreitman statistic, 151
Medicago (alfalfa), 44
MegaBACE 4000, 68
Melanie, 269
melanocytes, *244*
MEME program, 230
Mendelian disease genes, recombination mapping,
  152–155
mental retardation, 28
metabolic control analysis (MCA), 333–338
metabolic databases, 331–333

metabolic disorders, *39*
metabolic flux, 333, 334
metabolic profiling, 328–331
metabolic reconstruction, 340, *341*
metabolic systems, large-scale organization, 338–340
metabolomes
  defined, 324
  integrating with transcriptomic and proteomic
    data, 329–331
  in silico, 340, *341*
metabolomics
  analysis of cellular constituents, 325–328
  databases, 331–333
  defined, 325
  metabolic profiling, 328–331
metabonomics, 328
MetaCyc database, 331, *332,* 333
metagenomics, 59–60
metastatic tumors, 246
Metazoa, conservation of gene function, 124
Methanococcus jannaschii, 48–49
MIAME, 252
mice. *See* mouse
microarray data, mining, 220–224
Microarray Gene Expression Data Society, 252
microarrays
  analysis of cancer, 244–246
  analysis of differential transcription, 225–227
  analysis of yeast genome, 57
  applications, 192–194
  basic procedure, 191–192
  cDNA, 198–200
  ChIP chips, 225–227
  data mining, 220–224
  databases, 252–253
  DNA applications, 227, 230–231
  experimental designs, 195–198
  image processing, 211–212
  labeling and hybridization of cDNAs, 205–206
  long-oligonucleotide, 200–202
  major platforms, 194
  microbial transcriptomics, 239–242, *243*
  short-oligonucleotide, 202–204, 230–231
  See also protein microarrays
MicrobesOnline, 53
microbial genomes
  projects, 48–51
  representation of, *52*
  sequence completeness, 94
Microbial Genomes page, 52
microbial genomics
  funding for, 54
  metagenomics, 59–60
  parasites, 57–59
  projects and databases, 51–53
  range of organisms studied, *55*
  yeast, 54, 56–57
Microbial Sequencing Centers, 58
microbial transcriptomics, 239–242, *243*
Microcebus murinus (mouse lemur), 30
microinjection, 305
"microislands," 79
microRNAs, 107

microsatellites, 3, *92,* 110, *111*
Mimulus spp. (monkey flowers), 47
minimal genome, 48–51
"minimal" transcriptome, 243
minisatellites, 110
MIPS database, *332, 333*
MIRAGE database, *332, 333*
miRBase, 107
miRNAs, 107
MITEs, 42
model organisms
  biomedical, 34–35
  Human Genome Project and, *14*
  invertebrate, 36–39
  See also individual organisms
modifier screens, 311–312
molecular anthropology, 147, 150
monkey flowers (*Mimulus* spp.), 47
monoamine receptor knockouts, 308
monoclonal antibodies (MAbs), 276
monocytes, *244*
Morgan, Thomas Hunt, 5
mosquito (*Anopheles gambiae), 29, 57*
most recent common ancestor (MRCA), 148–150
motif detection, 228–230
mouse
  human synteny and sequence conservation, *32*
  monoamine receptor knockout, 308
  repetitive genome sequences, *108*
  transgenic, 301, 305
  See also rodent genome projects
Mouse Genome Database, 126
Mouse Genome Informatics (MGI), 31–32
mouse lemur (*Microcebus murinus),* 30
MS/MS. *See* tandem mass spectrometry
MULEs, 42
Multidimensional Protein Identification Technology (MudPIT), 274
multiple comparison problem, 164–165
Munich Information Center for Protein Sequences (MIPS), 54
mutagenesis
  gain-of-function, 306–307
  near-saturation, 4
  regional, 308–311
  saturation, 31, 295–299
  systematic, 56, 300–301, *302*
  targeted, 300–301, *302*
  transposon, 296–297
mutagens, 295–297
mutation(s)
  dominant, 297, *298*
  knockout, 301, *302*
  recessive, 298, 299
  SNPs and, 134
mutation rates, microsatellites and, 110
Mycobacterium, 246
Mycoplasma
  M. capricolum, 51
  M. genitalium, 48, 49, 50, 51, 292, *293*
  M. mycoides, 51
mycoplasmas, 51
myotonic dystrophy, *111*

Nanoarcheota, *55*
nanopore sequencing strategies, 82
NAPPA (nucleic-acid programmable protein arrays), 279–280
National Center for Biotechnology Information (NCBI)
  BLAST program, 23, 90–91, 113, 114, 115
  dbEST database, 96
  GenBank files, 26–27
  Web site, *24,* 25, 40
National Human Genome Research Institute (NHGRI), 18, 28
National Institute of Asthma and Infectious Disease, 58
National Institute of Health, 22, 23
NCBI. *See* National Center for Biotechnology Information
ncRNAs (non-protein coding RNAs), 104–107, *116*
Neanderthals, 30
near-saturation mutagenesis, 4
Needleman-Wunsch algorithm, 75–77
neighbor joining (NJ) method, 121
nematode worm. *See Caenorhabditis elegans*
neomorphs, 297
network theory, 283–285
neurological disorders, *39*
neutral theory of molecular evolution, 136
neutrality, tests of, 150–151
"next-generation" sequencing methods, 3, 79–83
Nimblegen, 204
NlaIII, 231
nodes, in networks, 283
non-protein coding genes, 104–107
non-protein coding RNAs (ncRNAs), 104–107, *116*
noncoding SNPs, 134
norm of reaction, 162
normalization, of cDNA microarray data, 208–210, 212
northern blots, 236–237
nuclear magnetic resonance (NMR) spectroscopy, 290–291, 327
nucleic-acid programmable protein arrays (NAPPA), 279–280
nucleotide diversity, *137, 138*
null hypothesis, 215

OAMB (octopamine receptor protein), *266*
obesity, 250–251
object-oriented databases, 40–41
oceans, metagenomics and, 59
octopamine receptor protein (OAMB), *266*
olfactory receptors, 17
oligonucleotide arrays
  bead arrays, 201–202
  long microarrays, 200–202
  purpose of, 191
  short microarrays, 202–204, 230–231
  tiling arrays, 227, 230
one-way ANOVA, 215–216
Online Mendelian Inheritance in Animals database, 35
Online Mendelian Inheritance in Man (OMIM) database, 25
open reading frames (ORFs), 287

Ornithorhynchus anatinus (platypus), *29*
ORNL site, 19
orphan genes, 113, 242
orthologs, 10, 117–119
Orthology browser tool, 34
Oryza sativa (rice), 44, *45*
Otolemur garnetti (bushbaby), 30
"out-of-Africa" theory, 147, 150
overcollapsed repeats, *92*
Overlapper, 89, *92*

P1 clones, 84, *85*
P-elements, 297, 304, *313*
P-loop motif, *116*
p-value, 91, 215
PAC clones, 84, *85*
Pacific Biosciences, 82
pairwise sequence alignment, 74–77, 90
palindromic sequences, 179
PAM scoring matrices, 91
Pan troglodytes (chimpanzee), 28, *29*, 30
PANTHER database, 128
parallel pyrosequencing, 79
paralogs, 10, 117–119
parasites
    gene expression profiling, 242
    genomics, 57–59
parsimony methods, 120–121
Pathema Microbial Resource Center, 58
Pathogenic Functional Genomics Resource Center, 58
Pathways database, 34
pattern formation modeling, 336–338
PAX6 gene, *115*
PCR-RFLP method, 179
pedigree-based analysis, 166–170
pedigree mapping, of disease genes, 152–153
penetrance, 160
PERL scripts, 41
Perlegen, 82
permutation testing, 166
PFAM database, *115*, 265
phage display, 278
phagemids, *85*, 88
pharmaceuticals, drug design and, 288
pharmacogenomics, 4, 325
.phd files, 72
phenocopies, 308
"phenomic" analysis, 31
PHI-BLAST, 117
phosphoglucose isomerase, 334
phrap assembler, 71, 72
phred program, 69, 72
phred score, 71
Phusion, 89
phylogenetic analysis, 119–122
phylogenetic footprinting, 103
phylogenetic shadowing, 103, *104*
phylogenomics, 122
phylogeny, 120
physical maps
    alignment with other maps, *10*
    Human Genome Project and, 13, *14*
    overview, 8
    synteny and, *10*

PipMaker, 103
plant genomes
    Arabidopsis thaliana, 40–44
    flowering plants, 46–48
    grasses and legumes, 44–46
    mapping, 6
plant nematodes, 58
"plant-specific" genes, 42
plants, transgenic, 304, 305
plasmid vectors, *85*
Plasmodium, 246
    P. falciparum, 242
platypus *(Ornithorhynchus anatinus)*, *29*
PLIER algorithm, 202
PLINK software, 175
point mutations, 296
poliovirus, 51
polonies, 81–82
polony sequencing, 81–82
polyclonal antibodies, 276
polymorphisms
    defined, 3
    haplotypes and, 135–136
    mutation and, 134
    See also single nucleotide polymorphisms
population-based case-control design, 162–166
population genetics
    coalescent theory, 148–150
    genetic mapping and, 7
    SNP technology and, 146–152
population stratification
    described, 161
    haplotype maps and, 144–145
Populus trichocarpa (black cottonwood), 47
positional cloning, 152, *163*
positive selection, 136
post-hoc tests, 216
posttranslational protein modification, 263
presenilin, 311
primates
    divergence in gene expression, *249*
    genome projects, 28, 30
    human divergence from, 152
principal component analysis (PCA), 174, 223–224, 328
print heads, 200
printing pin, 200
PRINTS, 261
privacy issues, 18
probability scores, 94
ProDom, 261
profile, 261, 264–265
profile plots, *223*
prokaryotes
    genome projects, 48–51
    horizontal gene transfer, 51
    number of genes defined in cellular processes, *49*
    See also microbial genomes; microbial genomics
promoter sequences, motif detection, 228–230
Prosite, 260, 261
PROSPECTOR algorithm, 292

prostate, *244*
protein annotation, 259–267
protein classification
    clusters of orthologous genes, 116–119
    domain clustering, 114–116
    See also functional annotation
Protein Data Bank (PDB), 286, 288, 290
protein domains
    classifying, 115–116
    databases, 260–261
    defined, 286
Protein Information Resource (PIR), 259
protein interaction maps, 280–286
protein kinase, *116*
Protein Lounge database, *332, 333*
protein microarrays, 277–280
protein separation
    affinity chromatography, 269–270
    2D-PAGE, 267–269
protein structure
    determination, 288–291
    overview, 261, 263
    prediction and threading, 291–294
Protein Structure Initiative (PSI), 287
proteins
    annotation, 259–267
    classification, 114–119
    E. coli production system, 278
    interaction maps, 280–286
    relationship of profiles to transcripts, 276
    separation, 267–270
    structure (*see* protein structure)
Proteobacteria, *55*
proteomics
    defined, 4, 259
    integrating with transcriptomic and metabolomic
        data, 329–331
    structural, 286–294
    See also functional proteomics
proteorhodopsin genes, 59
ProtScale, *266*
pseudocounts, 265
pseudogenes, 107
PSI-BLAST, 117
psychogenomics, 325
public health
    databases, 155
    disease control and, 58
PubMed, 23
pufferfish (*Tetraodon nigroviridis*), *29, 35*
purple sea urchin (*Stronglylocentrotus
        purpuratus*), *29*
pyrograms, *182*
pyrolysate, 326
pyrolysis, 325, 326
pyrosequencing, 79, *80, 182*–183

Q-PCR (quantitative PCR), 237–239
Q-Q plots, 171, 174
Q-RT-PCR (quantitative reverse-transcription PCR),
        237–239
QTL (quantitative trait loci), 136
QTL mapping, 155–158
QTN. *See* quantitative trait nucleotides

quantitative PCR (Q-PCR), 237–239
quantitative reverse-transcription PCR (Q-RT-PCR),
        237–239
quantitative trait loci (QTL), 136
quantitative trait nucleotides (QTN), 159
quantitative trait transcripts (QTT), 329
queries, to databases, 252–253
query languages, 41
query sequence, 90
qvalue, 219

R/Bioconductor open source software, 202
R genes, 42
race, 20
radiation hybrid (RH) mapping, 6
Raf serine kinase gene, 94
Raman spectrometry, 325
rats. *See* rodent genome projects
REACTOME database, *332*
"reads per kilobase of predicted exon per million
        total reads" (RPKM), 236
READSEQ, 40
recessive mutations, 298, 299
recombinant inbred lines (RILs), 6, 44, 155–156
recombinase, 314, *315*
recombination mapping, 152–155
recombination rates, 141
red jungle fowl (*Gallus gallus*), 34
reference sample analysis, 207–208
reference sample designs, 195
RefSeq (reference sequence), 98
regional mutagenesis, 308–311
regulatory elements, motif detection, 228–230
regulatory network analysis, 335, *336*
regulatory polymorphisms, 134
regulatory sequences, annotation, 103
relational databases, 40–41
relative risk, 160
renal diseases, *39*
repetitive sequences, 107–109
replacement polymorphisms, 134
Research Collaboratory for Structural Biology, 288
restriction fragment polymorphisms, 178–180
restriction profiling, 8
reticulocyte lysates, 278
reverse genetics, 294
reverse transcriptase, *116*
reversible terminator technology, *80*
rhesus macaques (*Macaca mulatta*), 28, *29*
rhodopsin-like GPCR, *116*
ribosomal RNA (rRNA), 105
ribosomes, ncRNA, 105
rice (*Oryza sativa*), 44, *45*
rice blast (*Magnaporthe grisea*), 54
RIKEN cDNA database, 98
river blindness, *57*
RNA-inducing silencing complex (RISC), 303
RNA-Seq, 234–236
RNA sequencing
    RNA-Seq, 234–236
    serial analysis of gene expression, 231–234
RNAi screens, 303, 306
RNase P, 106
Robust Multichip Analysis (RMA) algorithm, 202

rodent genome projects, 30–34
rooted trees, 120
Rosetta Stone approach, 282, 285
Roslin Institute, 35
RPKM, 236
Rrm domain, *116*
rRNA (ribosomal RNA), 105

Saccharomyces cerevisiae (budding yeast)
    comparative gene content, *17*
    genome conservation, 36
    genomics, 54
    integration of transcriptomic, proteomic, and
        metabolomic data, 329
    orthologs to genes from human disease, 39
    transcription factor analysis, *226, 227*
    transcriptome profiling, 239–240
    Y2H screens, 281
    See also yeast
Saccharomyces Genome Database (SGD), 54, 126
SAGE (serial analysis of gene expression), 3, 191,
    231–234, 252
San Diego Supercomputing Center, 288
sandfly, *57*
Sanger, Frederick, 2, 65
Sanger Institute, 23, 89
Sanger sequencing, 65–67, *80*
Sargasso Sea, 59
saturation forward genetics, 295–299
saturation mutagenesis, 295–299
saturation point, 295
saturation random mutagenesis, 31
saxophone gene, *38*
scaffolds
    of contigs, 92–94
    defined, 2
scale-free systems, 338–340
ScanAlyze, 211–212
schistosomiasis, *57*
Schizosaccharomyces pombe (fission yeast), 54, 240
SCOP, 292
scoring matrices, 91
Screener, 89
scripting languages, 41
sea squirt *(Ciona savignyi), 29*
seawater, metagenomics and, 59
segmental aneuploids, 310
segmental duplications, 42, *43,* 110
segregation distortion, 169
self-organizing maps (SOMs), 222, 223
SENTRA, 261
Sentrix Array Matrix (SAM), 184
separation chamber, 271–272
sequence alignment, pairwise, 74–77
sequence chromatograms, 70
sequence-contig scaffold, *86*
sequence identity, 74
sequence motifs, identification, 264–265
sequence scaffolds, 92–94
sequence-tagged sites (STSs), 8
sequence traces, reading, 68–71
sequencing by hybridization (SBH), 82, 175–176
Sequenom Corp., 82
    MassARRAY iPLEX Gold, 185

serial analysis of gene expression (SAGE), 3, 191,
    231–234, 252
serum uric acid, 329
short-oligonucleotide microarrays, 202–204, 230–231
short quantitative random RNA libraries (SQRLs),
    234, *235,* 236
shotgun sequencing, 2, 83, 88–89, 92–94
sibling-transmission disequilibrium test (S-TDT), 168
Significance Analysis of Microarrays (SAM) software,
    213
significance testing
    of cDNA microarray data, 213, 217–219
    overview, 215
simple nucleotide polymorphism, 134
simple sequence repeats (SSRs), 110
SINEs, *108,* 109
sing-feature polymorphisms (SFPs), 230
single-base extension (SBE), 181–182
single-channel microarrays, 211
single-gene analyses
    northern blots, 236–237
    quantitative PCR, 237–239
single-gene disorders, recombination mapping,
    152–155
single-molecule polony sequencing, 81–82
single-molecule real time (SMART) technology, 82
single nucleotide polymorphisms (SNPs)
    classifications, 133–136
    defined, 133–134
    discovery, 177–178
    distribution, 136–138
    genotyping (*see* SNP genotyping)
    haplotype maps, 144–146
    Human Genome Project and, 13, 17, 20
    identifying in sequence traces, 70, 71
    linkage and association, 160
    linkage disequilibrium, 138–142
    linkage disequilibrium mapping (*see* linkage dise-
        quilibrium mapping)
    population genetics and, 146–152
    QTL mapping and, 155–158
    recombination mapping and, 152–155
    sequencing, 180–183, 185–186
    significance, 3
    tagging, 144, *145*
single-sperm typing, 6
singular value decomposition, 223–224
SLC2A9 gene, 329
small nucleolar RNAs (snoRNAs), 105
SMART, 261
snapdragon *(Antirrhinum majus), 47*
SNaPshot Multiplex system, 184
Snip-SNPs method, 179
snoRNAs (small nucleolar RNAs), 105
SNP discovery, 177–178
SNP genotyping
    high-throughput genotyping platforms, 183–185
    restriction fragment polymorphism, 178–180
    SNP discovery, 177–178
SNP sequencing
    haplotype phasing methods, 185–186
    minisequencing methods, 180–183
SNPs. *See* single nucleotide polymorphisms